# Body Image and Nutritional Status Among Adolescents and Adults

# Body Image and Nutritional Status Among Adolescents and Adults

Guest Editors

**Emanuela Gualdi-Russo**
**Luciana Zaccagni**

MDPI

Basel • Beijing • Wuhan • Barcelona • Belgrade • Novi Sad • Cluj • Manchester

*Guest Editors*
Emanuela Gualdi-Russo
Department of Neuroscience
and Rehabilitation
University of Ferrara
Ferrara
Italy

Luciana Zaccagni
Department of Neuroscience
and Rehabilitation
University of Ferrara
Ferrara
Italy

*Editorial Office*
MDPI AG
Grosspeteranlage 5
4052 Basel, Switzerland

This is a reprint of the Special Issue, published open access by the journal *Nutrients* (ISSN 2072-6643), freely accessible at: https://www.mdpi.com/journal/nutrients/special_issues/7ON0C786D2.

For citation purposes, cite each article independently as indicated on the article page online and as indicated below:

| Lastname, A.A.; Lastname, B.B. Article Title. *Journal Name* **Year**, *Volume Number*, Page Range. |
|---|

**ISBN 978-3-7258-6342-6 (Hbk)**
**ISBN 978-3-7258-6343-3 (PDF)**
**https://doi.org/10.3390/books978-3-7258-6343-3**

# Contents

# About the Editors

**Emanuela Gualdi-Russo**

Emanuela Gualdi-Russo is currently an Eminent Scholar in the Department of Neuroscience and Rehabilitation at the University of Ferrara (Italy). Her areas of expertise are in the fields of Anthropology and Anthropometry, with a particular focus on body composition, nutritional status, and body image perception. She has authored over 150 original journal publications and numerous book chapters. Her current h-index is 31 (Scopus), and her Research Interest Score is 9337 (ResearchGate).

**Luciana Zaccagni**

Luciana Zaccagni is currently an Associate Professor of Methods and Didactics of Sports Activities in the Department of Neuroscience and Rehabilitation at the University of Ferrara (Italy). Her areas of expertise are in the field of sport sciences, with an emphasis on sports performance, kinanthropometry, and public health. She has authored over 60 original articles. Her current h-index is 24 (Scopus), and her Research Interest Score is 7339 (ResearchGate).

*nutrients*

MDPI

*Editorial*

# Body Image and Nutritional Status Among Adolescents and Adults

**Emanuela Gualdi-Russo [1,*] and Luciana Zaccagni [1,2]**

[1] Department of Neuroscience and Rehabilitation, Faculty of Medicine, Pharmacy and Prevention, University of Ferrara, 44121 Ferrara, Italy; luciana.zaccagni@unife.it

[2] Center for Exercise Science and Sports, University of Ferrara, 44123 Ferrara, Italy

* Correspondence: emanuela.gualdi@unife.it

## 1. Introduction

Body image perception is a composite construct that includes feelings, evaluations, and attitudes concerning the body. This perception may be independent of the body's actual appearance, and is strongly influenced by others' judgments. Body image dissatisfaction may arise from negative thoughts and feelings about one's own body [1].

Disturbances to body image, including dissatisfaction with and misperception of one's body, may be associated with eating disorders and unhealthy behaviors aimed at weight control [2].

The idealization of thinness is widely observed in Western societies, particularly among women and in those of increasingly younger ages. This can lead to dissatisfaction with one's body and low self-esteem, resulting in a distorted perception of body image [3,4].

Distorted body image is common and can cause anxiety regarding one's appearance, as well as depressive disorders and harmful eating behaviors, especially among adolescents and young adults. Furthermore, body dissatisfaction resulting from discrepancies between perceived and ideal body image can negatively affect individuals in terms of both physical and mental health.

This Special Issue of *Nutrients* collates twelve original research articles that discuss various aspects of body perception and eating disorders. The main aim of this Special Issue is to present an overview of the recent advances in this field, encompassing risk factors for eating disorders as well as the effects of physical activity on body image, from dissatisfaction with weight and body image to anorexia nervosa. Furthermore, it aims to explore differences in gender and age across a diverse range of contexts.

## 2. Eating Disorders, Risk Factors, and Assessment Tools

An article by Bozzola et al., forming contribution 1, shows that the experience of multiple adverse events, particularly during childhood and adolescence, can significantly increase the likelihood of an individual developing an eating disorder. Based on a sample of patients under 18 years of age hospitalized for anorexia, the authors hypothesized that considering the simultaneous presence of multiple potential risk factors, rather than a single risk factor, is crucial to understanding the severity of anorexia nervosa.

Especially in sports such as artistic and rhythmic gymnastics, in which success is associated with maintaining a lean body to meet esthetic and performance demands, participation is associated with a heightened risk of eating disorders. In this context, Donti et al. (contribution 2) demonstrate that adolescent artistic gymnasts exhibit symptoms of eating disorders and have limited knowledge of the key aspects of Relative Energy Deficiency in Sport (RED-S). Notably, a higher prevalence of eating disorders has been found among elite gymnasts compared to those at lower competitive levels. This suggests that reduced pressure on performance and appearance is associated with fewer eating disorders.

The incidence of anorexia nervosa was observed to have increased among children and adolescents during the pandemic. A study by Dylag et al. [contribution 3], conducted on a large sample of Polish patients aged 10 to 18, demonstrated the significant impact of the COVID-19 pandemic on the clinical course and hospitalization patterns of pediatric patients with anorexia nervosa. The results suggest that the pandemic may have worsened the disease's severity and resulted in alterations to treatment approaches, underscoring the need for increased awareness, early intervention, and targeted support strategies to ameliorate the impact of future pandemics on the incidence of eating disorders.

A diet rich in plant-based nutrients such as the Mediterranean diet has been shown to reduce plasma levels of pro-inflammatory cytokines, thus counteracting the biological mechanisms involved in the development of depression. In this context, Menghi et al. (contribution 4) found that the girls they examined exhibited higher levels of depression, anxiety, neuroticism, and food pickiness than the boys. They also observed that greater sensitivity to bitterness and astringency predicted higher levels of depressive symptoms in girls exclusively. This suggests that gender modulates the relationship between depressive symptoms, sensory perception, and eating habits in healthy adolescents.

The symptoms of eating disorders are complex and heterogeneous, making their clinical evaluation challenging. Despite the many validated assessment tools that have been developed, several demonstrated limited sensitivity to change, an inadequate coverage of associated factors, and excessive reliance on indicators such as BMI. Addressing this drawback, Golan et al. (contribution 5) examined the psychometric properties of the CONTASI-ED tool, demonstrating its strong reliability and that it has a high correlation with EAT-26. The scores acquired decreased significantly over the course of treatment, confirming the tool's sensitivity to changes in symptoms.

Furthermore, Gallagher et al. (contribution 6) investigated the Eating Disorders-Specific Nutrition-Focused Physical Examination (ED-NFPE) tool for use by registered dietitian nutritionists across different levels of emergency department care. Developed through expert consensus, this tool assesses nutrition-related complications in individuals with eating disorders. The results demonstrate that the ED-NFPE tool can enhance nutritional assessments and improve the overall patient experience.

## 3. Dissatisfaction with Body Image Perception and Weight

Girls, in particular, tend to be dissatisfied with their body weight. Fondell et al. (contribution 7) compared girls who wished to lose weight with those who intended to maintain their current weight in a nationally representative U.S. sample, examining eating habits, physical activity, vaping, alcohol consumption, sleep patterns, and screen time. Adolescent girls with the intention to lose weight exhibited various unhealthy behaviors, including skipping breakfast, increased screen time, and sleeping less. They also reported higher alcohol consumption and more vaping. Although weight loss intentions were more common among obese girls, they were also observed to have considerable prevalence among normal-weight girls. The authors suggested that girls would benefit from interventions targeting sleep, screen time, diet, physical activity, and substance use.

Negative body image perception is a hallmark symptom of anorexia nervosa. Affected individuals often demonstrate significantly distorted body size perception and high dissatisfaction with their body, including its size, shape, and appearance. A study by

Scarpina et al. (contribution 8) revealed that muscle mass and vitamin B6 concentrations are associated with the psychological manifestations of body image concerns in individuals with anorexia nervosa, particularly in the early stages of the disorder.

Adolescents generally tend to be dissatisfied with their bodies due to discrepancies between their current body image and prevalent beauty ideals. Through a longitudinal study of early adolescents, Gualdi-Russo et al. (contribution 9) found that perceptions of body image remained relatively stable over the three-year study period. Dissatisfaction was observed at similar rates in both genders, although different esthetic ideals were already noticeable even in early adolescence. Self-perception of weight status was not entirely consistent, with underweight adolescents tending to overestimate it and overweight/obese adolescents tending to underestimate it. Given increasing body dissatisfaction and the increasing misinterpretations of weight status, the promotion of a healthy lifestyle and a positive body image in schools is an advisable endeavor.

Relationships between body image and sociodemographic, morphological, and behavioral variables were examined by Knapik et al. (contribution 10) in a sample of adult women. The results indicate that women generally tend to be critical of their bodies. Excessive BMI negatively influences scores on the Body Esteem Scale, while BMI itself is less severely affected by sociodemographic variables. Engagement in physical activity had a particularly notable correlation with the "physical condition" domain of the Body Esteem Scale in young women.

Given that positive body image is linked to improved physical and mental well-being, Kriaučionienė et al. (contribution 11) examined the associations between body appreciation, body weight, lifestyle factors, and subjective health among undergraduate students. Their study highlighted strong links between body appreciation, healthy weight, positive lifestyle behaviors, and self-perceived health. The promotion of healthier habits may therefore foster an increased sense of body appreciation.

While numerous studies have explored body image and self-esteem in females, only a few have examined these aspects in adult men. McNeill's study (contribution 12) found that men are generally less influenced by celebrities or fashion trends in the development of their body image ideals and are largely satisfied with their appearance. The author found no significant influence of age on men's body image; instead, body image satisfaction appears to be shaped by other aspects of life satisfaction, such as relationship status and engagement in sports.

## 4. Conclusions

Altogether, the studies presented in this Special Issue emphasize the importance of body image perception and nutritional status for both physical and mental health. Furthermore, the availability of validated tools can substantially assist in clinical and nutritional assessments.

In conclusion, ongoing research on body image and weight perceptions, as well as the related eating disorders, continues to elucidate human variability and the factors that influence health maintenance. Nevertheless, further clinical, translational, nutritional, and anthropometric research is needed to address the many questions that remain unanswered.

**Author Contributions:** Conceptualization: E.G.-R. and L.Z.; writing—original draft, E.G.-R.; writing—review and editing, E.G.-R. and L.Z. All authors have read and agreed to the published version of the manuscript.

**Funding:** This research received no external funding.

**Conflicts of Interest:** The authors declare no conflicts of interest.

**List of Contributions**

1. Bozzola, E.; Cirillo, F.; Mascolo, C.; Antilici, L.; Raucci, U.; Guarnieri, B.; Ventricelli, A.; De Santis, E.; Spina, G.; Raponi, M.; et al. Predisposing Potential Risk Factors for Severe Anorexia Nervosa in Adolescents. *Nutrients* **2024**, *17*, 21. https://doi.org/10.3390/nu17010021.
2. Donti, A.; Maraki, M.I.; Psychountaki, M.; Donti, O. Eating Disorder Symptoms and Energy Deficiency Awareness in Adolescent Artistic Gymnasts: Evidence of a Knowledge Gap. *Nutrients* **2025**, *17*, 1699. https://doi.org/10.3390/nu17101699.
3. Dyląg, K.A.; Skowron, K.P.; Kurnik-Łucka, M.; Drzewiecki, Ł.; Przybyszewska, K.; Król-Dykas, M.; Dumnicka, P.; Gawlik, Z.; Gawlik, J.; Sikora, S.; et al. Anorexia Nervosa in Polish Children and Adolescents in the Context of the COVID-19 Pandemic-An Observational Single Centre Study. *Nutrients* **2024**, *16*, 4112. https://doi.org/10.3390/nu16234112.
4. Menghi, L.; Fontana, L.; Camarda, S.; Endrizzi, I.; Concas, M.P.; Gasparini, P.; Gasperi, F. Gender Moderates the Associations Between Responsiveness to Alarming Oral Sensations, Depressive Symptoms, and Dietary Habits in Adolescents. *Nutrients* **2025**, *17*, 1653. https://doi.org/10.3390/nu17101653.
5. Golan, M.; Sides, R.; Baum, K.; Arbib, R.; Abu Ahmad, W. The Conclusive and Continuous Tool to Assess Severity and Improvement of Eating Disorders (CONTASI-ED): Development and Psychometric Properties. *Nutrients* **2025**, *17*, 1790. https://doi.org/10.3390/nu17111790.
6. Gallagher, D.; Bailey, A.; Byham-Gray, L.; Rigassio-Radler, D.; Ziegler, J. Development of a Trauma-Informed, Culturally Sensitive Eating-Disorder-Specific Nutrition-Focused Physical Examination Tool: A Modified Delphi Study. *Nutrients* **2025**, *17*, 1449. https://doi.org/10.3390/nu17091449.
7. Fondell, E.; Vallejos, J.; Piazza, A.J.; Calle, M.C. Lifestyle Behaviors Associated with Weight Loss Intent in Adolescent Girls: Findings from the US 2021 National Youth Risk Behavior Survey. *Nutrients* **2025**, *17*, 1676. https://doi.org/10.3390/nu17101676.
8. Scarpina, F.; Cattaldo, S.; Prina, E.; Piterà, P.; Brusa, F.; Priano, L.; Mendolicchio, L.; Mauro, A. Muscle Mass and Vitamin B6 Are Linked to Negative Body Image in Women with Anorexia Nervosa: A Retrospective Study. *Nutrients* **2024**, *16*, 3902. https://doi.org/10.3390/nu16223902.
9. Gualdi-Russo, E.; Masotti, S.; Rinaldo, N.; De Luca, F.; Toselli, S.; Mazzoni, G.; Mandini, S.; Zaccagni, L. A Longitudinal Study on Body Image Perception and Size among Italian Early Adolescents: Changes over Time and Discrepancies between Genders. *Nutrients* **2024**, *16*, 3439. https://doi.org/10.3390/nu16203439.
10. Knapik, A.; Plinta, R.; Gnat, R. Association Between Body Image and Physical Activity, Sociodemographic, and Morphological Variables in Adult Women. *Nutrients* **2025**, *17*, 424. https://doi.org/10.3390/nu17030424.
11. Kriaučionienė, V.; Gajewska, D.; Raskilienė, A.; Myszkowska-Ryciak, J.; Ponichter, J.; Paulauskienė, L.; Petkevičienė, J. Associations Between Body Appreciation, Body Weight, Lifestyle Factors and Subjective Health Among Bachelor Students in Lithuania and Poland: Cross-Sectional Study. *Nutrients* **2024**, *16*, 3939. https://doi.org/10.3390/nu16223939.
12. McNeill, L.S. What Motivates Men to Improve Their Health? Understanding the Roles of Self-Esteem and Influential Others in Behaviour Change. *Nutrients* **2024**, *16*, 1916. https://doi.org/10.3390/nu16121916.

## References

1. Grogan, S. *Body Image: Understanding Body Dissatisfaction in Men, Women, and Children*; Routledge: Oxfordshire, UK, 2021.
2. McLean, S.A.; Paxton, S.J. Body Image in the Context of Eating Disorders. *Psychiatr. Clin. N. Am.* **2019**, *42*, 145–156. [CrossRef] [PubMed]

3. Swami, V.; Frederick, D.A.; Aavik, T.; Alcalay, L.; Allik, J.; Anderson, D.; Andrianto, S.; Arora, A.; Brännström, A.; Cunningham, J.; et al. The attractive female body weight and female body dissatisfaction in 26 countries across 10 world regions: Results of the international body project I. *Pers. Soc. Psychol. Bull.* **2010**, *36*, 309–325. [CrossRef] [PubMed]
4. Gualdi-Russo, E.; Rinaldo, N.; Masotti, S.; Bramanti, B.; Zaccagni, L. Sex Differences in Body Image Perception and Ideals: Analysis of Possible Determinants. *Int. J. Environ. Res. Public Health* **2022**, *19*, 2745. [CrossRef] [PubMed]

*nutrients*

MDPI

*Article*

# The Conclusive and Continuous Tool to Assess Severity and Improvement of Eating Disorders (CONTASI-ED): Development and Psychometric Properties

**Moria Golan [1,2,*], Roni Sides [2], Keren Baum [2], Rachel Arbib [2] and Wiessam Abu Ahmad [3]**

1 Department of Nutritional Sciences, Tel Hai College, North Galilee, Qiryat Shemona 1220800, Israel
2 SHAHAF—A Community-Based Facility for Treatment of Eating Disorders, Ganey Hadar 7683000, Israel; rachelarbib@gmail.com (R.A.)
3 Braun School of Public Health, Hebrew University-Hadassah, Jerusalem 91120, Israel; wiessam.abuahmad@mail.huji.ac.il
* Correspondence: moria.golan@mail.huji.ac.il

**Abstract: Background:** Accurately assessing eating disorder (ED) severity and treatment progress is essential for effective intervention. The Comprehensive and Continuous Tool to Assess Severity and Improvement of Eating Disorders (CONTASI-ED) was developed to address limitations in existing assessments by incorporating behavioral, cognitive, and physiological markers. **Objectives:** This study aimed to examine the psychometric properties and sensitivity to symptom changes of the CONTASI-ED in a community-based clinical sample of women with ED. **Methods:** Participants were 58 females diagnosed with EDs and 10 healthy controls. The CONTASI-ED assessments were conducted over multiple time points in outpatient and intensive treatment settings. We examined reliability, validity, and sensitivity to treatment-related change. The CONTASI-ED scores were compared with EAT-26, and multivariable analyses explored the effects of body mass index (BMI), age, and post-traumatic stress disorder (PTSD) on symptom trajectories. **Results:** The CONTASI-ED demonstrated strong reliability, with test–retest correlations between 0.72 and 0.90 and inter-rater reliability of 0.68–0.95. The tool effectively distinguished ED patients from healthy controls ($p < 0.001$) and correlated strongly with EAT-26. Significant reductions in the CONTASI-ED scores over time ($p < 0.001$) reflected treatment-related improvements—although temporary score increases highlighted greater self-awareness and symptom disclosure. BMI, age, and PTSD significantly influenced symptom severity and treatment response. **Conclusions:** The CONTASI-ED demonstrated strong reliability and validity in distinguishing clinical and non-clinical cases and in tracking treatment-related changes. However, the findings are based on a relatively small, all-female sample, underscoring the need for further validation in more diverse populations.

**Keywords:** eating disorders; assessment; severity; longitudinal; psychometric properties

## 1. Introduction

The clinical assessment of eating disorders (EDs) remains a significant challenge due to their complex, heterogeneous, and often fluctuating symptom presentations. EDs manifest in a wide range of physical, cognitive, and emotional symptoms, varying widely across individuals. Over the past decades, a wide range of validated assessment tools has been developed, each with distinct strengths and clinical applications. Existing screening tools have high specificity and sensitivity, yet many are impractical in primary care due to complexity, cultural adaptability issues, or administration demands [1,2]. Clinicians often rely on a combination of assessment tools to quantify illness severity over time or detect meaningful clinical changes in ED pathology [3–5].

Tools such as the Eating Disorder Inventory-3 (EDI-3) [6,7] and the Eating Disorder Examination (EDE) [8,9] have demonstrated diagnostic precision and contributed valuable insight into the cognitive and emotional correlates of EDs. The EDI-3 also provides valuable insight into psychological traits linked to EDs—such as perfectionism and low self-esteem [6,7]. However, it is time-consuming and better suited for diagnostic profiling than for ongoing outcome monitoring.

The Eating Disorder Examination (EDE), a well-validated interview protocol, is similarly robust. Yet, it requires specialized training and lengthy administration [8,9]. The EDE-Q offers a more practical self-report alternative but may be influenced by social desirability or lack of insight, particularly in individuals with poor insight or profound shame [10]. Several other tools focus on symptom tracking or life impact. For instance, the CHEDS [11] measures session-by-session fluctuations but lacks coverage of psychiatric comorbidities. The Munich ED-Quest [12] aligns with DSM-5 criteria but overlooks factors such as depression and trauma, which are crucial for predicting treatment outcomes [13].

Broader tools such as the Eating Disorder Quality of Life (EDQoL) Scale assess quality-of-life impacts but may not detect short-term symptom shifts, and thus are often used as supplementary measures [14]. Similarly, the CR-EAT captures long-term patterns but is less responsive to rapid clinical changes, limiting its utility for real-time decision-making [15].

Newer methods, such as the Ecological Momentary Assessment (EMA), use digital tracking to reduce recall bias, but lack standardization in scoring and interpretation, and are not designed to estimate overall illness severity or track treatment progress [16].

Common screening tools such as the SCOFF questionnaire [17], the EAT-26 [18], and the EDE-Q [19] are often based on cutoff scores and may miss atypical presentations—such as restrictive eating disorder (RED) patients who maintain normal weight despite severe metabolic dysregulation [20], or EDs in males, older adults, and culturally diverse populations [1,21]. For example, individuals from cultures where weight gain is perceived as a sign of health or prosperity may experience EDs differently and focus more on control or ritualistic eating behaviors than on body dissatisfaction [22,23]. These cultural and clinical nuances highlight the need for more inclusive, flexible tools that can account for diverse symptom profiles and trajectories. Moreover, the continued reliance on static indicators like BMI presents another limitation. While BMI is a relevant clinical parameter, it explains less than 15% of the variance in ED-related quality of life [24] and may obscure symptom severity in higher-weight individuals or those with normo-weight presentations.

To summarize, current assessment tools show limited sensitivity to change, insufficient coverage of comorbid factors, and overreliance on static indicators like BMI.

### 1.1. The Need for a Comprehensive and Continuous Assessment Tool

The Comprehensive and Continuous Tool to Assess Severity and Improvement of Eating Disorders (CONTASI-ED) was developed to address these challenges. Rather than replacing existing measures, the CONTASI-ED aims to complement them by integrating behavioral, cognitive, physiological, and psychosocial domains into a brief, structured format suitable for routine clinical use.

Grounded in contemporary evidence [25–27], The CONTASI-ED incorporates both state-based indicators (e.g., purging, somatic symptoms) and more stable trait-level contributors (e.g., compulsiveness, trauma history). This enables clinicians to track intra-individual changes over time, detect relapse risk, and support timely treatment adjustments. Its modular, transdiagnostic structure also increases applicability across ED subtypes and care settings.

To our knowledge, the CONTASI-ED is the first tool to combine this breadth of content with a scoring system sensitive enough for short-term progress tracking, while remaining feasible for real-world implementation.

### 1.2. Study Aims

This study aimed to evaluate the psychometric properties of the CONTASI-ED—including reliability, validity, and sensitivity to within-person symptom changes over time. Secondary analyses explored how the CONTASI-ED scores varied by BMI, age, and PTSD diagnosis, and how they compared with EAT-26, an established screening tool.

## 2. Methods

### 2.1. Development of CONTASI-ED

The development of the CONTASI-ED followed a structured, multi-phase process aimed at generating a clinically actionable instrument that captures the complexity of ED symptomatology and its evolution over time.

An initial item pool was generated based on clinical records, established ED instruments, and key domains identified in recent literature. Items were reviewed for redundancy, ambiguity, and contextual relevance, and were then organized into conceptual domains that reflected both theoretical frameworks and clinical utility. A multidisciplinary expert panel—including clinicians from psychiatry, psychology, nutrition, and research—refined the draft tool over two iterative rounds, focusing on clarity, scoring feasibility, and alignment with treatment decision-making. The tool was subsequently pilot tested across three clinical settings (inpatient, day program, and outpatient), leading to minor revisions in wording, item structure, and scoring anchors. The final tool was designed to be brief (completion time < 20 min), adaptable for self-report or clinician administration, and sensitive to both overt symptoms and underlying contributors to ED severity.

### 2.2. Item Generation and Tool Structure

Item content for the CONTASI-ED was informed by established ED assessment models and empirical literature [2,4,11,14,28–35]. The initial version included 78 items across eight candidate domains. During the refinement phase, concept saturation was monitored, and no new themes emerged, supporting content sufficiency. Following expert review, the final version included 61 items organized into six domains: Starting Point, Anthropometrics and Menstrual Cycle, Pathophysiology, Self-Care, Compulsiveness, and Obsessiveness.

The tool is structured as a checklist with standardized scoring anchors and a summed severity score. Item content includes behavioral symptoms (e.g., restriction, purging), cognitive-affective factors (e.g., intrusive thoughts), psychosocial history (e.g., trauma, treatment dropout), and a limited number of physiological markers (e.g., bradycardia, abnormal electrolyte or enzyme levels), which are commonly used in ED medical monitoring. These physiological items were included not as diagnostic thresholds, but as clinically relevant indicators that can support comprehensive severity profiling and facilitate interdisciplinary communication.

Higher total scores reflect greater illness severity and clinical complexity. Full scoring details and domain structure are provided in Table 1.

**Table 1.** Structure and scoring of the final version of the assessment tool.

| Domain | Number of Subcategories | Number of Items | Max Score | Weight Key Indicators * | Example of Factors Included |
|---|---|---|---|---|---|
| **Starting Point** | 6 | 11 | 22 | 22% | Previous treatments; occupation status; trauma history; substance use |
| **Anthropometrics and Menstrual Cycle** | 4 | 4 | 16 | 16% | Weight trajectory; weight stability; menstrual status |
| **Pathophysiology** | 3 | 19 | 13 | 13% | Bradycardia; orthostatic; hypotension; gastrointestinal symptoms; blood test |
| **Self-Care** | 4 | 4 | 16 | 16% | Sleep hygiene; general self-care behaviors; physical activity patterns; medication adherence |
| **Compulsiveness** | 4 | 22 | 16 | 16% | Eating restriction; over-controlling; other compulsive behaviors; use of laxatives or vomiting |
| **Obsessiveness** | 7 | 20 | 18 | 18% | Intrusive thoughts; cognitive rigidity |

* WKI = Max category score/max points.

### 2.3. Expert Panel Review and Scoring

An expert panel consisting of one psychiatrist, two clinical psychologists, four dietitians, and one statistician, all with substantial experience in eating disorder treatment—reviewed the tool for clinical relevance, clarity, and scoring feasibility.

The review process involved two formal consensus rounds. In Round 1, panelists rated each item on a 5-point Likert scale for clarity, clinical relevance, and contribution to illness severity and relapse risk. Items with high variability or low median ratings were flagged for further discussion. In Round 2, a structured Delphi method was employed to resolve disagreements and finalize the scoring schema. Consensus was defined as ≥75% agreement on each item's inclusion, phrasing, and weight.

Final weights were determined using a dual approach. Exploratory Factor Analysis (EFA) was used to identify item-level loadings, with those above 0.50 prioritized within their respective domains. These empirical results were supplemented with clinical judgment from the expert panel, particularly regarding each item's relevance to treatment decision-making and relapse prevention [3,4]. For instance, although some physiological items (e.g., bradycardia, CPK levels) showed modest loadings, they were retained at lower weights due to their clinical importance in medical risk assessment. Conversely, items reflecting obsessive thoughts or rigid behavioral patterns received higher weights, reflecting their consistent association with treatment resistance and relapse. Scoring anchors were standardized to improve consistency across raters and treatment contexts.

The tool was designed for flexible administration and can be used either as a clinician-administered interview or as a structured self-report questionnaire, depending on the setting and user preference. This flexibility enhances its usability across a wide range of clinical environments.

The final version of the CONTASI-ED includes six weighted domains, contributing to a total possible score of 101. These domains are outlined in Table 1, which details the number of subcategories, number of items, maximum scores, and examples of the key indicators included in each domain.

### 2.4. *Pilot Testing*

The study protocol was approved by the institutional review board of Tel Hai Academic College on 3.2021. Informed consent was obtained in writing from participants and their legal guardians.

This pilot study was conducted to evaluate item clarity, response feasibility, and clinical relevance. The preliminary version of the CONTASI-ED was pilot-tested by eight clinical dietitians across three different clinical settings: inpatient unit, intensive day treatment program, and outpatient clinics. During the pilot phase, clinicians completed the proposed tool alongside the EAT-26 questionnaire for all their patients and provided structured feedback through written comments and group debriefing meetings. Based on this feedback, 11 items were revised to enhance clarity, improve the comprehensibility of behavior descriptions, and align more closely with clinical language.

Key modifications included the addition of items related to previous treatment history and occupation; recoding of several categorial variables into continuous scales for greater sensitivity in capturing symptom severity; and inclusion of emotional and regulatory indicators commonly observed in comorbid presentations—such as mood disorders, anxiety, and impulse control disorders, which often underlie eating pathology.

These modifications aimed to improve the comprehensiveness and clinical interpretability of the tool. The revised version was subsequently evaluated for reliability and validity in a naturalistic clinical sample, where severity levels tend to vary continuously and do not align with categorical diagnostic threshold (Supplementary Materials S1).

### 2.5. *Statistical Analysis*

To evaluate the psychometric properties of the CONTASI-ED and its sensitivity in tracking longitudinal changes in illness severity and treatment outcomes, a series of statistical analyses were performed:

The dataset was reformatted from a wide to a long format to facilitate time-series analysis. Participants with at least two valid BMI measurements were included.

Each time point was modeled within a unified linear mixed-effects model, which accounts for repeated measures and estimates fixed effects jointly across the full-time series. As such, adjustment for multiple comparisons was not required, since the structure of the model controls for overall type I error without inflating significance at individual time points.

Additionally, multivariable models incorporated age and PTSD diagnosis as covariates to control for potential confounding effects. These adjustments improved model fit, as reflected in increased speculative marginal $R^2$ values, and enabled a more precise estimation of independent effects.

Although diagnostic categories (e.g., anorexia nervosa, bulimia nervosa, binge eating disorder) were available, analyses were conducted using BMI-based groups. This approach was chosen due to diagnostic fluidity across time points, particularly in patients transitioning between restrictive and binge-purge behaviors. Grouping by BMI allowed a clearer interpretation of symptom trajectories, minimized misclassification, and aligned with the tool's aim to track severity across a transdiagnostic spectrum. Thus, to investigate how the CONTASI-ED scores varied across different BMI categories, the BMI from the first three valid time points was calculated for each participant. Based on this, participants were categorized into three groups: Underweight, Normal Weight, and Overweight.

All analyses were conducted using R statistical software (version 3.5.0). Longitudinal data were analyzed using linear mixed-effects models, with random intercepts at the participant level, to account for repeated measurements. The model included Time as a fixed effect to assess the trajectory of symptom changes. An interaction term between Time and BMI group was introduced to determine whether the patterns of the CONTASI-ED and EAT-26 scores differed across BMI categories. Univariate and multivariate survival analyses were also performed to explore the predictive value of the CONTASI-ED scores for treatment outcomes. The CONTASI-ED and EAT26 scores were analyzed to examine changes over time and the effect of BMI group status on those changes.

To visualize the longitudinal trends, line plots were generated, displaying the mean CONTASI-ED and EAT-26 scores over time for each BMI group. To better understand the trend of EAT-26 measurements over time across BMI groups, we applied the locally estimated scatterplot smoothing (LOESS) method. Since the CONTASI-ED involved more frequent measurements than EAT-26, gaps in the data related to the latter made it difficult to observe a clear trend using traditional line plots. The LOESS method was used to smooth fluctuations and provide a continuous representation of the trajectory over time. This approach allowed us to visualize the overall trend of EAT26 scores while preserving local variations, ensuring a more accurate interpretation of changes within each BMI group.

### 2.6. Power Calculation

A post hoc power analysis was conducted to assess the adequacy of the sample size. Based on the observed effect sizes from the linear mixed-effects models, ranging from $-12.75$ to $-22.54$ units in STAT scores over time, and standard errors between 1.3 and 1.8, we estimated the statistical power with a two-sided alpha of 0.05 and an intra-class correlation (ICC) of 0.8. With 39 participants and 481 repeated observations, the power to detect a mean difference of 12.75 units was approximately 99%, and exceeded 99.9% for larger differences. These results support that the available sample size was sufficient to detect the longitudinal changes in STAT scores with high precision.

## 3. Results

### 3.1. Study Population

All participants were recruited and assessed in Israel. All procedures adhered to the Declaration of Helsinki. Assessments took place within community-based clinics, hospital-affiliated outpatient programs, and intensive day programs. This study included 58 Israeli female patients, who were admitted to community-based outpatient clinics for eating disorders during 2022–2023. Data from repeated assessments of 31 patients over 22 time points, and from the remaining 17 patients (who had not remained in treatment long enough) at 3 time points, were incorporated into the psychometric analysis of the CONTASI-ED. Ten other patients from this sample were treated in an intensive day program, with the CONTASI-ED assessments conducted over three months. There were no statistically significant differences in mean age or years of education among the three participant groups (Table 2). All participants provided written consent to have their data used for this study's analysis and publication.

**Table 2.** The sample demographic and clinical characteristics.

| Characteristic | Diagnosed with ED | | Healthy Participants |
|---|---|---|---|
| **Setting of treatment** | Outpatient clinic $N = 48$ | Intensive day clinic $N = 10$ | $N = 10$ |
| Age (mean ± SD) | 19.7 (SD = 6.4) | 24.0 (SD = 3.2) | 20 (SD = 5.5) |
| BMI at admission | 20.0 (SD = 5.7) | 19.66 (SD = 6.1) | 23.4 (SD = 6.8) |

**Table 2.** *Cont.*

| Characteristic | Diagnosed with ED | | Healthy Participants |
|---|---|---|---|
| Education | 11 | 12 | 13 |
| ED diagnosis at admission | Restrictive AN 24 (50%)<br>Binge Purge AN 5 (10%)<br>Atypical AN 7 (14.6%)<br>BN 7 (14.6%)<br>BED 5 (10.5%) | Restrictive AN 7 (70%)<br>Binge Purge AN 3 (30%) | – |
| PTSD | 32 (55%) | 2 (20%) | – |

### 3.2. *Psychometric Properties*

The psychometric evaluation of the CONTASI-ED revealed strong reliability and validity across multiple measures, as follows:

Test–retest reliability: Pearson's correlation coefficients (r) for test–retest reliability of the total CONTASI-ED score ranged from 0.72 to 0.90—indicating excellent stability over time.

Inter-rater reliability: Three clinicians independently interviewed and scored 22 patients. Fleiss's Kappa reliability (κ) ranged from 0.68 to 0.95—reflecting high agreement between different raters using the tool.

Patient–dietitian reliability: Spearman correlations between patient self-reports and dietitian-administered assessments ranged from 0.75 to 0.92—indicating strong alignment between the two modes of administration.

Criterion (discriminant) validity: The CONTASI-ED demonstrated strong criterion validity.

At baseline, scores were significantly higher among ED patients compared to healthy controls ($p < 0.001$; see Figure 1).

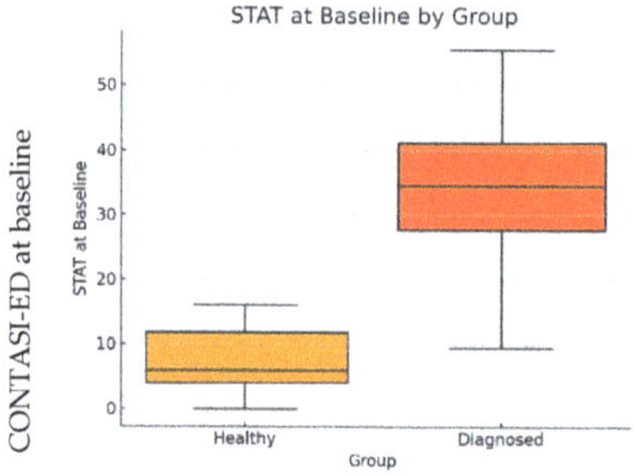


**Figure 1.** CONTASI-ED at baseline for healthy vs. ED-diagnosed patients.

A parallel analysis using the EAT-26 yielded a similar pattern, with both tools showing large effect sizes and clear discrimination between groups, as summarized in Table 3.

**Table 3.** Comparison of CONTASI-ED and EAT-26 scores between diagnosed and healthy participants.

| Variable | Overall $N = 58$ [1] | Diagnosed $N = 17$ [1] | Healthy $N = 41$ [1] | $p$-Value [2] |
|---|---|---|---|---|
| **STATT1** | 26.7 (14.9) | 7.6 (4.7) | 34.6 (9.4) | <0.001 |
| **Eat26T1** | 35.8 (19.0) | 5.7 (3.4) | 40.2 (16.0) | <0.001 |
| **Unknown** | 11 | 11 | 0 | |

[1] Mean (SD) and [2] Welch two sample *t*-test.

Convergent validity: Convergent validity was assessed by examining correlations between the CONTASI-ED and EAT-26 scores over the course of treatment. The CONTASI-ED scores were strongly correlated with EAT-26 scores throughout treatment (Figures 2 and 3). Initially, both measures were high, gradually decreasing over the course of treatment, until around 12 months—after which occasional regressions were observed. These regressions initially reflected the disclosure of previously hidden symptoms (such as restriction, purging or compulsive physical activity) and subsequent shifts in eating disorder subtypes (e.g., from restrictive anorexia nervosa to binge–purge anorexia or bulimia nervosa).

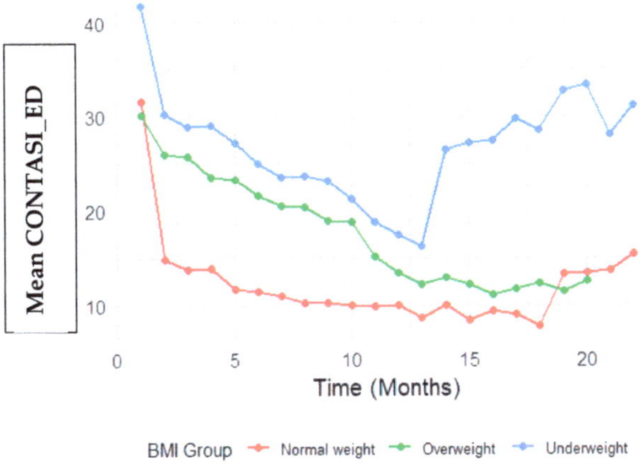


**Figure 2.** Differences in CONTASI-ED based on BMI group.

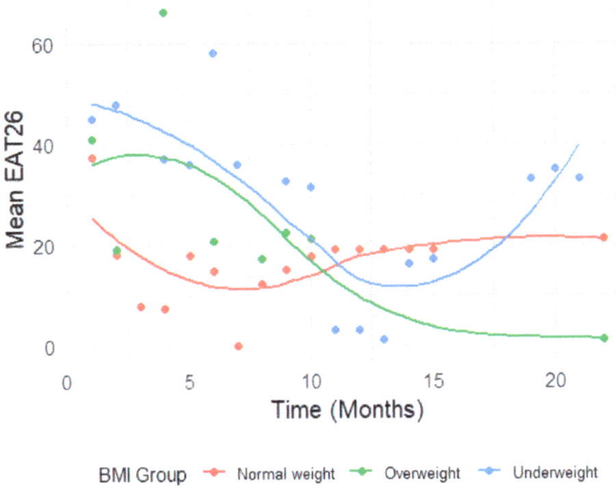


**Figure 3.** Differences in EAT-26 scores based on BMI group.

Internal consistency: Internal consistency was calculated using McDonald's Omega, accounting for different item loadings and variations in the strength of the items' associations with the construct being measured [36]. Apart from the "Start" category, all other categories exhibit omega coefficients above the acceptable threshold of 0.70—ranging between 0.72 and 0.9.

### 3.3. *Change in CONTASI-ED Scores over Time*

The CONTASI-ED scores decreased significantly over time. From Time 1 to Time 2, scores decreased by 12.75 units (B = −12.75, 95% CI: −15.12 to −10.39). By Time 13, the cumulative reduction reached 20.87 units (B = −20.87, 95% CI: −23.80 to −17.94).

Occasional score increases at later time points were observed, particularly among participants with extended treatment duration. While these patterns may reflect temporary symptom exacerbation, they could also indicate increased insight or willingness to disclose previously unreported symptoms. As this interpretation is speculative, further research is needed to clarify whether such score elevations reflect true clinical worsening or a shift in self-awareness and reporting.

BMI Group Differences

At baseline, Underweight participants had significantly higher CONTASI-ED scores than the Normal Weight group (B = 10.04, 95% CI: 2.86 to 17.21, $p = 0.006$). The Overweight group did not significantly differ from the Normal Weight group (B = −1.32, 95% CI: −10.51 to 7.87, $p = 0.778$) (Figure 2). Visual inspection of Figure 2 reveals distinct patterns of symptom reduction by BMI group. While all groups showed declining CONTASI-ED scores over time, the Underweight group exhibited a more gradual descent with a flatter slope in early treatment phases—suggesting potential resistance or slower response. In contrast, participants in the Overweight group demonstrated steeper initial declines, indicating a more rapid symptom reduction in the early weeks. The Normal Weight group displayed a more linear and consistent downward trend. These visual trends complement the model-based findings and underscore the clinical heterogeneity in treatment response trajectories.

Interaction effects indicated that Underweight participants showed a slower rate of improvement in the early stage of treatment. For example, at Time 2, the difference in reduction was significantly smaller (Time 2: B = 9.19, 95% CI: 3.98 to 14.39, $p = 0.001$). By Time 8, however, this difference diminished and was no longer statistically significant (B = 3.67, 95% CI: −1.67 to 9.01, $p = 0.177$).

Conversely, participants in the Overweight participants demonstrated greater reductions in the CONTASI-ED scores than those of the Normal Weight group, with significant interaction effects at multiple time points (e.g., Time 2: B = 11.08, 95% CI: 4.16 to 17.99, $p = 0.002$).

### 3.4. *Change in EAT-26 Scores over Time*

EAT-26 scores also showed significant reductions across time ($p < 0.001$). From baseline to Time 2, scores dropped by 14.77 units (B = −14.77, 95% CI: −21.01 to −8.54). By Time 10, the total reduction reached 22.13 units (B = −22.13, 95% CI: −28.79 to −15.46).

In the Underweight group, interaction effects indicated limited changes in EAT26 scores over time. While early time points suggested larger increases (e.g., Time 2: B = 9.85, 95% CI: −5.32 to 25.02, $p = 0.201$), these trends were not statistically significant and diminished further at subsequent time points (e.g., Time 10: B = 2.80, 95% CI: −14.01 to 19.62, $p = 0.742$).

In contrast, the Overweight group showed significant increases in EAT-26 scores at specific time points—notably at Time 4 (B = 33.54, 95% CI: 6.59 to 60.49, $p = 0.015$) (Figure 3). As shown in Figure 3, the Overweight group exhibited fluctuations in EAT-26 scores with notable spikes (e.g., at Time 4), possibly reflecting episodic changes in cognitive attitudes toward eating. The Underweight group showed relatively flat trajectories, with limited downward movement, indicating a potential disconnect between behavioral and cognitive symptom improvement. These visual patterns illustrate the differential sensitivity of EAT-26 across subgroups and reinforce the importance of multidimensional assessment tools.

However, these patterns were inconsistent over time, with many time points yielding non-significant interaction effects (e.g., Time 9: B = 4.86, 95% CI: −12.22 to 21.94, $p = 0.574$).

### 3.5. *Multivariable Analysis*

#### 3.5.1. Effect of Age on CONTASI-ED Scores

Age was significantly associated with higher CONTASI-ED scores in the Overweight group (B = 1.42, 95% CI: 0.36 to 2.49, $p = 0.010$), but only to a limited degree with Underweight and Normal Weight participants.

Including age as a covariate improved model fit for the Overweight group, with the marginal $R^2$ increasing from 0.437 to 0.604 (Supplementary Materials S2a).

#### 3.5.2. Effect of PTSD on CONTASI-ED Scores

Adjusting for PTSD had a more pronounced effect on the Normal Weight participants. In the Underweight group, PTSD had no significant impact (B = 1.42, 95% CI: −13.14 to 15.98, $p = 0.848$), and baseline CONTASI-ED scores remained stable (41.48, 95% CI: 34.94 to 48.02) (Supplementary Materials S2b).

Similarly, in the Overweight group, PTSD was not a significant predictor (B = 8.89, 95% CI: −1.85 to 19.63, $p = 0.102$), and its inclusion did not substantially change baseline CONTASI-ED estimates (30.12, 95% CI: 23.36 to 36.89). However, in the Normal Weight group, PTSD was strongly associated with lower CONTASI-ED scores (B = −16.81, 95% CI: −22.50 to −11.12, $p < 0.001$), and adjusting for PTSD increased the baseline estimate from 33.99 (95% CI: 28.29 to 39.70) to 31.45 (95% CI: 27.28 to 35.61). This stratified analysis was exploratory and intended to probe potential variation in comorbidity patterns, although the observed effects should be interpreted with caution due to limited subgroup sizes.

## 4. Discussion

This study provides initial validation of the CONTASI-ED as a psychometrically sound and clinically responsive tool for assessing eating disorder (ED) severity and tracking treatment progress. Designed to address key limitations of existing instruments, the CONTASI-ED integrates state-based indicators (e.g., physiological markers) with trait-level factors (e.g., trauma history, compulsivity), offering a holistic and nuanced assessment framework.

The CONTASI-ED demonstrated strong test–retest and inter-rater reliability, comparable to or exceeding those reported for established instruments such as the EDE-Q [19] and EPSI-CRV [3]. This suggests that the tool yields stable and consistent assessments across raters and time, even in varied clinical settings. Moreover, the strong alignment between clinician- and patient-rated scores reinforces its applicability as a dual-mode instrument, adaptable to clinician or self-report formats depending on patient needs.

The CONTASI-ED also demonstrated high internal consistency for most categories, supporting its structural coherence. As anticipated, the "Start" category yielded lower omega values due to its broader coverage of historical risk factors—consistent with psychometric literature suggesting reduced reliability in multidomain constructs [36].

Importantly, the CONTASI-ED effectively differentiated between ED patients and healthy controls providing evidence for its discriminant validity. This aligns with previous literature establishing large effect sizes for tools like EAT-26 in distinguishing clinical from non-clinical groups [18], but the present results suggest even greater sensitivity, potentially due to the inclusion of both behavioral and physiological domains.

One of the most notable findings was the CONTASI-ED's sensitivity to symptom change over time. Scores decreased significantly across treatment phases, with occasional increases interpreted as reflections of greater honesty and insight. These "blips" may reflect either temporary symptom exacerbation or increased trust in the therapeutic alliance, leading to greater disclosure of previously unreported symptoms. However, this interpretation remains speculative and should be further examined in future research before being considered indicative of 'authentic worsening' or deeper self-awareness—paralleling prior work suggesting that symptom disclosure can transiently elevate scores without indicating relapse [33]. In contrast, EAT-26 scores remained flatter, missing these nuanced shifts—suggesting that the CONTASI-ED may be better equipped to capture clinically meaningful fluctuations than static screening tools.

Group-based analyses offered further insight. Underweight participants had higher baseline scores and showed slower early improvement compared to those at normal weight, consistent with prior studies indicating that low BMI is associated with more entrenched cognitions and treatment resistance [26]. In contrast, participants with overweight status demonstrated greater symptom reduction, potentially reflecting responsiveness to interventions addressing binge-eating patterns [5].

Multivariable analyses revealed additional complexity. Older age was associated with higher severity in the Overweight group, suggesting that chronicity may moderate treatment response in this subgroup. PTSD was significantly associated with symptom severity only in the Normal Weight group—a finding that reinforces the importance of trauma-informed assessment tools. One possible explanation is that individuals in this weight category may experience fewer external indicators of illness, leading to an increased role of internalized emotional distress in symptom expression. PTSD-related hypervigilance, avoidance, or dissociation could interfere with treatment responsiveness or amplify underlying compulsive behaviors. These interpretations remain exploratory and warrant targeted investigation in future studies. This finding also echoes prior work on the prognostic role of comorbidity in ED outcomes [4,13].

In light of emerging research on emotion dysregulation as a central mechanism in eating pathology, it is notable that several CONTASI-ED domains—particularly Compulsiveness, Self-Care, and the inclusion of mood and anxiety indicators—may indirectly capture aspects of dysregulation. For example, patterns of impulsive behaviors, sleep disruption, and restrictive eating cycles often reflect difficulties in emotional containment. While the tool was not explicitly designed to quantify emotion regulation capacity, future work could explore how specific subdomains of the CONTASI-ED correlate with validated emotion dysregulation scales, thereby enriching its clinical interpretability and relevance to transdiagnostic processes, a finding that reinforces the importance of trauma-informed assessment tools and echoes prior work on the prognostic role of comorbidity in ED outcomes [4,13]. The CONTASI-ED also demonstrated clear advantages in routine clinical application. Its flexible design enables tracking of progress across settings—from inpatient to outpatient—without losing interpretive continuity. While not a focus of the present study, the modular structure and dual-mode administration of the CONTASI-ED may lend itself to future digital implementation. This possibility remains untested in our current data and is proposed solely as a direction for future research, particularly in the context of stepped-care models and remote monitoring strategies.

While the CONTASI-ED was designed with contextual adaptability in mind—including phrasing that allows for clinical and cultural tailoring—this flexibility has not yet been empirically tested. The current study was conducted in a culturally homogenous sample, limiting the ability to assess cross-cultural validity. Future studies should examine whether the tool performs equivalently across diverse populations and healthcare systems.

Unlike static tools with cutoff scores, the CONTASI-ED allows clinicians to monitor progress on a continuous scale, capturing meaningful intra-individual variation and informing real-time decisions. This approach is consistent with the rationale behind tools such as the CR-EAT, which emphasizes session-by-session tracking rather than fixed thresholds [15]. In our clinical observations, reductions in the CONTASI-ED scores below 10 coincided with substantial functional improvement and justified a step-down in treatment intensity—a pattern that warrants future empirical validation.

Finally, the inclusion of objective physiological markers such as bradycardia and electrolyte abnormalities further distinguishes the CONTASI-ED from existing tools. These markers not only improve face validity and clinical confidence but also facilitate communication across interdisciplinary teams (e.g., psychiatry, internal medicine, nutrition)—a key requirement in ED care settings [28].

*Limitations and Future Research Directions*

While the present findings support the CONTASI-ED's strong psychometric properties and clinical applicability, several limitations warrant consideration. First, the sample was predominantly female and relatively homogenous in terms of cultural background, which limits the generalizability of the findings to males, non-binary individuals, and more diverse populations. This limitation is common in ED research but highlights the importance of validating assessment tools in underrepresented groups [32] Further studies should include broader demographic samples to assess measurement invariance across sex and cultural background.

Second, although the CONTASI-ED was designed to capture a wide range of behavioral, cognitive, and physiological indicators, certain contextual and protective factors—such as social connectedness, body appreciation, or self-compassion—were not included. These factors may play a critical role in moderating ED severity and predicting recovery trajectories [37], and their exclusion may limit the tool's comprehensiveness. Incorporating such elements into future revisions or as parallel modules may enhance its predictive value. Moreover, future studies may want to validate each subdomain with specific tools—particularly constructs such as emotion regulation and compulsivity.

Third, although domain weights were informed by both expert consensus and factor analysis, their empirical grounding remains preliminary. For example, obsessive cognitions were weighted more heavily than some physiological markers, based on the clinical assumption that they are more resistant to change and more strongly associated with relapse risk. While this reflects established clinical reasoning [3], it may also introduce a theoretical risk that multidimensional scoring may inadvertently over-pathologize certain subgroups—especially when psychosocial variables are weighted alongside physiological indicators. Future large-scale predictive modeling could help refine the weighting schema and ensure that clinical relevance does not come at the expense of diagnostic inflation.

Additionally, the relatively modest sample size and the naturalistic setting of the study, while ecologically valid, may limit statistical power and preclude certain subgroups or longitudinal comparisons. Larger samples across multiple sites would allow for more robust analyses of measurement invariance, predictive validity, and sensitivity to treatment phase or subtype transitions.

Finally, it is important to note that the "Social and/or mentalizing impairment" item category was added after the primary analyses were conducted. Thus, it was excluded from all quantitative analyses reported in this study and is not reflected in any of the psychometric evaluations. While this domain reflects a growing interest in social-cognitive factors in EDs [23], its psychometric contribution and predictive utility should be evaluated in future studies before being integrated into the core scoring algorithm.

## 5. Conclusions

The CONTASI-ED represents a novel, multidimensional assessment tool that bridges key gaps in ED evaluation—combining clinical depth, time efficiency, and sensitivity to therapeutic change. Its adaptability across care settings, robust psychometric properties, and incorporation of physiological and psychosocial indicators position it as a valuable addition to the measurement-based care landscape.

As mental health systems move toward more personalized, data-informed practices, tools like the CONTASI-ED offer a scalable and clinically meaningful way to guide decision-making and improve treatment outcomes across the diverse spectrum of eating disorders.

**Supplementary Materials:** The following supporting information can be downloaded at: https://www.mdpi.com/article/10.3390/nu17111790/s1, Supplementary Materials S1—the questionnaire; Supplementary Materials S2—table of statistics.

**Author Contributions:** Conceptualization, M.G.; methodology, M.G. and W.A.A.; software, W.A.A.; validation, M.G., R.S., R.A. and W.A.A.; formal analysis, W.A.A.; investigation, M.G.; resources, K.B., R.S. and R.A.; data curation, R.S. and W.A.A.; writing—original draft preparation, M.G. and W.A.A.; writing—review, editing and visualization, W.A.A.; supervision, M.G.; project administration, R.A. All authors have read and agreed to the published version of the manuscript.

**Funding:** This research received no external funding.

**Institutional Review Board Statement:** The study was conducted in accordance with the Declaration of Helsinki. The psychometric properties analysis did not require ethical approval, which is approved by Tel Hai Institutional Review Board (March 2021).

**Informed Consent Statement:** Written informed consent was obtained from all subjects involved in this study and their legal guardians. Written informed consent has been obtained from the patient (s) to publish this paper.

**Data Availability Statement:** The dataset supporting the conclusions of this article is available from the corresponding author upon reasonable request. Due to concerns about potential re-identification despite anonymization, and to protect participant confidentiality, the dataset will not be publicly uploaded at this stage.

**Conflicts of Interest:** The authors declare no conflicts of interest.

## References

1. Alexander, T.; Burnette, C.B.; Cory, H.; McHale, S.; Simone, M. The need for more inclusive measurement to advance equity in eating disorders prevention. *Eat. Disord.* **2024**, *32*, 798–816. [CrossRef] [PubMed]
2. Tournayre, L.; Alencar, M.; Balbinotti, A.; Monthuy-Blanc, J. Some hope for a dimensional assessment? A critical review of psychometric validated (semi-) structured interview to assess eating disorders. *Eur. Eat. Disord. Rev.* **2024**, *32*, 1138–1156. [CrossRef]
3. Forbush, K.T.; Bohrer, B.K.; Hagan, K.E.; Chapa, D.A.N.; Gould, S.R.; Petkus, J.M. Development and initial validation of the Eating Pathology Symptoms Inventory-Clinician-Rated Version (EPSI-CRV). *Psychol. Assess.* **2020**, *32*, 943–955. [CrossRef]
4. Treasure, J.; Schmidt, U. The cognitive-interpersonal maintenance model of anorexia nervosa revisited: A summary of the evidence for cognitive, socio-emotional and interpersonal predisposing and perpetuating factors. *J. Eat. Disord.* **2013**, *1*, 13. [CrossRef] [PubMed]
5. Austin, A.; De Silva, U.; Ilesanmi, C.; Likitabhorn, T.; Miller, I.; Fialho, M.D.L.S.; Austin, S.B.; Caldwell, B.; Chew, C.S.E.; Chua, S.N.; et al. International consensus on patient-centred outcomes in eating disorders. *Lancet Psychiatry* **2023**, *10*, 966–973. [CrossRef]

6. Garner, D.M.; Olmsted, M.P.; Polivy, J. Development and validation of a multidimensional eating disorder inventory for anorexia nervosa and bulimia. *Int. J. Eat. Disord.* **1983**, *2*, 15–34. [CrossRef]
7. Clausen, L.; Rosenvinge, J.H.; Friborg, O.; Rokkedal, K. Validating the Eating Disorder Inventory-3 (EDI-3): A comparison between 561 female eating disorders patients and 878 females from the general population. *J. Psychopathol. Behav. Assess.* **2011**, *33*, 101–110. [CrossRef]
8. Fairburn, C.G.; Cooper, Z. The Eating Disorder Examination (12th ed.). In *Binge Eating: Nature, Assessment, and Treatment*; Fairburn, C.G., Wilson, G.T., Eds.; Guildford Press: New York, NY, USA, 1993; pp. 317–360.
9. Thomas, J.J.; Roberto, C.A.; Berg, K.C. The Eating Disorder Examination: A semi-structured interview for the assessment of the specific psychopathology of eating disorders. *Adv. Eat. Disord.* **2014**, *2*, 190–203. [CrossRef]
10. Luce, K.H.; Crowther, J.H.; Pole, M. Eating Disorder Examination Questionnaire (EDE-Q): Norms for undergraduate women. *Int. J. Eat. Disord.* **2008**, *41*, 273–276. [CrossRef] [PubMed]
11. Spangler, D.L. The change in eating disorder symptoms scale: Scale development and psychometric properties. *Eat. Behav.* **2010**, *11*, 131–137. [CrossRef]
12. Fichter, M.M.; Quadflieg, N.; Gierk, B.; Voderholzer, U.; Heuser, J. The Munich Eating and Feeding Disorder Questionnaire (Munich ED-Quest) DSM-5/ICD-10: Validity, reliability, sensitivity to change and norms. *Eur. Eat. Disord. Rev.* **2015**, *23*, 229–240. [CrossRef] [PubMed]
13. Weissman, R.S.; Becker, A.E.; Bulik, C.M.; Frank, G.K.; Klump, K.L.; Steiger, H.; Strober, M.; Thomas, J.; Waller, G.; Walsh, B.T. Speaking of that: Terms to avoid or reconsider in the eating disorders field. *Int. J. Eat. Disord.* **2021**, *54*, 1501–1507. [CrossRef] [PubMed]
14. Adair, C.E.; Marcoux, G.C.; Cram, B.S.; Ewashen, C.J.; Chafe, J.; Cassin, S.E.; Pinzon, J.; Gusella, J.L.; Geller, J.; Scattolon, Y.; et al. Development and multi-site validation of a new condition-specific quality of life measure for eating disorders. *Health Qual. Life Outcomes* **2007**, *5*, 23. [CrossRef]
15. Moessner, M.; Fassnacht, D.B.; Bauer, S. Online assessment of eating disorders: The clinical and research inventory for eating disorders (CR-EAT). *Ment. Health Prev.* **2015**, *3*, 170–177. [CrossRef]
16. Heron, K.E.; Everhart, R.S.; McHale, S.M.; Smyth, J.M. Using mobile-technology-based ecological momentary assessment (EMA) methods with youth: A systematic review and recommendations. *J. Pediatr. Psychol.* **2017**, *42*, 1087–1107. [CrossRef]
17. Morgan, J.F.; Reid, F.; Lacey, J.H. The SCOFF questionnaire: Assessment of a new screening tool for eating disorders. *Br. Med. J.* **1999**, *319*, 1467–1468. [CrossRef]
18. Garner, D.M.; Olmsted, M.P.; Bohr, Y.; Garfinkel, P.E. The Eating Attitudes Test: Psychometric features and clinical correlates. *Psychol. Med.* **1982**, *12*, 871–878. [CrossRef] [PubMed]
19. Mond, J.M.; Hay, P.J.; Rodgers, B.; Owen, C.; Beumont, P.J.V. Validity of the Eating Disorder Examination Questionnaire (EDE-Q) in screening for eating disorders in community samples. *Behav. Res. Ther.* **2004**, *42*, 551–567. [CrossRef]
20. Dufour, R.; Steiger, H.; Booij, L. Examining dimensionality and item-quality of the Eating Disorder Examination Questionnaire in individuals with eating disorders using item response theory analysis. *Int. J. Eat. Disord.* **2025**, *58*, 349–361. [CrossRef]
21. Zhang, J.; Cui, S.; Xu, Y.; Cui, T.; Barnhart, W.R.; Ji, F.; Nagata, J.M.; He, J. Introducing diagnostic classification modelling as an unsupervised method for screening probable eating disorders. *Assessment* **2024**, *32*, 405–416. [CrossRef]
22. Mishra, K.; Harrop, E. Teaching how to avoid overreliance on BMI in diagnosing and caring for patients with eating disorders. *AMA J. Ethics* **2023**, *25*, E507–E513. [CrossRef] [PubMed]
23. Schmidt, U.; Sharpe, H.; Bartholdy, S.; Bonin, E.-M.; Davies, H.; Easter, A.; Goddard, E.; Hibbs, R.; House, J.; Keyes, A.; et al. Treatment of anorexia nervosa: A multimethod investigation translating experimental neuroscience into clinical practice. *Program Grants Appl. Res.* **2025**, *5*, 1–258. [CrossRef]
24. American Psychiatric Association. *Diagnostic Statistical Manual of Mental Disorders*, 5th ed.; DSM-5; American Psychiatric Association: Washington, DC, USA, 2013. [CrossRef]
25. Ouellette, A.S.; Rodrigue, C.; Lemieux, S.; Tchernof, A.; Biertho, L.; Bégin, C. An examination of the mechanisms and personality traits underlying food addiction among individuals with severe obesity awaiting bariatric surgery. *Eat. Weight. Disord. Stud. Anorex. Bulim. Obes.* **2018**, *23*, 1–8. [CrossRef]
26. Lydecker, J.A.; White, M.A.; Grilo, C.M. Examining the clinical utility of the DSM-5 severity specifier for binge-eating disorder. *Int. J. Eat. Disord.* **2021**, *54*, 1455–1460. [CrossRef]
27. Koehler, J.M.; Breithaupt, L.; Waring, M.E.; Bulik, C.M.; Matheson, B.E. A systematic review of smartphone applications for eating disorder prevention. *Eur. Eat. Disord. Rev.* **2021**, *29*, 808–824.
28. Schmidt, U.H.; Claudino, A.; Fernández-Aranda, F.; Giel, K.E.; Griffiths, J.; Hay, P.J.; Kim, Y.R.; Marshall, J.; Micali, N.; Schaefer, L.M.; et al. A systematic review of instruments for the assessment of eating disorders among adults. *Curr. Opin. Psychiatry* **2021**, *34*, 543–562. [CrossRef] [PubMed] [PubMed Central]

29. Gleaves, D.H.; Graham, E.C.; Ambwani, S. Measuring "Orthorexia": Development of the Eating Habits Questionnaire. *Int. J. Eat. Disord.* **2013**, *46*, 276–287. [CrossRef]
30. De Young, K.P.; Anderson, D.A. Prevalence and correlates of exercise motivated by negative affect. *Int. J. Eat. Disord.* **2017**, *50*, 52–56. [CrossRef] [PubMed]
31. Bachner-Melman, R.; Lev-Ari, L.; Zohar, A.H.; Lev, S.L. Can we predict who will be more successful in a cognitive behavioral therapy for eating disorders? *J. Clin. Psychol.* **2018**, *74*, 1612–1625. [CrossRef]
32. Kennedy, S.F.; Kovan, J.; Werner, E.; Egan, A.M.; Becker, C.B.; Becker, A.E. Examining eating disorder screening instruments for use in transgender and gender diverse populations. *J. Eat. Disord.* **2021**, *9*, 1–11.
33. Schaefer, L.M.; Crosby, R.D.; Machado, P.P. Defining remission, recovery, and relapse in eating disorders: A systematic review. *Int. J. Eat. Disord.* **2021**, *54*, 2079–2098.
34. Hatoum, A.S.; Burton, J.D.; Berry, J.D.; Keller, M.C.; Munn-Chernoff, M.A. Genome-wide association study of disordered eating in a large twin sample. *Mol Psychiatry.* **2022**, *27*, 485–493.
35. Jurich, P.D.; Madison, M.J. Measuring item influence for diagnostic classification models. *Educ. Assess.* **2023**, *28*, 229–245. [CrossRef]
36. McNeish, D. Thanks coefficient alpha, we'll take it from here. *Psychol. Methods* **2018**, *23*, 412–433. [CrossRef]
37. Tylka, T.L.; Wood-Barcalow, N.L. The Body Appreciation Scale-2: Item refinement and psychometric evaluation. *Body Image* **2015**, *12*, 53–67. [CrossRef]

*nutrients*

MDPI

*Article*

# Eating Disorder Symptoms and Energy Deficiency Awareness in Adolescent Artistic Gymnasts: Evidence of a Knowledge Gap

**Anastasia Donti, Maria I. Maraki, Maria Psychountaki and Olyvia Donti ***

School of Physical Education & Sport Science, National & Kapodistrian University of Athens, 17237 Athens, Greece; adonti@phed.uoa.gr (A.D.); mmaraki@phed.uoa.gr (M.I.M.); mpsychou@phed.uoa.gr (M.P.)

* Correspondence: odonti@phed.uoa.gr; Tel.: +30-2107276109

**Abstract: Background/Objectives**: Pressure to stay lean may lead adolescent athletes to dietary restraint and disordered eating. Lack of nutrition awareness can also contribute to suboptimal dietary habits, increasing the risk of eating disorders and Relative Energy Deficiency in Sport [RED-S], though evidence in competitive athletes is limited. This study explored eating disorder symptoms and RED-S knowledge in adolescent artistic gymnasts. **Methods:** Eighty-four female artistic gymnasts, thirty-nine international and national level gymnasts [high-level; 14 [14, 15] y] and forty-five recreational and club level gymnasts [low-level; 14 [13, 15] y] completed the Eating Disorder Examination Questionnaire [EDE-Q 6.0], the RED-S knowledge Questionnaire and provided training details. **Results:** Seventeen gymnasts (20.2%) scored above the cutoff point on the EDE-Q. In addition, high-level gymnasts scored higher than low-level on EDE-Q [2.21 $\pm$ 1.37 (35.9%) vs. 1.19 $\pm$ 0.79 (6.7%), respectively; $p < 0.001$] and on its subscales: *Restraint, Eating Concerns, Weight Concerns,* and *Shape Concerns* [$p < 0.001$ to 0.009], thus indicating more severe disordered eating symptoms. No group differences were found in binge eating and compensatory behaviors. An important percentage of gymnasts reported at least one episode of binge eating and excessive training [39.3–58.3%], while four gymnasts reported self-induced vomiting. RED-S knowledge did not differ between groups. On average, gymnasts were unaware of correct answers related to RED-S [51.5%], its definitions [79.8–92.9%], and its association with menstrual disturbances and bone health [54.8–86.9%]. However, gymnasts reported better awareness of the impact of food restriction on illness and performance [47.6–84.5%]. **Conclusions:** Elite artistic gymnasts exhibited a higher prevalence of eating disorder symptoms than lower-level peers. Gymnasts at all levels demonstrated limited knowledge of the effects of RED-S on menstrual and bone health. Failure to recognize these risks may influence gymnasts' eating behaviors and delay RED-S detection and management.

**Keywords:** energy deficit; eating behaviors; adolescence; gymnastics; energy availability; aesthetic sports

## 1. Introduction

Eating disorders are complex psychiatric disorders that may negatively affect health, development, and daily life [1]. They are often accompanied by other mental health disorders, their exact causes remain unclear, and their treatment is challenging, costly, and frequently ineffective [2]. Eating behaviors exist along a continuum, ranging from optimal nutritional practices to clinically diagnosable eating disorders, with disordered eating representing a midpoint [1]. Anorexia Nervosa and Bulimia Nervosa represent the most severe forms of disordered eating [3]. Subclinical behaviors—such as body image preoccupation, food restriction, bingeing, purging, and laxative or diuretic misuse—can negatively impact health and increase the risk of developing clinical eating disorders [3]. Intensive food restriction, and, as a result, low energy availability are the primary causes of Relative Energy Deficiency Syndrome (RED-S), a clinical syndrome characterized by an energy imbalance in which an athlete's energy intake is insufficient to meet the demands of exercise, growth, and normal physiological functioning [4,5].

Previous studies have shown that participation in sports with strict demands on body weight, such as gymnastics, is associated with an increased risk of eating disorders in female athletes [6–9]. Success in these sports often necessitates maintaining a lean, sometimes prepubescent body, to meet sport-specific aesthetic demands and optimize performance [6,7,10,11]. Furthermore, high-level athletes frequently exhibit elevated rates of disordered eating compared to their lower-level or non-athlete counterparts, driven by performance pressures, appearance demands, and specific personality traits [12–15].

Eating disorders predominantly affect female adolescents and tend to follow a chronic course with significant impairment in social functioning [16]. Early diagnosis and timely therapeutic intervention are essential for improving long-term outcomes [17]. Adolescent athletes in aesthetic sports, such as artistic and rhythmic gymnastics, are often exposed to excessive training and competition loads during a critical developmental stage [18,19], when performance pressures coincide with significant physical changes [20,21]. The drive for a lean physique may lead to food restriction, and initial performance gains can reinforce this behavior, increasing the risk of eating disorders [22,23].

Importantly, a lack of understanding of the nutritional requirements of sport may influence athletes' dietary habits and increase the risks of eating disorders and RED-S [4]. Research on RED-S has typically been performed among adolescent female athletes; however, emerging research demonstrates that male athletes are also affected, exhibiting clinical indicators of compromised bone health [24]. RED-S can result in a variety of adverse outcomes, including disruptions in hormonal regulation, growth retardation, impaired bone health, decreased immune function, and alterations in metabolic processes [25,26]. To lower the risk of RED-S, athletes, their families, and their support teams must be aware of its signs, symptoms, and consequences on sports performance, injury risk, recovery, and overall health. Failure to recognize these risks may not only influence gymnasts' eating behavior but also delay early RED-S detection and management [27,28]. However, despite its significance, evidence on athletes' awareness of the impact of RED-S on health and performance remains scarce. Thus, the aim of this study was to examine eating disorder symptoms and RED-S knowledge in a population of adolescent artistic gymnasts. It was hypothesized that: (a) a higher percentage of disturbed eating attitudes and behaviors would be observed in high-compared to low-level artistic gymnasts, and (b) a high percentage of gymnasts would be unaware of RED-S.

## 2. Materials and Methods

### 2.1. Participants

Eighty-four female artistic gymnasts, aged 12–18 years, participated in this research. Male athletes were excluded from the study due to an insufficient sample size, which would have limited the statistical power, prevented meaningful sex-based comparisons, and minimized confounding factors associated with sex-based physiological, behavioral, and nutritional differences.

The study questionnaire consisted of two parts: (a) a questionnaire which included age, training and competition details such as training experience (i.e., the years of systematic training and competing in artistic gymnastics), number of training sessions/hours per week and number of competitions per year, and (b) the Eating Disorder Examination Questionnaire (EDE-Q) [29,30], and the Knowledge of Relative Energy Deficiency Syndrome Questionnaire (RED-S Knowledge) [31]. The questionnaires were distributed to the participants in their training facilities, one hour before training. Athletes were gathered in small groups in a separate room, with no coaches or parents present. A member of the research team was present during questionnaire completion to provide clarification if needed and to ensure that participants remained focused and completed the inventories independently. Furthermore, athletes were informed about the confidentiality and anonymity of their responses and their right to withdraw from the study at any time without being required to provide any explanation for their decision. Participants responded to the questionnaire on paper, and a researcher was available to answer questions and collect the completed questionnaires.

The eighty-four artistic gymnasts included a population of international and national level gymnasts (High-level gymnasts; n = 39) and club and recreational level gymnasts (Low-level gymnasts; n = 45). The sample of high-level gymnasts (n = 39) comprised all the national team members of the country, trained six times per week, for approximately 3–4.5 h per session, and participated in competitions 3–4 times a year. Low-level artistic gymnasts trained 2–4 times per week, for 1.5–2.5 h per session, and participated in contests, festivals, or club competitions 2–3 times a year. The high-level artistic gymnasts were recruited in collaboration with the Hellenic Gymnastics Federation. The club and recreational level artistic gymnasts were randomly recruited from three different gymnastics clubs.

Participants' Body Mass Index (BMI) was calculated as the ratio of body weight to the squared standing height ($kg/m^2$). The characteristics of the participants are shown in Table 1. Prior to the study, the athletes and their parents were fully informed about the purpose and procedures of this study and gave written informed consent. Instructions to the participants included a reminder to respond to all items and a statement that there were no right or wrong answers. Procedures were approved by the local Institutional Ethics Review Committee (number 1645/15-05-2024) and followed the Code of Ethics set by the World Medical Association.

**Table 1.** Age, competitive experience, and anthropometric characteristics of the participants [1].

| | High Level Gymnasts (n = 39) | Low Level Gymnasts (n = 45) | *p*-Value |
|---|---|---|---|
| Age (y) | 14 (14, 15) | 14 (13, 15) | 0.061 |
| Training experience (y) | 8 (6, 8) | 6 (3, 8) | <0.001 |
| Height (cm) | 159 (157, 162) | 160 (156, 163) | 0.815 |
| Body mass (Kg) | 47 (44, 50) | 50 (47, 55) | 0.019 |
| Body Mass Index ($kg/m^2$) | 18.7 (17.7, 19.7) | 20.1 (18.4, 21.4) | 0.006 |

[1] Note: Values are expressed as median (Q1, Q3). *p*-values derived from Mann–Whitney rank test. Statistical significance was set at the 5% level.

### 2.2. *Study Measures*

#### 2.2.1. Eating Disorder Examination Questionnaire (EDE-Q)

The Eating Disorder Examination Questionnaire (EDE-Q) [29,30] is a 28-item self-report questionnaire designed to assess the range and severity of features associated with a diagnosis of an eating disorder and is derived from the Eating Disorder Examination interview [32]. The EDE-Q focuses on the past 28 days and is scored using a 7-point, forced-choice rating scheme. Subscale scores, relating to *Dietary Restraint*, *Eating Concerns*, *Concerns about Shape*, and *Concerns about Weight*, and a global score, are derived from the 22 items addressing attitudinal aspects of eating-disorder psychopathology. The total score of the inventory is an average of the subscales. A global EDE-Q score of 2.612 for the Greek population yields an optimal trade-off to determine eating disorder diagnosis [30]. Questions 13 to 18 (6 items) of EDE-Q are individually examined to ascertain the frequency of binge eating and compensatory behaviors occurring during the past four weeks (28 days) [29,33]. These items do not contribute to the subscale scores. The questionnaire also includes three more questions, which are not numbered. The two questions ask the respondent to estimate his/her weight and height. The third question concerns the frequency of menstruation for the last 3–4 months, as well as the use of contraceptive pills. Cronbach's $\alpha$ values for the total score of the questionnaire as well as its subscales were 0.840, 0.808, 0.817, 0.880, and 0.811, for EDE-Q total score, *Dietary Restraint*, *Eating Concerns*, *Concerns about Shape*, and *Concerns about Weight*, respectively.

#### 2.2.2. Knowledge of Relative Energy Deficiency Syndrome

This comprehensive 18-item questionnaire assesses theoretical and practical understanding of the signs, symptoms, and consequences of RED-S, including menstrual irregularities, poor bone health, impaired immunity, and decreased neuromuscular performance [31]. The domains of the inventory that assessed RED-S knowledge are labelled as follows: (1) Awareness and definition of Low Energy Availability (LEA), Athletic Triad, and RED-S (items 1–3); (2) knowledge of RED-S signs and symptoms (items 4–8); and (3) knowledge of health and performance consequences of RED-S (items 9–18). The questionnaire has high content as well as construct validity [31]. Total scores are calculated by summing the scores of 18-items. Preliminary analyses of the RED-S knowledge questionnaire's reliability in the Greek population indicate acceptable reliability, with Cronbach's alpha values around 0.70.

### *2.3. Statistical Analysis*

The normality of distribution of continuous variables was assessed using the Shapiro–Wilks test and Q-Q plots. For normally distributed data, data are presented as means $\pm$ SD, and differences between high- and low-level artistic gymnasts were determined using independent samples *t*-tests. Data that were not normally distributed are presented as medians (Q1, Q3), while data on categorical variables are presented as frequencies (n, %); differences between high- and low-level artistic gymnasts were determined using Mann–Whitney rank or Chi-Square tests, respectively. Statistical significance was set at the 5% level ($p$-value < 0.05). All hypotheses tested were two-tailed. All analyses were performed using SPSS (version 29, SPSS Inc. Chicago, IL, USA).

## 3. Results

### *3.1. Eating Disorder Examination Questionnaire*

A total of 84 artistic gymnasts provided data for the variables of interest and were included in the analyses for the present study. High-level artistic gymnasts had more training experience, lower body mass, and BMI than low-level artistic gymnasts ($p < 0.05$, Table 1).

High-level artistic gymnasts scored higher than low-level in EDE-Q global score and its subscales Restraint, Eating Concerns, Weight Concerns, and Shape Concerns ($p < 0.05$; Table 2).

**Table 2.** Participants' values of the EDE-6 Questionnaire [1].

| | High Level Gymnasts (n = 39) | Low Level Gymnasts (n = 45) | *p*-Value |
|---|---|---|---|
| EDE-Q total score | 2.07 (1.12, 3.13) | 1.17 (0.44, 1.73) | <0.001 |
| Subscales | | | |
| *Restrain* | 1.40 (0.60, 3.60) | 0.60 (0.20, 1.70) | 0.014 |
| *Eating Concern* | 1.20 (0.40, 2.60) | 0.60 (0.20, 1.20) | 0.009 |
| *Weight Concern* | 2.00 (1.60, 4.20) | 1.00 (0.40, 2.20) | <0.001 |
| *Shape concern* | 2.38 (1.63, 3.75) | 1.25 (0.50, 2.31) | <0.001 |
| RED-S total score | 8.74 ± 3.04 | 8.71 ± 2.98 | 0.961 |

[1] Note: Values are expressed as median (Q1, Q3) for EDE-Q and its subscales and as means ± SD for REDS knowledge score. *p*-values derived from Mann–Whitney rank test and from *t*-test. Statistical significance was set at the 5% level.

The six items of EDE were transformed into dichotomous variables, with 0 indicating the absence of binge eating and compensatory behaviors, and 1 indicating the presence of at least one episode. No differences were found between groups across these six EDE-Q items ($p > 0.05$). In contrast, more high-level artistic gymnasts had missed menstrual cycles in the last 3–4 months ($p < 0.001$), while no differences were observed between groups in contraceptive use ($p = 0.283$). Furthermore, when frequencies were computed for the six EDE-Q items for all the artistic gymnasts combined, it was found that 49 girls (58.3%) were concerned about the amount of food (EDE13) and 48 girls (57.2%) lost control of their eating at least once (EDE14). Forty-two (50%) experienced at least one subjective or objective binge-eating episode (EDE15). Four athletes (4.8%) reported self-induced vomiting as compensatory behavior after binge eating (EDE16). No gymnast reported laxative or diuretic misuse (EDE 17). Thirty-three athletes (39.3%) reported additional excessive training as compensatory behavior to avoid feeling guilty about eating (EDE18).

The percentage of artistic gymnasts scoring higher than the cut-off of 2.612 in EDE-Q was 20.2% (n = 17; 35.9% in high and 6.7% in low-level gymnasts). Those gymnasts scored significantly higher in all EDE-Q subscales ($p < 0.001$), with no significant differences in main characteristics (age, height, weight, BMI, training experience and RED-S knowledge score ($p > 0.05$); however, they were more likely to train at high than low level, compared to those scoring lower than 2.612 ($p < 0.001$).

### *3.2. RelativeED-S Knowledge*

No differences were observed between groups in the total scores of the RED-S knowledge questionnaire ($p = 0.961$) (Table 2). Athletes' responses to each question in this questionnaire are presented in Figure 1 and in Supplementary Table S1. In the domain assessing awareness and definition of LEA, Triad, and RED-S, 79.8 to 92.9% of the participants reported unawareness of these definitions. In the domain assessing knowledge of RED-S signs and symptoms, 54.8 to 86.9% of all artistic gymnasts failed to recognize known RED-S signs and symptoms and in the domain knowledge of the health and performance consequences of RED-S, 15.5 to 52.4% of the participants exhibited limited understanding of the impact of food restriction on their health and performance (Figure 1 and Supplementary Table S1). On average, 51.5% of artistic gymnasts provided incorrect answers related to RED-S.

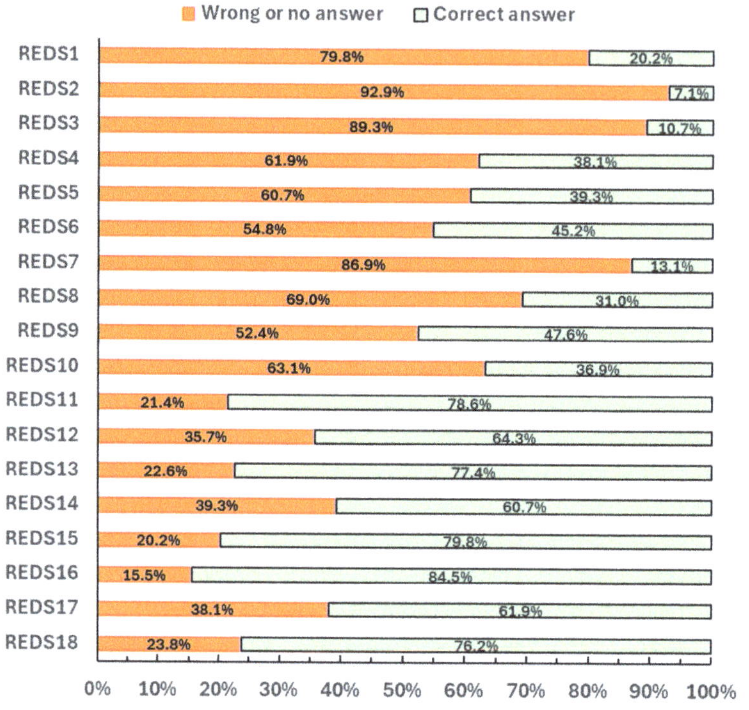


**Figure 1.** Percentage of correct and incorrect responses to the RED-S knowledge questionnaire among gymnasts (n = 84).

## 4. Discussion

This study examined eating disorder symptoms and RED-S knowledge in adolescent artistic gymnasts. A higher percentage of eating disorder symptoms and higher total scores in EDE-Q and its subscales were found in high-level compared to lower-level artistic gymnasts. An important percentage of artistic gymnasts reported at least one episode of binge eating and excessive training. RED-S knowledge did not differ between groups, with most artistic gymnasts unaware of the main aspects of Relative Energy Deficiency in Sports.

It is well-established that eating disorders have a greater prevalence among high-performance athletes as compared with non-elite athletes and the general population, especially in disciplines that emphasize leanness, like gymnastics [10,22,34]. In a previous study with adolescent artistic gymnasts, Tan et al. [22] also reported that disordered eating among elite female artistic gymnasts was higher than that observed in lower-level gymnasts, and the same was found for international-level adolescent rhythmic gymnasts [35]. The finding that a large percentage (35.9%) of high-level adolescent artistic gymnasts in this study demonstrated eating disorder symptoms-based on their EDE-Q scores, is in line with previous research highlighting the need for psychological support and treatment in high-level sport environments [6,12]. In contrast, 6.7% of club and recreational artistic gymnasts scored higher than 2.612 in EDE-Q, a percentage almost similar to that found in Greek adolescents (9.8%) [36]. This shows that moderate exercise and lower performance and appearance demands are linked to reduced occurrence of disturbed eating attitudes compared to high-level competitive athletes (35.9% in the present study), while remaining comparable to the general adolescent population. High-level artistic gymnasts also scored higher than low-level in the subscales *Restraint*, *Eating Concerns*, *Weight Concerns*, and *Shape*

*Concerns.* Artistic gymnasts are required to execute complex technical skills during flight while maintaining difficult body shapes. Thus, the ratio of force and power generation to body mass is a critical performance determinant in gymnastics from an early age, potentially leading youth gymnasts to engage in food restriction [37,38]. Notably, despite having a lower BMI and more extensive training experience, high-level artistic gymnasts in this study demonstrated greater concerns regarding weight and body shape compared to lower-level peers. Heightened preoccupation with weight and a drive for thinness are significant risk factors for the development of eating disorders [37,39,40].

No differences were found between groups regarding binge eating and compensatory behaviors. Over 50% of artistic gymnasts reported overeating with a loss of control and binge eating at least once in the past 28 days, highlighting the potential influence of the gymnastics environment (e.g., negative comments, performance anxiety) in triggering binge eating as a coping mechanism for emotional distress, guilt, or shame [39,41]. Regarding compensatory behaviors, fewer gymnasts reported excessive training or self-induced vomiting, with approximately 4% engaging in self-induced vomiting. Similar percentages of self-vomiting, a symptom associated with severe health, psychological, and, in some cases, life-threatening consequences, have also been reported in previous studies in elite athletes [42,43]. While the athletes' responses in these areas are concerning, these findings must be interpreted with caution. Previous research involving Greek adolescents has highlighted discrepancies between the six items of the EDE-Q and objectively assessed compensatory behaviors during clinical evaluations [30]. Similarly, Mond et al. [44] found that about 42% of the participants who reported self-induced vomiting or laxative misuse on a questionnaire denied these behaviors when they were questioned in a face-to-face interview. Thus, these findings highlight the need to interpret results from self-report inventories, such as the EDE-Q, within the framework of clinical evaluation. While inventories offer practical benefits, including ease of use and preliminary insights, accurate diagnosis of eating disorders requires validation through clinical interviews and comprehensive diagnostic assessments.

RED-S is common among female athletes at different ages and performance levels and results in serious health and performance consequences [45–47]. Exposure to low energy availability is the main underlying etiological factor thus, awareness of the effects of food restriction on metabolic rate, menstrual, bone, cardiovascular, and gastroenterological health, immunity, and overall well-being is of paramount importance and especially for adolescent athletes [48]. The results of this study highlight an interesting finding: irrespective of performance level, most artistic gymnasts reported unawareness of the RED-S definition, and over 50% of them failed to recognize known associations between food restriction and menstrual and bone health. This lack of awareness was not specific to high or low performance levels but rather reflects a broader, systemic gap in knowledge among athletes, and possibly coaches, or parents. Limited understanding of RED-S and its health implications may delay recognition and treatment of serious conditions, placing athletes at risk [49].

The finding that 86.9% of gymnasts responded "No or I don't know" to the question about whether amenorrhea increases the risk of fractures highlights a substantial educational gap among athletes, coaches, and parents. This lack of awareness is especially concerning given the well-established association between menstrual dysfunction and impaired bone health, which significantly increases the risk of stress fractures. Ackerman et al. [50] reported that adolescent athletes with amenorrhea had a markedly higher lifetime incidence of fractures (47%) compared to their eumenorrheic peers (25.7%) and non-athletes (12.5%). In artistic gymnastics, the prevalence of stress fractures is further amplified by the sport's high mechanical loads imposed on skeletally immature athletes. A recent systematic review found that women's gymnastics has one of the highest rates of stress fractures among collegiate sports, with an incidence of 25.58 per 100,000 athlete-exposures [51]. In line with this finding, a previous survey of 712 adolescent and young adult runners, dancers, and figure skaters found that only 12% had heard of the female athlete triad and only 7% were able to name two of the three components of the triad [52]. Anecdotally, over 60% of the artistic gymnasts in this study reported that it is normal for female athletes to miss their period when they are following intensive training regimes. This result aligns with the finding that high-level artistic gymnasts reported a greater number of missed menstrual cycles compared to low-level peers over the past 3–4 months. At the elite level, misconceptions regarding menstruation are prevalent, with evidence suggesting that 28–56% of female adolescent athletes believe that the absence of menstruation is a normal consequence of intense athletic training [53,54]. Several studies on female athletes highlight a culture that encourages the pursuit of the ideal body through restrictive eating and excessive training, cultivating a "lighter is better" mentality from the early stages of their careers, while often neglecting the significance of menstrual disturbances [54,55]. Nevertheless, artistic gymnasts in this study demonstrated a better understanding of the effects of severe food restriction on immunity and neuromuscular performance, likely influenced by feedback from coaches and parents or their own personal experiences. It is plausible that malnourished athletes face performance difficulties due to weakness and energy depletion.

Although this study is the first to examine awareness of energy deficiency and eating disorder symptoms in adolescent artistic gymnasts, there are some limitations that should be acknowledged. EDE-Q, when using its global score, is a proper screening tool for assessing the core psychopathology of eating disorders [30,33]. However, the assessment of pathological behaviors using the six items of the EDE-Q has limitations within the Greek population, as adolescents may misinterpret terms such as "large amount of food" and "loss of control" or misunderstand the motivation behind excessive exercise [30,33]. They may perceive the term as referring to additional exercise for physical fitness rather than compensatory behavior. Therefore, it is crucial to recognize that the results obtained from these self-report questions should be cross-validated with findings from face-to-face interviews. A second limitation is that this study relied solely on self-report instruments. Thus, a comprehensive clinical evaluation is necessary to accurately assess the presence of eating disorders. In addition, data on menstruation were derived solely from self-reported responses to non-standardized items included in the EDE-Q. These items, which asked about menstrual frequency and contraceptive use over the past 3–4 months, may be subject to recall bias or misreporting, potentially affecting the accuracy and reliability of the menstrual data collected. A final limitation is the small sample size of high-level adolescent artistic gymnasts and unequal recruitment methods, which restrict the statistical power to detect small or medium effect sizes and may introduce bias. However, it should be reported that the sample of high-level gymnasts (n = 39) comprised all female artistic gymnasts in Greece who were members of the national team and trained at the two high-performance centers located in Athens and Thessaloniki. For comparison purposes, a demographically similar group of club and recreational-level artistic gymnasts (n = 45) was randomly selected from three different gymnastics clubs. The inclusion of a large sample of elite athletes was crucial for investigating sensitive issues like eating disorders and RED-S awareness.

Nevertheless, the findings of this study highlight the need for the gymnastics community to educate athletes and coaches on the importance of maintaining adequate energy availability to optimize both health and performance. It is crucial to teach athletes effective fueling strategies for different training durations and intensities, as well as for growth [26]. Adolescent aesthetic athletes who fail to recognize potential risks may gradually develop pathological behaviors, eventually losing control over their actions, which can compromise both their health and performance. The treatment of RED-S and low energy availability necessitates addressing the underlying causes [25]. Researchers and specialists should focus on strengthening protective factors (e.g., enhancing self-esteem and a positive body image, encouraging acceptance of physical changes during adolescence, and improving media literacy) while reducing risk factors (e.g., internalization of the stereotype of the "ideal body", peer pressure, and fat shaming), involving the entire athlete support system [25,26]. In addition, long-term coach education programs, especially those offered at the university level or through national sport federations, should include modules on RED-S and training load management. To further promote awareness, workshops and digital resources should also be made available to parents and athletes through National Olympic Committees and sport federations. Increasing knowledge in these areas can help identify additional educational needs and support athletes, coaches, and sports medicine professionals in making informed decisions about training, competition, and recovery. Failure to recognize these risks may not only impact artistic gymnasts' eating behaviors but also hinder the early detection and management of RED-S, where timely intervention may be required.

## 5. Conclusions

A higher prevalence of eating disorder symptoms was observed among high-level artistic gymnasts compared to their lower-level counterparts, suggesting that reduced performance and appearance-related pressures may be associated with fewer disordered eating behaviors. The finding that RED-S knowledge did not differ between groups—and that most artistic gymnasts were unaware of the core aspects of Relative Energy Deficiency in Sport—is both novel and concerning. This widespread lack of awareness was not limited to athletes of a specific performance level, but instead highlights a broader, systemic gap in understanding that may extend to coaches and parents as well.

**Supplementary Materials:** The following supporting information can be downloaded at: https://www.mdpi.com/article/10.3390/nu17101699/s1. Table S1: Gymnasts' responses to the RED-S knowledge questionnaire (n = 84).

**Author Contributions:** Conceptualization, A.D., O.D., M.I.M. and M.P.; methodology, O.D., M.I.M. and M.P.; investigation, A.D.; data curation, O.D., M.I.M., A.D. and M.P.; writing—original draft preparation, A.D.; writing—review and editing, O.D. and M.I.M.; supervision, O.D. All authors have read and agreed to the published version of the manuscript.

**Funding:** This research received no external funding.

**Institutional Review Board Statement:** The study was conducted in accordance with the Declaration of Helsinki and approved by the Institutional Ethics Committee of School of Physical Education & Sport Science, National & Kapodistrian University of Athens, Greece [registration number: 1645/15-05-2024, approval date: 15 May 2024].

**Informed Consent Statement:** Written informed consent was obtained from the parents and the adolescent gymnasts involved in the study.

**Data Availability Statement:** All data will be made available on request to the corresponding author.

**Acknowledgments:** The authors would like to thank the athletes and their parents for their dedicated time and commitment throughout this study.

**Conflicts of Interest:** The authors declare no conflicts of interest.

## Abbreviations

The following abbreviations are used in this manuscript:

| | |
|---|---|
| RED-S | Relative Energy Deficiency Syndrome |
| EDE-Q | Eating Disorder Examination Questionnaire |
| LEA | Low energy availability |

## References


1. American Psychiatric Association. *Diagnostic and Statistical Manual of Mental Disorders*, 5th ed.; American Psychiatric Association: Washington, DC, USA, 2013. [CrossRef]
2. Fisher, M. Treatment of eating disorders in children, adolescents, and young adults. *Pediatr. Rev.* **2006**, *27*, 5–16. [CrossRef] [PubMed]
3. Garner, D. Measurement of eating disorder psychopathology. In *Eating Disorders and Obesity: A Comprehensive Handbook*, 2nd ed.; Fairburn, C.G., Brownell, K.D., Eds.; Guilford Press: New York, NY, USA, 2002; pp. 141–146.
4. Mountjoy, M.; Ackerman, K.E.; Bailey, D.M.; Burke, L.M.; Constantini, N.; Hackney, A.C.; Heikura, I.A.; Melin, A.; Pensgaard, A.M.; Stellingwerff, T.; et al. International Olympic Committee's (IOC) consensus statement on relative energy deficiency in sport (REDs). *Br. J. Sports Med.* **2023**, *57*, 1073–1098. [CrossRef] [PubMed]
5. Angelidi, A.M.; Stefanakis, K.; Chou, S.H.; Valenzuela-Vallejo, L.; Dipla, K.; Boutari, C.; Ntoskas, K.; Tokmakidis, P.; Kokkinos, A.; Goulis, D.G.; et al. Relative energy deficiency in sport (REDs): Endocrine manifestations, pathophysiology and treatments. *Endocr. Rev.* **2024**, *45*, 676–708. [CrossRef] [PubMed]
6. Sundgot-Borgen, J. Eating disorders, energy intake, training volume, and menstrual function in high-level modern rhythmic gymnasts. *Int. J. Sport Nutr.* **1996**, *6*, 100–109. [CrossRef]
7. Okano, G.; Holmes, R.A.; Mu, Z.; Yang, P.; Lin, Z.; Nakai, Y. Disordered eating in Japanese and Chinese female runners, rhythmic gymnasts, and gymnasts. *Int. J. Sports Med.* **2005**, *26*, 486–491. [CrossRef]
8. Anderson, C.; Petrie, T.A. Prevalence of disordered eating and pathogenic weight control behaviors among NCAA Division I female collegiate gymnasts and swimmers. *Res. Q. Exerc. Sport* **2012**, *83*, 120–124. [CrossRef]
9. de Bruin, A.P.; Oudejans, R.R.D.; Bakker, F.C. Dieting and body image in aesthetic sports: A comparison of Dutch female gymnasts and non-aesthetic sport participants. *Psychol. Sport Exerc.* **2007**, *8*, 507–520. [CrossRef]
10. Van Durme, K.; Goossens, L.; Braet, C. Adolescent aesthetic athletes: A group at risk for eating pathology? *Eat. Behav.* **2012**, *13*, 119–122. [CrossRef]
11. Kontele, I.; Vassilakou, T.; Donti, O. Weight pressures and eating disorder symptoms among adolescent female gymnasts of different performance levels in Greece. *Children* **2022**, *9*, 254. [CrossRef]
12. Reardon, C.L.; Hainline, B.; Aron, C.M.; Baron, D.; Baum, A.L.; Bindra, A.; Budgett, R.; Campriani, N.; Castaldelli-Maia, J.M.; Currie, A.; et al. Mental health in elite athletes: International Olympic Committee consensus statement. *Br. J. Sports Med.* **2019**, *53*, 667–699. [CrossRef]
13. Byrne, S.; McLean, N. Elite athletes: Effects of the pressure to be thin. *J. Sci. Med. Sport* **2002**, *5*, 80–94. [CrossRef] [PubMed]
14. Sundgot-Borgen, J.; Torstveit, M.K. Prevalence of eating disorders in elite athletes is higher than in the general population. *Clin. J. Sport Med.* **2004**, *14*, 25–32. [CrossRef] [PubMed]
15. Coelho, G.M.; Soares Ede, A.; Ribeiro, B.G. Are female athletes at increased risk for disordered eating and its complications? *Appetite* **2010**, *55*, 379–387. [CrossRef]
16. Golden, N.H.; Schneider, M.; Wood, C. Preventing obesity and eating disorders in adolescents. *Pediatrics* **2016**, *138*, e20161649. [CrossRef]
17. Ward, Z.J.; Rodriguez, P.; Wright, D.R.; Austin, S.B.; Long, M.W. Estimation of eating disorders prevalence by age and associations with mortality in a simulated nationally representative US cohort. *JAMA Netw. Open* **2019**, *2*, e1912925. [CrossRef]
18. Malina, R.M.; Baxter-Jones, A.D.G.; Armstrong, N.; Beunen, G.P.; Caine, D.; Daly, R.M.; Lewis, R.D.; Rogol, A.D.; Russell, K. Role of intensive training in the growth and maturation of artistic gymnasts. *Sports Med.* **2013**, *43*, 783–802. [CrossRef]
19. Donti, O.; Bogdanis, G.C.; Kritikou, M.; Donti, A.; Theodorakou, K. The relative contribution of physical fitness to the technical execution score in youth rhythmic gymnastics. *J. Hum. Kinet.* **2016**, *51*, 143–152. [CrossRef]
20. Desbrow, B.; McCormack, J.; Burke, L.M.; Cox, G.R.; Fallon, K.; Hislop, M.; Logan, R.; Marino, N.; Sawyer, S.M.; Shaw, G.; et al. Sports Dietitians Australia position statement: Sports nutrition for the adolescent athlete. *Int. J. Sport Nutr. Exerc. Metab.* **2014**, *24*, 570–584. [CrossRef]

21. Bingham, M.E.; Borkan, M.; Quatromoni, P. Sports nutrition advice for adolescent athletes: A time to focus on food. *Am. J. Lifestyle Med.* **2015**, *9*, 398–402. [CrossRef]
22. Tan, J.O.A.; Calitri, R.; Bloodworth, A.; McNamee, M.J. Understanding eating disorders in elite gymnastics: Ethical and conceptual challenges. *Clin. Sports Med.* **2016**, *35*, 275–292. [CrossRef]
23. Jagim, A.R.; Fields, J.; Magee, M.K.; Kerksick, C.M.; Jones, M.T. Contributing factors to low energy availability in female athletes: A narrative review of energy availability, training demands, nutrition barriers, body image, and disordered eating. *Nutrients* **2022**, *14*, 986. [CrossRef] [PubMed]
24. Stenqvist, T.B.; Melin, A.K.; Torstveit, M.K. Relative Energy Deficiency in Sport (REDs) indicators in male adolescent endurance athletes: A 3-year longitudinal study. *Nutrients* **2023**, *15*, 5086. [CrossRef] [PubMed]
25. Torstveit, M.K.; Ackerman, K.E.; Constantini, N.; Holtzman, B.; Koehler, K.; Mountjoy, M.L.; Sundgot-Borgen, J.; Melin, A. Primary, secondary and tertiary prevention of Relative Energy Deficiency in Sport (REDs): A narrative review by a subgroup of the IOC consensus on REDs. *Br. J. Sports Med.* **2023**, *57*, 1119–1128. [CrossRef]
26. Torstveit, M.K. The Female Athlete Triad in Norwegian Elite Athletes and Non-Athletic Controls: Identification and Prevalence of Disordered Eating, Menstrual Dysfunction, and Osteoporosis. Doctoral Dissertation, Norwegian School of Sport Sciences, Oslo, Norway, 2004.
27. Burke, L.M.; Lundy, B.; Fahrenholtz, I.L.; Melin, A.K. Pitfalls of conducting and interpreting estimates of energy availability in free-living athletes. *Int. J. Sport Nutr. Exerc. Metab.* **2018**, *28*, 350–363. [CrossRef]
28. Fostervold-Mathisen, T.; Ackland, T.; Burke, L.M.; Constantini, N.; Haudum, J.; Macnaughton, L.S.; Meyer, N.L.; Mountjoy, M.; Slater, G.; Sungot-Borgen, J. Best practice recommendations for body composition considerations in sport to reduce health and performance risks: A critical review, original survey and expert opinion by a subgroup of the IOC consensus on Relative Energy Deficiency in Sport (REDs). *Br. J. Sports Med.* **2023**, *57*. [CrossRef]
29. Fairburn, C.G.; Beglin, S. Eating Disorder Examination Questionnaire (EDE-Q 6.0). In *Cognitive Behavior Therapy and Eating Disorders*; Fairburn, C.G., Ed.; Guilford Press: New York, NY, USA, 2008; pp. 309–314.
30. Pliatskidou, S.; Samakouri, M.; Kalamara, E.; Papageorgiou, E.; Koutrouvi, K.; Goulemtzakis, C.; Nikolaou, E.; Livaditis, M. Validity of the Greek Eating Disorder Examination Questionnaire 6.0 (EDE-Q-6.0) among Greek adolescents. *Psychiatriki* **2015**, *26*, 204–216.
31. Pai, N.N.; Brown, R.C.; Black, K.E. The development and validation of a questionnaire to assess relative energy deficiency in sport (RED-S) knowledge. *J. Sci. Med. Sport* **2022**, *25*, 794–799. [CrossRef]
32. Fairburn, C.G.; Cooper, P.J. The Eating Disorder Examination. In *Binge Eating: Nature, Assessment And treatment*, 12th ed.; Fairburn, C.G., Wilson, G.T., Eds.; Guilford Press: New York, NY, USA, 1993; pp. 317–360.
33. Pliatskidou, S.; Samakouri, M.; Kalamara, E.; Goulemtsakis, C.; Koutrouvi, K.; Papageorgiou, E.; Livadites, M. Reliability of the Greek version of the Eating Disorder Examination Questionnaire (EDE-Q) in a sample of adolescent students. *Psychiatriki* **2012**, *23*, 295–301.
34. Bratland-Sanda, S.; Sundgot-Borgen, J. Eating disorders in athletes: Overview of prevalence, risk factors, and recommendations for prevention and treatment. *Eur. J. Sport Sci.* **2013**, *13*, 499–508. [CrossRef]
35. Donti, O.; Donti, A.; Gaspari, V.; Pleksida, P.; Psychountaki, M. Are they too perfect to eat healthy? Association between eating disorder symptoms and perfectionism in adolescent rhythmic gymnasts. *Eat. Behav.* **2021**, *41*, 101514. [CrossRef]
36. Kyrgios, I.; Papageorgiou, V.; Kotanidou, E.; Kokka, P.; Kleisarchaki, A.; Mouzaki, K.; Tsara, I.; Efstratiou, E.; Haidich, A.-B.; Galli-Tsinopoulou, A. Eating Disorders in Greek Adolescents: Frequency and Characteristics. In Proceedings of the 54th Annual ESPE, Barcelone, Spain, 1–3 October 2015; Volume 84.
37. Bacciotti, S.; Baxter-Jones, A.; Gaya, A.; Maia, J. The physique of elite female artistic gymnasts: A systematic review. *J. Hum. Kinet.* **2017**, *58*, 247–259. [CrossRef] [PubMed]
38. Claessens, A.L.; Lefevre, J.; Beunen, G.; Malina, R.M. The contribution of anthropometric characteristics to performance scores in elite female gymnasts. *J. Sports Med. Phys. Fit.* **1999**, *39*, 355–360.
39. de Bruin, A.P.K.; Oudejans, R.R.D.; Bakker, F.C.; Woertman, L. Contextual body image and athletes' disordered eating: The contribution of athletic body image to disordered eating in high-performance women athletes. *Eur. Eat. Disord. Rev.* **2011**, *19*, 201–215. [CrossRef]
40. Donti, O.; Theodorakou, K.; Kambiotis, S.; Donti, A. Self-esteem and trait anxiety in girls practicing competitive and recreational gymnastics. *Sci. Gymnast. J.* **2012**, *4*, 6–12. [CrossRef]
41. Kristjánsdóttir, H.; Sigurðardóttir, P.; Jónsdóttir, S.; Þorsteinsdóttir, G.; Saavedra, J. Body image concern and eating disorder symptoms among elite Icelandic athletes. *Int. J. Environ. Res. Public Health* **2019**, *16*, 2728. [CrossRef]
42. Killen, J.D.; Taylor, C.B.; Telch, M.J.; Saylor, K.E.; Maron, D.J.; Robinson, T.N. Self-induced vomiting and laxative and diuretic use among teenagers. *J. Am. Med. Assoc.* **1986**, *255*, 1447–1449. [CrossRef]

43. Rosen, L.W.; McKeag, D.B.; Hough, D.O.; Curley, V. Pathogenic weight-control behavior in female athletes. *Physician Sportsmed.* **1986**, *14*, 79–86. [CrossRef]
44. Mond, J.M.; Hay, P.J.; Rodgers, B.; Owen, C.; Mitchell, J.E. Correlates of self-induced vomiting and laxative misuse in a community sample of women. *J. Nerv. Ment. Dis.* **2006**, *194*, 40–46. [CrossRef]
45. Wasserfurth, P.; Palmowski, J.; Hahn, A.; Krüger, K. Reasons for and consequences of low energy availability in female and male athletes: Social environment, adaptations, and prevention. *Sports Med. Open* **2020**, *6*, 44. [CrossRef]
46. Torstveit, M.K.; Fahrenholtz, I.L.; Lichtenstein, M.B.; Stenqvist, T.B.; Melin, A.K. Exercise dependence, eating disorder symptoms, and biomarkers of relative energy deficiency in sports (RED-S) among male endurance athletes. *BMJ Open Sport Exerc. Med.* **2019**, *5*, e000439. [CrossRef]
47. Villa, M.; Villa-Vicente, J.G.; Seco-Calvo, J.; Mielgo-Ayuso, J.; Collado, P.S. Body composition, dietary intake, and the risk of low energy availability in elite-level competitive rhythmic gymnasts. *Nutrients* **2021**, *13*, 2083. [CrossRef] [PubMed]
48. Logue, D.M.; Madigan, S.M.; Melin, A.; Delahunt, E.; Heinen, M.; Donnell, S.-J.M.; Corish, C.A. Low energy availability in athletes 2020: An updated narrative review of prevalence, risk, within-day energy balance, knowledge, and impact on sports performance. *Nutrients* **2020**, *12*, 835. [CrossRef] [PubMed]
49. Mountjoy, M.; Sundgot-Borgen, J.; Burke, L.; Carter, S.; Constantini, N.; Lebrun, C.; Meyer, N.; Sherman, R.; Steffen, K.; Budgett, R.; et al. The IOC consensus statement: Beyond the female athlete triad—Relative energy deficiency in sport (RED-S). *Br. J. Sports Med.* **2014**, *48*, 491–497. [CrossRef] [PubMed]
50. Ackerman, K.E.; Sokoloff, N.C.; Maffazioli, G.D.N.; Clarke, H.; Lee, H.; Misra, M. Fractures in relation to menstrual status and bone parameters in young athletes. *Med. Sci. Sports Exerc.* **2015**, *47*, 1577. [CrossRef]
51. Campbell, R.A.; Bradshaw, E.J.; Ball, N.B.; Pease, D.L.; Spratford, W. Injury epidemiology and risk factors in competitive artistic gymnasts: A systematic review. *Br. J. Sports Med.* **2019**, *53*, 1056–1069. [CrossRef]
52. Tosi, M.; Maslyanskaya, S.; Dodson, N.A.; Coupey, S.M. The female athlete triad: A comparison of knowledge and risk in adolescent and young adult figure skaters, dancers, and runners. *J. Pediatr. Adolesc. Gynecol.* **2019**, *32*, 165–169. [CrossRef]
53. Torstveit, M.K.; Sundgot-Borgen, J. Participation in leanness sports but not training volume is associated with menstrual dysfunction: A national survey of 1276 elite athletes and controls. *Br. J. Sports Med.* **2005**, *39*, 141–147. [CrossRef]
54. Calcaterra, V.; Vandoni, M.; Bianchi, A.; Pirazzi, A.; Tiranini, L.; Baldassarre, P.; Diotti, M.; Cavallo, C.; Nappi, R.E.; Zuccotti, G. Menstrual dysfunction in adolescent female athletes. *Sports* **2024**, *12*, 245. [CrossRef]
55. Sherman, R.T.; Thompson, R.A.; Rose, J.S. Body mass index and athletic performance in elite female gymnasts. *J. Sport Behav.* **1996**, *19*, 338–346.

*nutrients*

MDPI

*Article*

# Lifestyle Behaviors Associated with Weight Loss Intent in Adolescent Girls: Findings from the US 2021 National Youth Risk Behavior Survey

**Elinor Fondell *, Jaime Vallejos, Andrew J. Piazza and Mariana C. Calle**

Department of Health Sciences, Worcester State University, Worcester, MA 01602, USA; mcalle@worcester.edu (M.C.C.)
* Correspondence: efondell@worcester.edu

**Abstract: Background**: External social influences on body image affect females differently than males, and adolescent girls are more likely to want to change their weight status. Understanding the healthy and unhealthy habits of adolescent girls is vital for developing effective and targeted health promotions and interventions. **Methods**: Using data from the 2021 Youth Risk Behavior Surveillance System (YRBS) survey, this cross-sectional study compares dietary habits, physical activity, vaping, alcohol use, sleep, and screen time in females (9th–12th grades) who intend to lose weight versus those who want to maintain their current weight. **Results**: The sample consisted of 4362 females, of which 56.7% reported an intent to lose weight. The average BMI percentile was 64.1 compared to 75.4 among those trying to lose weight and 50.1 among those not trying to lose weight. Adolescent girls intending to lose weight also reported less frequent breakfast consumption (OR 0.52; 0.40–0.69), less sleep (OR 0.72; 0.59–0.89), more screen time (OR 1.27; 1.02–1.58), engaging in muscle toning exercises (OR 1.30; 1.07–1.57), vaping (OR 1.22; 1.01–1.47), and alcohol use (OR 1.61; 1.32–1.98) compared to those not intending to lose weight. **Conclusions**: Adolescent girls trying to lose weight would likely benefit from interventions to help them improve sleep, reduce screen time, improve dietary and exercise habits, and monitor alcohol and vaping use.

**Keywords:** weight loss; weight control; weight dissatisfaction; female adolescents; dietary behaviors; sleep; screen time; physical activity; breakfast; girls' lifestyle

## 1. Introduction

Obesity prevalence reached 22.2% among adolescents aged 12–19, based on the latest National Health Report released in 2021 [1]. Around 80% of obese adolescents carry it into adulthood, leading to health risks such as metabolic tissue dysfunction, low-grade systemic inflammation, cardiovascular diseases, and type 2 diabetes [2,3].

External social influences on body image affect females differently than males [4]. Weight dissatisfaction is particularly prevalent among girls, ranging from 19.2 to 83.4% among 10–19-year-old girls in an international systematic review [5]. Perception of body weight plays a role in the motivation to lose weight in young adults, and physically active adolescents have reported a better body perception [6,7]. However, there is less information regarding physical activity in adolescent girls intending to lose weight. The PA Guidelines for Americans recommend 60 min (1 h) or more of moderate-to-vigorous physical activity daily, including aerobic, muscle-strengthening and bone-strengthening exercise for children and adolescents [8]. Unfortunately the adherence to these recommendations in adolescents is low, approximately 23.9% adherence overall, and significantly lower for girls at 15.7% [9].

The Healthy Eating Index (HEI) score for boys and girls aged 14–18 is 51/100 [10]. The average vegetable intake for girls that age is less than 1 serving per day. Adhering to the Dietary Guidelines for Americans increases the likelihood of maintaining body weight since it encompasses balance, moderation, variety, and calorie control.

The American Academy of Sleep Medicine recommends that teenagers aged 13 to 18 sleep 8 to 10 h per 24 h regularly to promote optimal health [11]. There is a consistent association between screen time and sleep in adolescents [12]. Data from the longitudinal, U.S.-based Adolescent Brain Cognitive Development (ABCD) study showed that females recorded about 1 h more average daily app and smartphone use than males [13]. Coincidentally, the Centers for Disease Control and Prevention (CDC) reported that the percentage of high school students who do not get enough sleep is higher among girls (80%) compared to boys (75%) [14]. Furthermore, inadequate sleep (including both short duration and poor quality) is associated with overweight and obesity in adolescents [15]. Adolescents are also susceptible to direct pulmonary injury, addiction, and other health risks from e-cigarettes [16]. According to the Monitoring the Future survey report, by the National Institute of Health, Institute on Drug Abuse, 27% of 12th grade students had vaped nicotine within the last 12 months in 2022, while 52% of students had used alcohol within the last 12 months in 2022 [17].

While there have been some descriptive studies on adolescents' habits [9,13,18–20], to date, a comprehensive analysis of lifestyle habits among girls intending to lose weight using a representative US sample has been limited [21–23].

Understanding the health-related behavior of girls intending to reduce their weight can inform strategies for addressing body weight management in this population. Using a national representative sample, this cross-sectional study compares dietary habits, physical activity, vaping, alcohol use, sleep, and screen time in females (9th–12th grades) who intend to lose weight versus those who want to maintain their current weight. Additionally, this study describes the relationship between actual BMI (using self-reported weight) and weight perception between the two groups of girls (weight loss intent vs. maintaining weight).

## 2. Materials and Methods

### 2.1. Study Design

The 2021 Youth Risk Behavior Survey (YRBS) is part of a biennial cross-sectional survey conducted by the CDC since 1991 to monitor the prevalence of health risk behaviors among 9th through 12th grade students in public and private schools in the United States. YRBS utilizes a three-stage cluster sample design to generate a representative sample. Each student record was assigned a weighting factor based on student sex, race, ethnicity, and grade to adjust for nonresponse and the oversampling of black and Hispanic students in the sample. The weighted count of students equals the total sample size, and the weighted proportions match the national population proportions.

The 2021 YRBS covered 152 schools, received 17,232 usable questionnaires, and had an overall response rate of 57.5% (72.7% school response rate and 79.1% student response rate) [24]. Of the 87 questions covered in the questionnaire, only 58 were required, and schools could customize the survey. Question Q67 regarding weight loss was an optional question and was completed by 9273 students, of which 4362 were girls. The test–retest reliability has been shown to be strong for the questions in YRBS [25]. The institutional review board at the CDC approved the protocol for the YRBS. The YRBS has been described in more detail here [24]. In addition, the present authors obtained IRB approval at their institution to conduct this secondary data analysis.

### 2.2. YRBS Measures

The dependent variable, intent to lose weight, was measured as "Which of the following are you trying to do about your weight" (lose weight, gain weight, stay the same weight, I am not trying to do anything about my weight). For the main analyses, this question was dichotomized into (trying to lose weight, not trying to lose weight).

A healthy diet was measured by looking at vegetable intake, fruit intake, breakfast consumption, soda, and milk intake. Dichotomized versions were used for the variables fruit, soda, milk, and breakfast consumption (yes or no regarding 1 or more fruit per day during the past 7 days, 1 or more soda [not including diet soda] per day during the last 7 days, 1 or more glass of milk per day during the previous 7 days, eating breakfast on all 7 days before the survey).

We used two measures to assess physical activity: muscle strengthening ("During the past 7 days, on how many days did you exercise to strengthen or tone your muscles, such as push-ups, sit-ups, or weight lifting?") and being physically active for 60 min per day ("During the past 7 days, on how many days were you physically active for a total of at least 60 min per day?"). Screen time was measured as time spent in front of a TV, computer, smartphone, or other electronic device watching shows or videos, playing games, accessing the Internet, or using social media on an average school day, not counting time spent doing schoolwork. This variable was dichotomized into spending 3 or more hours per day on screen time (yes, no). Sleep was dichotomized into getting 8 h or more of sleep on an average school night (yes, no). Alcohol use was dichotomized as drinking at least one drink on at least 1 day during the last 30 days before the survey (yes, no). Similarly, vaping was assessed as using electronic vapor products (including e-cigarettes, vapes, vape pens, e-cigars, e-hookahs, hookah pens, and mods) on at least 1 day during the 30 days before the survey. The BMI percentile was available in the YRBS data set and calculated using the 2000 CDC growth charts [26].

Body size perception was derived from a question regarding how they view their weight (very underweight, slightly underweight, about the right weight, slightly overweight, very overweight) and comparing it to their BMI percentile (overweight/obese $\geq$ 85th BMI percentile). Body size perception was then categorized into underestimator (those who perceive themselves lighter than their actual weight), accurate estimator (their weight perception is close to their actual weight classification), and overestimator (their body weight perception is above their actual weight).

### 2.3. Data Analysis

All variables were weighted to adjust for nonresponse and oversampling. SPSS complex samples were used to account for the YRBS survey design and weighting. Logistic regression was used to assess the association between dietary variables, physical activity variables, sleep, screen time, alcohol intake, vaping, BMI percentile, and weight loss intent. The fit of the model was assessed by examining ROC curves (AUC = 0.78), which indicated a good fit and good discrimination between groups.

To assess whether associations differed among girls with good versus poor body perception, we conducted analyses stratified by body size perception.

SPSS v29 was used for all statistical analyses (IBM Corp. Released 2023. IBM SPSS Statistics for Windows, Version 29.0.2.0 Armonk, NY, USA: IBM Corp). Statistical significance was determined at the 5% level (2-sided).

## 3. Results

### 3.1. Participant Characteristics

The mean BMI percentile among the girls in our sample was 64.2 (95% CI 62.4–65.9). In total, 55.5% (95% CI 53.6–57.4) wanted to lose weight. This intent was the highest among girls of Hispanic (66.3%) or multiple-Hispanic descent (59.9%) and lowest among girls of Asian descent (50.5%) or 'Other' descent (50.0%) (Table 1).

**Table 1.** Demographic characteristics of adolescent girls by weight loss intent status (2021 YRBS).

| | Total | | Trying to Lose Weight | | Not Trying to Lose Weight | |
|---|---|---|---|---|---|---|
| | Unweighted *N* | Weighted % (95% CI) Mean (SE) | Unweighted *N* | Weighted % (95% CI) Mean (SE) | Unweighted *N* | Weighted % (95% CI) Mean (SE) |
| Age | | | | | | |
| <14 years old | 819 | 20.4% (18.8–22.1) | 448 | 20.2% (18.1–22.4) | 371 | 20.7% (19.1–22.4) |
| 15–16 years old | 2192 | 49.8% (48.4–51.2) | 1229 | 50.2% (48.2–52.1) | 963 | 49.4% (46.6–52.1) |
| >17 years old | 1347 | 29.5% (28.3–31.3) | 769 | 29.7% (27.6–31.8) | 578 | 30.0% (27.4–32.7) |
| Grade | | | | | | |
| 9th | 1058 | 24.8% (23.4–26.2) | 586 | 24.9% (23.0–26.9) | 472 | 24.6% (22.9–26.7) |
| 10th | 1127 | 25.2% (23.2–27.3) | 619 | 24.6% (22.1–27.2) | 508 | 26.0% (23.5–28.7) |
| 11th | 1068 | 25.5% (23.9–27.2) | 595 | 25.4% (22.6–28.5) | 473 | 25.6% (23.3–28.1) |
| 12th | 1102 | 24.5% (23.1–25.9) | 644 | 24.9% (23.2–26.7) | 458 | 23.9% (21.6–26.5) |
| Race/Ethnicity | | | | | | |
| Asian | 259 | 5.6% (3.2–9.8) | 130 | 5.0% (3.0–8.1) | 129 | 6.4% (3.3–12.3) |
| Black | 713 | 12.2% (8.6–16.9) | 372 | 11.4% (7.9–16.3) | 341 | 13.1% (9.4–18.0) |
| White | 2033 | 48.5% (43.4–53.7) | 1142 | 48.2% (43.2–53.1) | 891 | 49.0% (43.1–54.9) |
| Hispanic | 346 | 9.8% 7.2–13.2) | 227 | 11.4% (8.3–15.5) | 119 | 7.7% (5.7–10.2) |
| Multiple Hispanic | 637 | 17.1% (14.9–19.5) | 383 | 18.0% (15.8–20.5) | 254 | 15.9% (13.1–19.1) |
| Other (American Indian, Alaska Native, Native Hawaii, Multiple-Non-Hispanic) | 325 | 6.8% (5.1–9.1) | 172 | 6.0% (4.3–8.3) | 153 | 7.9% (5.9–10.5) |

### 3.2. Weight Loss Intent and Lifestyle Behaviors

Results from the weighted logistic regression models are presented in Table 2 and were adjusted for race/ethnicity, grade level, BMI percentile, vegetable consumption, muscle strengthening, sleep, screen time, alcohol intake, and vaping. Girls trying to lose weight were more likely to report engaging in muscle-strengthening exercises (OR 1.30, 95% CI 1.07–1.57), spending 3 h or more per day using screens (OR 1.27; 1.02–1.58), drinking alcohol (OR 1.61; 1.32–1.98), and vaping (OR 1.22; 1.01–1.47). Girls trying to lose weight were less likely to report eating breakfast daily (OR 0.52; 0.40–0.69), eating other vegetables (not including potatoes, salad, or carrots) (OR 0.78; 0.67–0.90), and getting 8 h of sleep (OR 0.72; 0.59–0.89) (Table 2).

**Table 2.** Adjusted odds ratios for lifestyle behaviors associated with weight loss intent among adolescent girls (2021 YRBS).

| | Weighted *N* | Weighted % | Adjusted Odds Ratio (OR) [1] | 95% Confidence Interval (CI) |
|---|---|---|---|---|
| Dietary intake | | | | |
| No daily breakfast intake | 1951 | 44.9% | 1.00 | Reference |
| Daily breakfast intake | 2398 | 55.1% | 0.52 | 0.40–0.69 |
| No daily vegetable intake (green salad, carrots, or other vegetables excluding potatoes) | 3295 | 76.0% | 1.00 | Reference |
| Daily vegetable intake (green salad, carrots, or other vegetables excluding potatoes) | 1039 | 24.0% | 0.91 | 0.77–1.08 |
| No daily green salad intake | 4036 | 92.9% | 1.00 | Reference |
| Daily green salad intake | 309 | 7.1% | 1.31 | 0.73–2.35 |
| No daily carrot intake | 4158 | 96.2% | 1.00 | Reference |
| Daily carrot intake | 166 | 3.8% | 1.01 | 0.61–1.66 |
| No daily intake of other vegetables (not including salad, carrots, or potatoes) | 3500 | 80.4% | 1.00 | Reference |
| Daily other vegetable intake (not including salad, carrots, or potatoes) | 854 | 19.6% | 0.78 | 0.67–0.90 |
| No daily fruit intake | 3104 | 71.3% | 1.00 | Reference |
| Daily fruit intake | 1247 | 28.7% | 0.91 | 0.74–1.12 |
| No daily milk intake | 722 | 16.6% | 1.00 | Reference |
| Daily milk intake | 3626 | 83.4% | 0.90 | 0.61–1.32 |
| No daily soda intake (not including diet) | 3835 | 88.1% | 1.00 | Reference |
| Daily soda intake (not including diet) | 517 | 11.9% | 0.99 | 0.79–1.23 |
| Physical activity | | | | |
| Less than 3 days per week of muscle-strengthening activities | 2939 | 67.6% | 1.00 | Reference |
| Engage in muscle-strengthening activities on 3 days or more per week | 1411 | 32.4% | 1.30 | 1.07–1.57 |
| Less than 5 days per week of 60-min physical activity | 2783 | 62.8% | 1.00 | Reference |
| Physically active for 60 min or more on 5 or more days per week | 1652 | 37.2% | 1.09 | 0.82–1.45 |
| Did not play on a sports team | 2257 | 51% | 1.00 | Reference |
| Played on 1 or more sports team during the last 12 months | 2169 | 49% | 1.11 | 0.97–1.26 |
| Screen time | | | | |
| Did not spend 3 or more hours per day using screens | 893 | 20.5% | 1.00 | Reference |
| Spent 3 h or more using screens per school day (not including schoolwork) (yes vs. no) | 3457 | 79.5% | 1.27 | 1.02–1.58 |
| Sleep | | | | |
| Slept less than 8 h | 3449 | 79.3% | 1.00 | Reference |
| Slept 8 h or more on an average school night (yes vs. no) | 901 | 20.7% | 0.72 | 0.59–0.89 |
| Substance use | | | | |
| Did not drink alcohol during the last 30 days | 3197 | 73.5% | 1.00 | Reference |
| 1 or more drinks of alcohol on one or more days during the last 30 days (yes vs. no) | 1153 | 26.5% | 1.61 | 1.32–1.98 |
| No binge drinking | 3872 | 87.8% | 1.00 | Reference |
| Binge drinking (4 or more drinks in a row) on one or more days during the last 30 days (yes vs. no) | 539 | 12.2% | 1.29 | 0.97–1.70 |
| Did not use vapor products during the last 30 days | 3414 | 78.5% | 1.00 | Reference |
| Use of electronic vapor product on 1 or more days during the last 30 days (yes vs. no) | 936 | 21.5% | 1.22 | 1.01–1.47 |
| Did not smoke in the last 30 days | 4524 | 96.4% | 1.00 | Reference |
| Smoked cigarettes on at least 1 day during the last 30 days | 167 | 3.6% | 1.89 | 0.96–3.72 |

[1] Weighted multiple logistic regression models, adjusted for grade level, race/ethnicity, BMI percentile, breakfast intake, sleep, screen time, muscle strengthening activities, alcohol intake, and vaping.

### 3.3. Weight Loss Intent and BMI

A large proportion of girls within the normal weight category indicated an intent to lose weight despite being within the normal BMI percentile (43.4%). The intent to lose weight was the highest among obese girls (87.8%). (Figure 1) As previously shown in Table 1, Hispanic and multiple-Hispanic girls were more likely to want to lose weight when considering the entire sample of girls. When considering girls within normal weight only, this body size dissatisfaction was the largest among white girls and girls of Asian descent (47.5% and 45.7%, respectively).

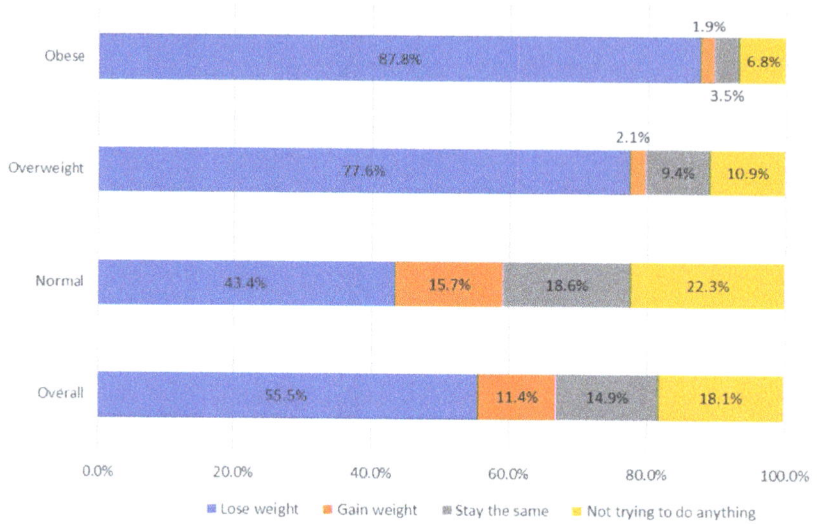


**Figure 1.** Proportion of adolescent girls reporting weight loss intent by BMI percentile categories (Normal Weight, Overweight, Obese) in the 2021 YRBS Survey: BMI, Body Mass Index. Normal weight BMI% < 85, overweight BMI% 85–<95, obese BMI% ≥ 95.

## 4. Discussion

This study examined the associations between sleep, screen time, physical activity, dietary-related variables, alcohol, vaping, and weight loss intent among adolescent girls. The relationship between BMI percentile and weight loss intent is also described. We found that girls intending to lose weight were associated with several unhealthy lifestyle behaviors. Girls intending to lose weight were more likely to report skipping breakfast, more screen time, and less sleep. Furthermore, a weight loss intent was associated with a higher consumption of alcohol and vaping use. The only healthy lifestyle behavior associated with weight loss intent was muscle-strengthening exercises. Due to the cross-sectional nature of the data, we cannot know if girls engaging in these unhealthy lifestyle activities are more likely to develop an intent to lose weight or if girls intending to lose weight are more likely to develop listed unhealthy lifestyle habits. Lastly, we found that weight-loss intent was most common among obese girls. However, many girls within a normal BMI range also reported an intent to lose weight.

### 4.1. *Physical Activity*

There were no differences in following the PA recommendation of 60 min. per day among the groups. Our results show that girls trying to lose weight were more likely to do strength training 1–3 times per week. These results were similar to data from the 2017 YRBS [22]. Engaging in muscle-strengthening activities can increase body weight, making it harder to fit the slim female ideal often seen in the media. Therefore, it is possible that girls who engage in strength training do so with the intention of losing weight, rather than the training itself leading to weight loss.

Physical activity can help regulate body weight [27] and dramatically lower the risk of many chronic illnesses [28]. Promoting increased physical activity levels could improve health in all adolescents regardless of weight loss intent. Conversely, the correlation between weight loss intent and muscle-strengthening exercise should be studied further to identify potential needs to improve body image among girls in sports. Low energy availability is highly prevalent among female athletes, up to 90% in some sports like ballet [29]. It is crucial to ensure proper food intake in these girls because under-fueling in sports can not only reduce sports performance but also impair immune function, bone health, cardiovascular health, and menstrual function [29].

### 4.2. Diet

In the present study, those trying to lose weight were less likely to eat other vegetables than those who were not trying to lose weight. Other vegetables refer to any vegetable, excluding potatoes, carrots, and salad greens. The YRBS survey only includes questions on the frequency of consumption of fruits and vegetables, and it does not specify the amounts. Although we did not see any significant differences in the consumption frequency of fruits and vegetables, aside from other vegetables, we could not explore a difference in the portion sizes of fruits and vegetables consumed.

Intervention studies targeting increased vegetable intake in this group can improve eating behaviors. However, as previously stated, the frequency of fruit and vegetable intake was low across this population, indicating that both groups of girls would benefit from increasing fruit and vegetable consumption. Adolescents have limited freedom to choose their diet as they follow their family's eating habits and grocery shopping. In this context, school cafeterias could play a pivotal role in offering healthier options to provide opportunities to increase intake.

Our results indicate that skipping breakfast was correlated with a desire to lose weight, confirming results from a previous study [21]. Skipping breakfast has also been positively associated with obesity [30] and negative body image [31] in girls. The association between breakfast and weight can also be attributed to other factors, such as sleep deprivation. According to data from Project EAT (Eating and Activity in Teens and Young Adults) [18], those who went to bed after 12:30 a.m. were more likely to skip breakfast.

A crossover study using block randomization of breakfast type in children showed that low-glycemic index foods eaten at breakfast had a significant impact on food intake at lunch [32]. Consuming breakfast increases satiety and contributes to higher fiber intake in adolescent girls [33]. Common barriers expressed by adolescents to have breakfast are lack of time and not being hungry [21]. This could be associated with eating late, which is usually associated with being awake late at night. Early school start times are also a likely contributor to poor breakfast habits and delaying school start time has been shown to improve breakfast consumption among highschoolers [34]. Insufficient sleep changes melatonin levels that in turn affects energy homeostasis related to the melanocortin system in the hypothalamus [35]. Additionally, sleeping duration regulates ghrelin, a hormone that stimulates appetite and feeding behavior [36].

Given the health consequences of skipping breakfast, providing adolescents with strategies to incorporate breakfast in their daily routine is paramount. An environment conducive to healthy food choices is key to implementing effective behavioral changes. An example is to provide incentives for participation in the school breakfast program, offering high-quality nutrients and breakfast options that are low in saturated fat and sugar. Education is only one aspect of health promotion. A more comprehensive approach needs to be applied to change health behaviors.

### 4.3. Sleep and Screen Time

Sleep is another lifestyle factor affecting body weight. Adolescents are chronically sleep-deprived [37]. The percentage of high school students who do not get enough sleep is higher among girls (80%) [24]. Regularly sleeping fewer than the recommended hours is associated with attention, behavior, and learning problems [11]. In the present study, girls trying to lose weight were less likely to sleep 8 h or more. Lack of sleep in adolescents affects their cognitive functioning and mental health. A recent review on sleep deprivation reported associations between sleepiness and subjective perception of depression, anxiety, and antisocial behavior [38]. Improvements in sleep need to be addressed in all adolescents, but specific strategies could be tailored to girls trying to lose weight.

Screen time also affects sleep [39]. The percentage of girls spending more than 3 h on screen time was high overall, and it was higher in those girls who intend to lose weight (81%) versus those who do not (75%). These results are consistent with reported data on screen time and body weight dissatisfaction [23], and girls who want to lose weight are more likely to spend more than 3 h per day using screens (not counting school work). Consistently, girls with higher than median screen time over a two-year period rated their body image lower than girls with below median screen time screen [40]. These studies showed the negative role of screen time on body image. The association between screen time and suicidality was 4.67% mediated by overweight/obesity (observed only in female adolescents) and 9.66% mediated by self-perceived overweight (both male and females) [41]. There is likely a bidirectional association between screen time and weight loss intent. Reducing social media use has been shown to improve body satisfaction in adolescents [42]. Therefore, limiting screen time in girls may be highly important to improve mental health and body image. A 7-month school-based intervention using social cognitive theory reduced screen time in overweight and obese adolescent girls [43]. Results from a meta-analysis showed that interventions targeting screen time are effective in reducing total screen time and television time in children and adolescents [44]. Hence, the implementation of these interventions in this group of girls is possible and warranted.

### 4.4. Vaping and Alcohol Use

Preventing primary use in teens and young adults is the most important step that can be taken to reduce the long-term complications of nicotine exposure [16]. In the present analysis, girls who intend to lose weight are more likely to vape. Consistent with our results, Mohapatra et al. reported in their recent systematic review that the high rates of vaping seemed to correlate with increased weight concerns, particularly among females. Girls facing body image pressures may see vaping as a weight loss or weight control strategy [45].

Alcohol intake was higher in females intending to lose weight in this sample. These results are consistent with those from the COMPASS, which is a longitudinal cohort study of secondary school students (Grades 9–12) in Canada [46]. Restrictive eating behavior has also been associated with binge drinking [47], suggesting that girls intending to lose weight may be more prone alcohol consumption. Risky behaviors, such as vaping and drinking alcohol, may be used as coping mechanisms [48]. Therefore, this group may benefit from increased support from school counselors to address mental health and prevent addictive behaviors.

Overall, vaping and alcohol are public health issues that need to be addressed in youth and understanding motivation can help to better target prevention strategies. Existing campaigns, such as the U.S. Food and Drug Administration's (FDA's) Real Cost Campaign [49], have been shown to influence youth beliefs related to vaping and can be useful tools to help students avoid or quit vaping [50]. Tackling one of the potential factors triggering these addictive behaviors can be a cost-effective measure. Thus, offering programs targeting girls promoting healthier strategies to manage and accept their body weight is an appropriate preventative measure.

### *4.5. BMI and Weight Loss Intent*

Despite having a normal BMI, more than half of these girls are actively trying to change their weight—revealing a significant degree of body image dissatisfaction. Notably, 43% are attempting to lose weight, while 16% are striving to gain weight.

This corresponds well with previous years of YRBS. Approximately 80% of the girls within BMI of 85–95% intend to lose weight in our sample; this is consistent with another representative sample of American adolescents where 86.9% of the ones in the overweight category reported that they intend to lose weight [51].

Culture impacts body image perception. Additionally, the immediate environment, such as family and friends, plays a role in how adolescents perceive themselves. A recent study reported that frequent negative familial weight talk was associated with higher weight bias internalization across gender in non-Hispanic White 10–15-year-old children living in MA [52]. In the present analysis, weight loss intent was the highest among white girls and girls of Asian descent. Similarly, according to data from The Ningbo Youth Risk Behavior Survey, self-perception of overweight and obesity was positively associated with lower-calorie diets and increased levels of PA in Chinese adolescents [53]. This information can be used to specifically target more vulnerable girls and develop culturally appropriate programs for these groups.

### *4.6. Limitations*

The general limitations of the 2021 YRBS survey were previously reported [24]. Limitations specific to this study are outlined below. Given the cross-sectional nature of the YRBS, it is impossible to determine the temporal directionality of associations or assess the directionality of correlations. Future studies should use a longitudinal design to establish cause-and-effect relationships. Nevertheless, cross-sectional research can help identify subpopulations that may be more vulnerable and help provide targeted interventions. As mentioned earlier, the YRBS survey includes questions on the frequency of food consumption, not specific amounts. This limited our analyses and we were unable to assess portion sizes. All questions in YRBS were assessed through self-reports, which can lead to overestimation or underestimation of the variables measured. The 2021 survey was conducted during COVID-19, which may have influenced dietary behaviors and access to food. Despite any limitations COVID-19 might have introduced, it will be useful to have these results recorded in the literature so comparisons can be made as future versions of the YRBS are administered.

## 5. Conclusions

School interventions may focus on offering healthier lunch options and providing breakfast programs, emphasizing high-quality nutrients that can enhance eating habits among all youth. Parents and caregivers should limit screen time to promote better sleep patterns, improve mental health, and foster positive body image. Programs designed to prevent substance abuse among youth should specifically target girls who are attempting to lose weight, as they are at a higher risk.

**Author Contributions:** Conceptualization, E.F., M.C.C., J.V. and A.J.P.; methodology and formal analysis, E.F.; writing—original draft preparation, E.F. and M.C.C.; writing—review and editing, M.C.C., A.J.P. and J.V. All authors have read and agreed to the published version of the manuscript.

**Funding:** This research received no external funding.

**Institutional Review Board Statement:** The YRBS was conducted in accordance with the Declaration of Helsinki, and approved by the Institutional Review Board of Worcester State University 2425-0001, 7 October 2024.

**Informed Consent Statement:** As this study involved secondary analysis of existing data, the requirement for informed consent was waived.

**Data Availability Statement:** The data can be downloaded from https://www.cdc.gov/healthyyouth/data/yrbs/data.htm/, accessed on 11 May 2025.

**Conflicts of Interest:** The authors declare no conflicts of interest.

## Abbreviations

The following abbreviations are used in this manuscript:

| | |
|---|---|
| BMI | Body Mass Index |
| YRBS | Youth Risk Behavior Surveillance System |
| CDC | Centers for Disease Control and Prevention |
| FDA | U.S. Food and Drug Administration |
| OR | Odds Ratio |
| CI | Confidence Interval |

## References


1. Stierman, B.; Afful, J.; Carroll, M.D.; Chen, T.-C.; Davy, O.; Fink, S.; Fryar, C.D.; Gu, Q.; Hales, C.M.; Hughes, J.P.; et al. National Health and Nutrition Examination Survey 2017-March 2020 Prepandemic Data Files-Development of Files and Prevalence Estimates for Selected Health Outcomes. *Natl. Health Stat. Rep.* **2021**, *158*, 15610–15620. [CrossRef]
2. Calle, M.C.; Andersen, C.J. Assessment of Dietary Patterns Represents a Potential, Yet Variable, Measure of Inflammatory Status: A Review and Update. *Dis. Markers* **2019**, *2019*, 3102870. [CrossRef] [PubMed]
3. Simmonds, M.; Llewellyn, A.; Owen, C.G.; Woolacott, N. Predicting Adult Obesity from Childhood Obesity: A Systematic Review and Meta-analysis. *Obes. Rev.* **2016**, *17*, 95–107. [CrossRef]
4. Toselli, S.; Zaccagni, L.; Rinaldo, N.; Mauro, M.; Grigoletto, A.; Maietta Latessa, P.; Marini, S. Body Image Perception in High School Students: The Relationship with Gender, Weight Status, and Physical Activity. *Children* **2023**, *10*, 137. [CrossRef] [PubMed]
5. Martini, M.C.S.; Assumpção, D.D.; Barros, M.B.D.A.; Mattei, J.; Barros Filho, A.D.A. Prevalence of Body Weight Dissatisfaction among Adolescents: A Systematic Review. *Rev. Paul. Pediatr.* **2023**, *41*, e2021204. [CrossRef] [PubMed]
6. Wharton, C.M.; Adams, T.; Hampl, J.S. Weight Loss Practices and Body Weight Perceptions Among US College Students. *J. Am. Coll. Health* **2008**, *56*, 579–584. [CrossRef]
7. Gualdi-Russo, E.; Rinaldo, N.; Zaccagni, L. Physical Activity and Body Image Perception in Adolescents: A Systematic Review. *Int. J. Environ. Res. Public Health* **2022**, *19*, 13190. [CrossRef]
8. Piercy, K.L.; Troiano, R.P.; Ballard, R.M.; Carlson, S.A.; Fulton, J.E.; Galuska, D.A.; George, S.M.; Olson, R.D. The Physical Activity Guidelines for Americans. *JAMA* **2018**, *320*, 2020. [CrossRef]
9. Michael, S.L.; Jones, S.E.; Merlo, C.L.; Sliwa, S.A.; Lee, S.M.; Cornett, K.; Brener, N.D.; Chen, T.J.; Ashley, C.L.; Park, S. Dietary and Physical Activity Behaviors in 2021 and Changes from 2019 to 2021 Among High School Students—Youth Risk Behavior Survey, United States, 2021. *MMWR Suppl.* **2023**, *72*, 75–83. [CrossRef]
10. U.S. Department of Agriculture, Food and Nutrition Service, Center for Nutrition Policy and Promotion. *Average Healthy Eating Index-2020 Scores for the U.S. Population—Total Ages 2 and Older and by Age Groups, WWEIA, NHANES 2017–2018*; U.S. Department of Agriculture, Food and Nutrition Service: Alexandria, VA, USA, 2023.
11. Paruthi, S.; Brooks, L.J.; D'Ambrosio, C.; Hall, W.A.; Kotagal, S.; Lloyd, R.M.; Malow, B.A.; Maski, K.; Nichols, C.; Quan, S.F.; et al. Consensus Statement of the American Academy of Sleep Medicine on the Recommended Amount of Sleep for Healthy Children: Methodology and Discussion. *J. Clin. Sleep Med.* **2016**, *12*, 1549–1561. [CrossRef]
12. Hale, L.; Guan, S. Screen Time and Sleep among School-Aged Children and Adolescents: A Systematic Literature Review. *Sleep Med. Rev.* **2015**, *21*, 50–58. [CrossRef] [PubMed]
13. Alexander, J.D.; Linkersdörfer, J.; Toda-Thorne, K.; Sullivan, R.M.; Cummins, K.M.; Tomko, R.L.; Allen, N.B.; Bagot, K.S.; Baker, F.C.; Fuemmeler, B.F.; et al. Passively Sensing Smartphone Use in Teens with Rates of Use by Sex and across Operating Systems. *Sci. Rep.* **2024**, *14*, 17982. [CrossRef] [PubMed]

14. Centers for Disease Control and Prevention. *Youth Risk Behavior Survey Data Summary & Trends Report for Dietary. Physical Activity, and Sleep Behaviors: 2013–2023*; Centers for Disease Control and Prevention: Atlanta, GA, USA, 2024.
15. Fatima, Y.; Doi, S.A.R.; Mamun, A.A. Sleep Quality and Obesity in Young Subjects: A Meta-analysis. *Obes. Rev.* **2016**, *17*, 1154–1166. [CrossRef]
16. Overbeek, D.L.; Kass, A.P.; Chiel, L.E.; Boyer, E.W.; Casey, A.M.H. A Review of Toxic Effects of Electronic Cigarettes/Vaping in Adolescents and Young Adults. *Crit. Rev. Toxicol.* **2020**, *50*, 531–538. [CrossRef] [PubMed]
17. Miech, R.A.; Johnston, L.D.; Patrick, M.E.; O'Malley, P.M.; Bachman, J.G.; Schulenberg, J.E. *Monitoring the Future National Survey Results on Drug Use, 1975–2022: Secondary School Students*; Monitoring the Future Monograph Series; University of Michigan, Institute for Social Research: Ann Arbor, MI, USA, 2023.
18. Ogilvie, R.P.; Lutsey, P.L.; Widome, R.; Laska, M.N.; Larson, N.; Neumark-Sztainer, D. Sleep Indices and Eating Behaviours in Young Adults: Findings from Project EAT. *Public Health Nutr.* **2018**, *21*, 689–701. [CrossRef]
19. Nagata, J.M.; Weinstein, S.; Alsamman, S.; Lee, C.M.; Dooley, E.E.; Ganson, K.T.; Testa, A.; Gooding, H.C.; Kiss, O.; Baker, F.C.; et al. Association of Physical Activity and Screen Time with Cardiovascular Disease Risk in the Adolescent Brain Cognitive Development Study. *BMC Public Health* **2024**, *24*, 1346. [CrossRef]
20. Sanchez, R.; Ranjit, N.; Kelder, S.H.; Gill, M.; Hoelscher, D.M. Intention to Lose Weight and Use of Electronic Cigarettes among Adolescents. *Prev. Med. Rep.* **2021**, *23*, 101406. [CrossRef]
21. Godin, K.M.; Patte, K.A.; Leatherdale, S.T. Examining Predictors of Breakfast Skipping and Breakfast Program Use Among Secondary School Students in the COMPASS Study. *J. Sch. Health* **2018**, *88*, 150–158. [CrossRef]
22. Burns, R.D. Energy Balance-Related Factors Associating with Adolescent Weight Loss Intent: Evidence from the 2017 National Youth Risk Behavior Survey. *BMC Public Health* **2019**, *19*, 1206. [CrossRef]
23. Bordeleau, M.; Gilbert, J.-A.; Alméras, N.; Monthuy-Blanc, J.; Gagnon, J.; Mathieu, M.-È.; Drapeau, V. Body Image and Health-Related Behaviors among Fitspirit Participants. *BMC Public Health* **2022**, *22*, 1764. [CrossRef]
24. Mpofu, J.J.; Underwood, J.M.; Thornton, J.E.; Brener, N.D.; Rico, A.; Kilmer, G.; Harris, W.A.; Leon-Nguyen, M.; Chyen, D.; Lim, C.; et al. Overview and Methods for the Youth Risk Behavior Surveillance System—United States, 2021. *MMWR Suppl.* **2023**, *72*, 1–12. [CrossRef] [PubMed]
25. Brener, N.; Kann, L.; Mcmanus, T.; Kinchen, S.; Sundberg, E.; Ross, J. Reliability of the 1999 Youth Risk Behavior Survey Questionnaire. *J. Adolesc. Health* **2002**, *31*, 336–342. [CrossRef]
26. Kuczmarski, R.J.; Ogden, C.L.; Guo, S.S.; Grummer-Strawn, L.M.; Flegal, K.M.; Mei, Z.; Wei, R.; Curtin, L.R.; Roche, A.F.; Johnson, C.L. 2000 CDC Growth Charts for the United States: Methods and Development. *Vital Health Stat. 11* **2002**, *246*, 1–190.
27. Chaput, J.-P.; Klingenberg, L.; Rosenkilde, M.; Gilbert, J.-A.; Tremblay, A.; Sjödin, A. Physical Activity Plays an Important Role in Body Weight Regulation. *J. Obes.* **2011**, *2011*, 360257. [CrossRef] [PubMed]
28. Anderson, E.; Durstine, J.L. Physical Activity, Exercise, and Chronic Diseases: A Brief Review. *Sports Med. Health Sci.* **2019**, *1*, 3–10. [CrossRef]
29. Jagim, A.R.; Fields, J.; Magee, M.K.; Kerksick, C.M.; Jones, M.T. Contributing Factors to Low Energy Availability in Female Athletes: A Narrative Review of Energy Availability, Training Demands, Nutrition Barriers, Body Image, and Disordered Eating. *Nutrients* **2022**, *14*, 986. [CrossRef]
30. Wang, K.; Niu, Y.; Lu, Z.; Duo, B.; Effah, C.Y.; Guan, L. The Effect of Breakfast on Childhood Obesity: A Systematic Review and Meta-Analysis. *Front. Nutr.* **2023**, *10*, 1222536. [CrossRef]
31. Ben Ayed, H.; Yaich, S.; Ben Jemaa, M.; Ben Hmida, M.; Trigui, M.; Jedidi, J.; Sboui, I.; Karray, R.; Feki, H.; Mejdoub, Y.; et al. What Are the Correlates of Body Image Distortion and Dissatisfaction among School-Adolescents? *Int. J. Adolesc. Med. Health* **2021**, *33*, 20180279. [CrossRef]
32. Warren, J.M.; Henry, C.J.K.; Simonite, V. Low Glycemic Index Breakfasts and Reduced Food Intake in Preadolescent Children. *Pediatrics* **2003**, *112*, e414. [CrossRef]
33. Zakrzewski-Fruer, J.K.; Seall, C.; Tolfrey, K. Breakfast Consumption Suppresses Appetite but Does Not Increase Daily Energy Intake or Physical Activity Energy Expenditure When Compared with Breakfast Omission in Adolescent Girls Who Habitually Skip Breakfast: A 7-Day Randomised Crossover Trial. *Nutrients* **2021**, *13*, 4261. [CrossRef]
34. Widome, R.; Erickson, D.J.; Laska, M.N.; Berger, A.T.; Lenk, K.M.; Iber, C.; Kilian, G.; Lammert, S.; Wahlstrom, K.L. Impact of Delaying High School Start Times on Weight and Related Behaviors—The START Study. *Prev. Med.* **2023**, *172*, 107548. [CrossRef] [PubMed]
35. Ahluwalia, M.K. Chrononutrition—When We Eat Is of the Essence in Tackling Obesity. *Nutrients* **2022**, *14*, 5080. [CrossRef]
36. Sovetkina, A.; Nadir, R.; Fung, J.N.M.; Nadjarpour, A.; Beddoe, B. The Physiological Role of Ghrelin in the Regulation of Energy and Glucose Homeostasis. *Cureus* **2020**, *12*, e7941. [CrossRef]

37. Owens, J.; Adolescent Sleep Working Group; Committee on Adolescence; Au, R.; Carskadon, M.; Millman, R.; Wolfson, A.; Braverman, P.K.; Adelman, W.P.; Breuner, C.C.; et al. Insufficient Sleep in Adolescents and Young Adults: An Update on Causes and Consequences. *Pediatrics* **2014**, *134*, e921–e932. [CrossRef]
38. Davidson-Urbain, W.; Servot, S.; Godbout, R.; Montplaisir, J.-Y.; Touchette, E. La somnolence chez les adolescents: Étiologie et conséquences multiples. *L'Encéphale* **2023**, *49*, 87–93. [CrossRef] [PubMed]
39. Priftis, N.; Panagiotakos, D. Screen Time and Its Health Consequences in Children and Adolescents. *Children* **2023**, *10*, 1665. [CrossRef] [PubMed]
40. Hrafnkelsdottir, S.M.; Brychta, R.J.; Rognvaldsdottir, V.; Chen, K.Y.; Johannsson, E.; Guðmundsdottir, S.L.; Arngrimsson, S.A. Screen Time and Body Image in Icelandic Adolescents: Sex-Specific Cross-Sectional and Longitudinal Associations. *Int. J. Environ. Res. Public Health* **2022**, *19*, 1308. [CrossRef]
41. Gao, H.; Wang, Y.; Wang, X.; Gao, M. Mediation of the Association between Screen Time and Suicidality by Overweight/Obesity and Perceived Overweight: Results from the Youth Risk Behavior Surveillance System of the United States. *Front. Psychiatry* **2024**, *15*, 1287021. [CrossRef]
42. Thai, H.; Davis, C.G.; Mahboob, W.; Perry, S.; Adams, A.; Goldfield, G.S. Reducing Social Media Use Improves Appearance and Weight Esteem in Youth with Emotional Distress. *Psychol. Pop. Media* **2024**, *13*, 162–169. [CrossRef]
43. Bagherniya, M.; Mostafavi Darani, F.; Sharma, M.; Maracy, M.R.; Allipour Birgani, R.; Ranjbar, G.; Taghipour, A.; Safraian, M.; Keshavarz, S.A. Assessment of the Efficacy of Physical Activity Level and Lifestyle Behavior Interventions Applying Social Cognitive Theory for Overweight and Obese Girl Adolescents. *J. Res. Health Sci.* **2018**, *18*, e00409.
44. Zhang, P.; Tang, X.; Peng, X.; Hao, G.; Luo, S.; Liang, X. Effect of Screen Time Intervention on Obesity among Children and Adolescent: A Meta-Analysis of Randomized Controlled Studies. *Prev. Med.* **2022**, *157*, 107014. [CrossRef] [PubMed]
45. Mohapatra, S.; Wisidagama, S.; Schifano, F. Exploring Vaping Patterns and Weight Management-Related Concerns among Adolescents and Young Adults: A Systematic Review. *J. Clin. Med.* **2024**, *13*, 2896. [CrossRef] [PubMed]
46. Patte, K.A.; Leatherdale, S.T. A Cross-Sectional Analysis Examining the Association between Dieting Behaviours and Alcohol Use among Secondary School Students in the COMPASS Study. *J. Public Health* **2016**, *39*, fdw034. [CrossRef]
47. Leatherdale, S.T. An Examination of the Co-Occurrence of Modifiable Risk Factors Associated with Chronic Disease among Youth in the COMPASS Study. *Cancer Causes Control* **2015**, *26*, 519–528. [CrossRef] [PubMed]
48. Magee, K.E.; Connell, A.M. The Role of Substance Use Coping in Linking Depression and Alcohol Use from Late Adolescence through Early Adulthood. *Exp. Clin. Psychopharmacol.* **2021**, *29*, 659–669. [CrossRef]
49. Santiago, S.; Mahoney, C.; Murray, M.P.; Benoza, G. "The Real Cost": Reaching At-Risk Youth in a Fragmented Media Environment. *Am. J. Prev. Med.* **2019**, *56*, S49–S56. [CrossRef]
50. MacMonegle, A.; Bennett, M.; Speer, J.L.; O'Brien, E.K.; Pitzer, L.; Jaarsma, A.; Nguyen Zarndt, A.; Duke, J. Evaluating the Real Cost Digital and Social Media Campaign: Longitudinal Effects of Campaign Exposure on E-Cigarette Beliefs. *Nicotine Tob. Res.* **2024**, *26*, S19–S26. [CrossRef]
51. Kakinami, L.; Houle-Johnson, S.A.; Demissie, Z.; Santosa, S.; Fulton, J.E. Meeting Fruit and Vegetable Consumption and Physical Activity Recommendations among Adolescents Intending to Lose Weight. *Prev. Med. Rep.* **2019**, *13*, 11–15. [CrossRef]
52. Rancaño, K.M.; Puhl, R.; Skeer, M.; Eliasziw, M.; Must, A. Negative Familial Weight Talk and Weight Bias Internalization in a US Sample of Children and Adolescents. *Pediatr. Obes.* **2024**, *19*, e13108. [CrossRef]
53. Liu, X.; Liu, M.; Wu, Z.; Liang, J. Secular Trends in the Prevalence of Meeting 24-Hour Movement Guidelines among U.S. Adolescents: Evidence from NHANES 2007–2016. *Front. Public Health* **2024**, *12*, 1362718. [CrossRef]

nutrients

MDPI

Article

# Gender Moderates the Associations Between Responsiveness to Alarming Oral Sensations, Depressive Symptoms, and Dietary Habits in Adolescents

**Leonardo Menghi [1,2,*], Lara Fontana [3], Silvia Camarda [4], Isabella Endrizzi [3], Maria Pina Concas [5], Paolo Gasparini [4,5] and Flavia Gasperi [1,3]**

1 Center Agriculture Food Environment, University of Trento, Via Mach 1, 38098 San Michele all'Adige, Italy; flavia.gasperi@unitn.it
2 Department of Green Technology, University of Southern Denmark, Campusvej 55, 5230 Odense, Denmark
3 Research and Innovation Centre, Edmund Mach Foundation, Via Mach 1, 38098 San Michele all'Adige, Italy; lara.fontana@fmach.it (L.F.); isabella.endrizzi@fmach.it (I.E.)
4 Department of Medicine, Surgery and Health Sciences, University of Trieste, Strada di Fiume 447, 34149 Trieste, Italy; silvia.camarda@phd.units.it (S.C.); paolo.gasparini@burlo.trieste.it (P.G.)
5 Institute for Maternal and Child Health, I.R.C.C.S. "Burlo Garofolo", Via dell'Istria, 65/1, 34137 Trieste, Italy; mariapina.concas@burlo.trieste.it
* Correspondence: leom@igt.sdu.dk

**Abstract: Background/Objectives**: As a peripheral effect of depression-related traits, sensory responses may predispose individuals to depressive symptoms by prompting suboptimal dietary patterns with long-term effects on mood. Mood disturbances in adolescence are strong predictors of adult mental illness, making it crucial to identify factors that may shift transient mood fluctuations into more severe mental health issues during this vulnerable period. Given the substantial gender differences in susceptibility to comorbidities of depression, we examined whether the link between sensory perception and depressive symptoms in nonclinical adolescents varied by gender and was related to dietary habits. **Methods**: In this cross-sectional study, 232 healthy adolescents (41.8% girls, aged 13–17) reported their diet over the past year using the EPIC Food Frequency Questionnaire and rated their liking and perceived intensity of oral sensations from four grapefruit juices and dark chocolate puddings with varying sucrose levels. Additionally, participants completed assessments of anxiety, neuroticism, pickiness, body dissatisfaction, and the Patient Health Questionnaire (PHQ-9) to evaluate depressive symptoms. **Results**: We found that girls exhibited higher levels of depression, anxiety, neuroticism, and pickiness compared to boys (Wilcoxon Rank Sum Test), and that greater responsiveness to bitterness (e.g., $\beta = 0.264$, $p = 0.037$) and astringency ($\beta = 0.269$, $p = 0.029$) predicted higher depressive symptoms exclusively in girls. PHQ-9 scores were positively associated with alcohol use in both girls ($\rho = 0.176$, $p = 0.003$) and boys ($\rho = 0.148$, $p = 0.004$) and inversely related to the intake of beneficial nutrients (e.g., fiber, polyunsaturated fats), particularly in girls. Intriguingly, moderation analyses suggested that associations between nutrient intake and acuity for alarming oral sensations were largely moderated by depression-related traits in girls, but not in boys. **Conclusions**: Our findings suggest that gender moderates the links between depressive symptoms, sensory perception, and dietary habits in healthy adolescents, possibly reflecting gender-specific coping strategies for comorbidities of depression.

**Keywords:** adolescents; gender; depressive symptoms; taste; oral responsiveness; diet

## 1. Introduction

Adolescence is a transient stage of life marked by profound physiological and psychosocial changes that can increase vulnerability to mood disorders. As emotional challenges intensify, mood disturbances may escalate to more significant mental health conditions, including depression. Characterized by persistent loss of pleasure or interest in activities and depressed mood [1], depressive disorders are one of the leading causes of morbidity globally among adolescents, contributing to 15% of the disease burden in this age group [2]. Approximately 34% of adolescents aged 10–19 are thought to experience intense depressive symptoms [3], with prevalence doubling during the first year of the COVID-19 pandemic [4] and increasing by 14% from the first (24%) to the second (38%) decade of the 21st century [3]. In particular, girls are twice as likely as boys to experience severe depressive symptoms, e.g., [5–7]. This gap typically emerges during preadolescence and peaks at ages 15–18, e.g., [6], likely due to factors such as earlier onset of puberty, increased exposure to severe stressors, intensified societal and academic pressures, and heightened susceptibility to the psychosocial correlates of depression (for review, see [7]). This is inevitably a cause for concern, as severe mood disturbances during adolescence are a strong predictor of mental illness in adulthood [8], and early symptoms often go undetected and untreated [2,9]. Therefore, identifying factors that may facilitate the progression from short-term mood fluctuations to significant mental health problems during adolescence is a critical public health priority.

In this context, changes in dietary habits are increasingly recognized as one of the most promising modifiable risk factors for preventing depression (for review, see [10]). In particular, an adequate intake of vegetables, fruits, vegetable oils, fish, and whole grains, along with limited consumption of refined grains, simple sugars, red and processed meats, and full-fat dairy products, has been consistently associated with a reduced risk of depression among adolescents worldwide, e.g., [11–13]. Indeed, diets rich in plant-based nutrients, such as the Mediterranean Diet, have been shown to reduce plasma levels of pro-inflammatory cytokines (e.g., IL-6, INF-$\gamma$, TNF-$\alpha$), thereby counteracting biological mechanisms implicated in the etiology of depression, including systemic oxidative stress and dysregulation in the production of glucocorticoids, monoamines, and brain-derived neurotrophic factor (for review, see [14]). Therefore, uncovering barriers to adherence to dietary patterns beneficial for mood is one of the essential steps in addressing the rising rates of depression among youth.

Despite greater independence compared to childhood and significant familial and social influences, the sensory properties of food (taste, smell, flavor, texture, appearance) remain one of the key determinants of dietary choices during this developmental stage (for review, see [15]). Adolescents generally favor more intensely sweet tastes relative to adults [16] and tend to avoid foods that evoke inherently disliked oral sensations, such as bitterness [15]. However, there are substantial interpersonal differences in taste perception, which can influence adherence to or deviation from healthy dietary patterns, e.g., [17,18]. An example is the genetic ability to perceive bitter thiourea compounds, such as phenylthiocarbamide or 6-n-propylthiouracil (PROP), which can have downstream effects on food preferences (for review, see [18]). While PROP perception is a continuous trait, individuals are typically classified as non-tasters (NTs), medium-tasters (MTs), or super-tasters (STs) based on cut-offs that reflect varying levels of perceived bitterness, from null to extreme, with haplotypes in the *TAS2R38* gene largely responsible for this phenotypic variation [18,19].

In adolescents, PROP status appears to influence food liking more than actual intake. For instance, NTs rate cruciferous vegetables as more pleasant than STs [20], whereas STs find animal-based foods (e.g., bacon, fried chicken, and herring), sauces, condiments [21], and sweetened foods more appealing [22]. While these findings might suggest that STs are more likely to adopt suboptimal dietary patterns, previous reports yielded mixed results. Only one study reported a higher sugar intake among STs relative to NTs [22], while others failed to link PROP phenotypes and/or genotypes to diet [20,21,23]. Thus, although the evidence remains insufficient to draw definitive conclusions due to small sample sizes and varying methods of operationalizing PROP status and dietary habits [24], other factors underlying variations in taste perception may contribute more to undesired dietary patterns during this phase of life.

In addition to PROP, notable gender differences in taste perception have been documented across the lifespan [22,25]. Girls at various stages of adolescence generally show greater responsiveness to oral stimuli than boys, e.g., [22,26], and are more likely to be STs [19]. However, whether these differences are primarily psychological or physiological remains debated [27]. In fact, a large body of literature suggests that sensory perception can also be influenced by mood states and personality traits (for reviews, see [17,28]). In nonclinical adults, anxiety and exposure to stressors (e.g., airhorn sounds, cold pressor test) tend to express higher intensity ratings to sweet, salty, and bitter stimuli [29–31], and being neurotic [32] or a picky eater [33] has been positively associated with acuity for bitter and for sweet and bitter aqueous solutions, respectively.

Interestingly, similar findings have been reported for depression. Dess and Chapman [34] found a positive correlation between depressive symptoms and bitterness ratings of quinine solutions, and Platte et al. [31] extended the same findings to the sweet taste elicited by varying levels of sucrose in water. Nevertheless, a link between greater responsiveness to oral stimuli and pronounced internalizing symptoms (anxiety, depression, stress) has not always been supported [35–37], and the opposite has often been observed in clinically depressed adults (for review, see [28]). Although findings are inconclusive and largely based on small adult samples, there is evidence that vulnerability to correlates of depression may drive variations in taste perception, potentially shaping dietary choices with long-term negative effects on mood.

Building on this, adolescents worldwide often cite the unpleasant sensory properties of healthy foods as a key barrier to healthy eating, leading them to opt for less nutritious options [15]. This is evident because prototypical healthy foods (e.g., vegetables, fruits, nuts, vegetable oils), whether raw or cooked, can evoke a range of sensory qualities such as bitterness, sourness, or astringency (hereafter referred to as alarming oral sensations), which elicit psychobiological states (i.e., arousal) that enhance alertness and facilitate rejection due to their perceived dangerousness [38]. However, the impact of arousal on food choices varies widely among individuals and depends on how well a sensation aligns with their optimal level of activation [38]. Consequently, an individual's susceptibility to sources of arousal in foods (stimulus intensity, novelty, and complexity) is contingent upon their responsiveness to exogenous stimuli and mood state (for review, see [39]). Individuals with traits that heighten responses to external inputs, such as those related to depression, may thus be more reactive to negative arousal from food and potentially reinforce undesired dietary choices with long-term effects on mood. Given the increased vulnerability of girls to depression and its comorbidities [3,7], this holds the potential to yield new insights into gender-based antecedents of depressive symptoms and warrants further investigation.

While taste plays a crucial role in shaping food habits among adolescents [15], and diet may prevent depressive disorders [10], no studies have yet simultaneously examined the links between sensory perception, depressive symptoms, and dietary habits in healthy adolescents, particularly concerning variations in vulnerability to correlates of depression. Additionally, the current knowledge has focused on either small adult samples, e.g., [31,34,35], or documented inconclusive results on the impact of taste on dietary habits due to a paucity of studies [24]. Also, the majority of previous research has relied on detection or recognition thresholds of diluted tastants in simple solutions, which have shown minimal correlations with everyday perceptions at suprathreshold levels [40]. To address this gap, the use of actual or model foods with varying tastant concentrations could offer more ecologically relevant insights, e.g., [41].

Against this backdrop, we examined whether the link between sensory perception and depressive symptoms in nonclinical adolescents varied by gender and was related to dietary habits. Additionally, we explored whether these associations were moderated by common psychosocial correlates of depression, such as anxiety [42], neuroticism [43], picky eating [33], and body dissatisfaction [44].

In this study, 232 nonclinical adolescents provided data on PROP responsiveness, as well as hedonic and psychophysical ratings to oral stimuli elicited by four variants of two food models with varying sucrose levels. Participants also completed the European Prospective Investigation into Cancer and Nutrition Food Frequency Questionnaire (EPIC-FFQ) [45] to monitor their habitual diet, a series of questionnaires assessing the aforementioned correlates of depression, and reported their depressive symptoms over the previous two weeks using the nine-item Patient Health Questionnaire (PHQ-9) [46].

## 2. Materials and Methods

### 2.1. Participants

This work builds on a broader project aimed at elucidating the genetic, non-genetic, and psychosocial determinants of individual variations in sensory perception among adolescents and their impact on diet. Due to the lack of comparable studies, sample size was estimated based on a previous study from our group involving healthy young adults (aged 18–30) from the same location (Autonomous Province of Trento, Italy) and engaged in similar experimental tasks [47]. In that study, we found significant differences (Cohen's d = 0.402) in responsiveness to bitterness and sourness elicited by commercially available foods between groups with comparable salivary microbial profiles. Accordingly, we conducted a power analysis using the *pwr.t.test* R function [48], which indicated that 198 participants would be needed to detect such an effect with 80% power at $\alpha = 0.05$ (two-tailed). To account for deviations from normality and potential dropouts, the target sample was increased by 15%, resulting in an expected sample size of 228 participants.

This was later accommodated with a final sample of 232 healthy adolescents (41.8% girls, 13–17 years; mean age ± SD = 14.5 ± 0.6 years; mean BMI ± SD = 21.0 ± 3.2 kg/m$^2$), who were recruited via a series of promotional events targeting both potential participants and their parents from two high schools in the Autonomous Province of Trento, Italy.

No formal exclusion criteria were applied. Nonetheless, participants reported no current diagnosis of major depressive disorder, nor were they undergoing treatment or had used medications in the past 6 months that could affect mood or taste function. Furthermore, girls and boys did not differ ($p > 0.05$) in terms of age, BMI, smoking habits, weekly alcohol consumption, food allergies, and habitual diet. In contrast, boys engaged in more physical activity than girls (t = 2.532, $p = 0.012$). A full demographic description of our sample is provided in Table S1.

Before data collection, parents or legal guardians provided written consent, and adolescents gave written assent. The study was approved by the Research Ethics Committee of the University of Trento (n° prot. 2023-047, approved on 28 September 2023) and followed the principles of the Declaration of Helsinki (as amended in Fortaleza, Brazil, 2013).

*2.2. Overview of Data Collection*

To align with the project objectives, a 7-day cross-sectional protocol was designed to collect a broad range of sensory, psychometric, demographic, health-related, and dietary data, as well as biological samples (saliva and tongue swabs) for metagenomic and genetic analyses. Key variables used in this study are highlighted in bold in the graphical overview of the experimental design (Figure 1), while Tables 1 and 2 provide details on tasks and measures used during data collection.

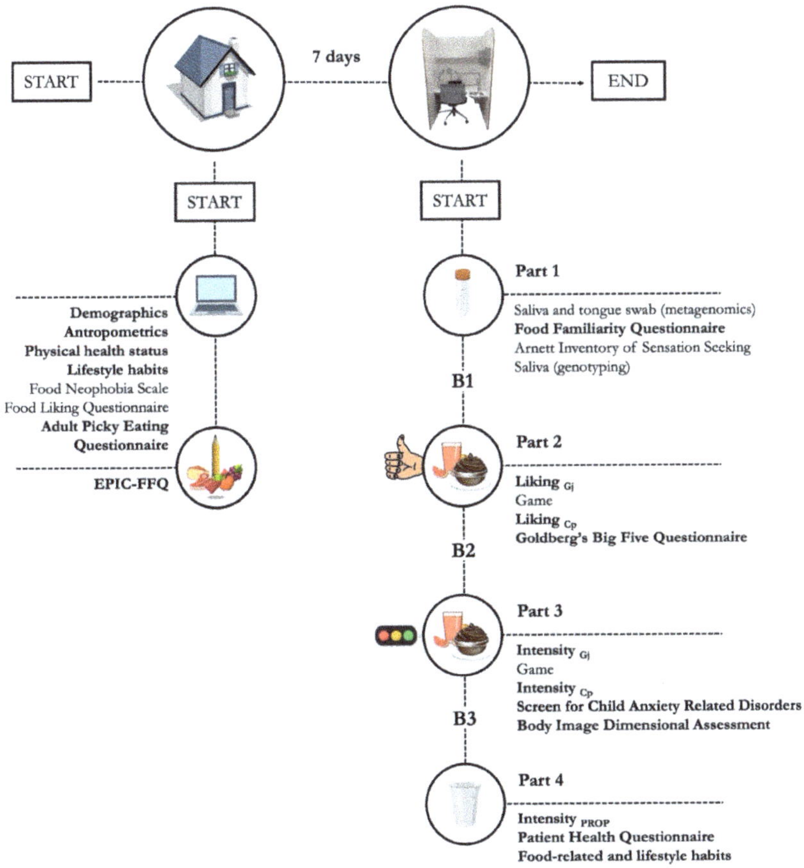


**Figure 1.** Graphical overview of data collection, with measures employed in this study highlighted in bold. Gj: grapefruit juice series; Cp: dark chocolate pudding series.

To optimize logistics, participants took part in two distinct work sessions: one remote and one conducted in an ISO 8589:2007 [49] compliant sensory lab. At least 7 days before the lab session, a researcher visited participants during school hours to provide detailed instructions for the home tasks (Table 1). These included an online questionnaire assessing demographics, anthropometrics, physical health status, lifestyle habits, stated preferences for a list of 57 food items, and selective/avoidant feeding attitudes using the Food Neophobia Scale and the Adult Picky Eating Questionnaire. Additionally, to assess their habitual diet, participants completed a paper version of the EPIC-FFQ, which they returned at the beginning of the lab session.

**Table 1.** Data collected during the remote work session. Relevant questionnaires, response options, number of items, rating scales, and references are listed. † Data collected in paper form.

| Questionnaire | Output | Options (Scale) | References |
| --- | --- | --- | --- |
| Demographics | Age | Years old at the moment of the test | |
| | Gender | Girl/Boy | |
| Anthropometrics | Weight | 40–180 kg | |
| | Height | 100–220 cm | |
| Physical health status (ongoing diagnosis or within the last 6 months) | Chronic diseases | Asthma; COVID-19; Type I or II diabetes; Celiac disease; Crohn's disease; Gastroesophageal reflux; Hiatal hernia; Irritable bowel syndrome | |
| | Oral diseases | Aphthous ulcers; Halitosis; Oral candidiasis; Gingivitis; Periodontitis; Xerostomia | |
| | Psychiatric disorders | Anorexia nervosa; Bulimia nervosa; Binge-eating disorder; Generalized anxiety disorder; Major depressive disorder; Autism spectrum disorder | |
| | Taste and smell disorders | Ageusia; Anosmia; Hypogeusia; Hyposmia | |
| | Food allergies | Yes/No [if yes, details asked] | |
| | Food intolerances | Yes/No [if yes, details asked] | |
| Medications and supplements use (past 6 month) | Antibiotics use | Yes/No | |
| | Medicines use | Yes/No [if yes, details asked] | |
| | Probiotics use | Yes/No | |
| Lifestyle habits | Diet | 10 items | [50] |
| | International Physical Activity Questionnaire (IPAQ) | 7 items | [51] |
| | Oral Hygiene Behavior Questionnaire | 8 items | [52] |
| Food Liking Questionnaire | Stated liking for a range of foods | 57 items rated (9-point hedonic scale; 1 = extremely disliked, 9 – extremely liked) | Adapted from [53] |
| Food Neophobia Scale | Selective and avoidant feeding attitudes | 10 items (7-point Likert scale; 1 = strongly disagree; 7 = strongly agree) | [54] |
| Adult Picky Eating Questionnaire | | 20 items (5-point Likert scale; 1 = never; 5 = always) | [55] |
| Dietary habits | EPIC Food Frequency Questionnaire † | 163 items | [45] |

After the remote session, participants were instructed to refrain from eating, drinking (except water), smoking, and brushing their teeth for at least 2 h before visiting the sensory lab at the Edmund Mach Foundation (San Michele all'Adige, Trento, Italy). Due to the extensive data collection, the lab session was divided into four slots (Parts 1–4) to minimize participants' burden and ensure data quality (Figure 1).

**Table 2.** Data collected during the four task slots (Part) designed for the lab session. Relevant tasks, expected outputs, number of items, rating scales, and references are listed. Gj: grapefruit juice series; Cp: dark chocolate pudding series.

| Part | Task | Output | Items | Scale | References |
|---|---|---|---|---|---|
| 1 | Saliva and tongue swab (metagenomics) | One salivary and one tongue swab sample for shotgun metagenomics | | | |
| | Food familiarity questionnaire | Familiarity with a range of foods | 57 items | 5-point Likert scale (1 = I do not recognize it; 5 = I regularly eat it) | Adapted from [53] |
| | Arnett Inventory of Sensation Seeking questionnaire | Inclination for novel, varied, and intense experiences | 20 items | 4-point Likert scale (1 = describes me very well; 4 = does not describe me at all) | [56] |
| | Saliva (genotyping) | One salivary sample for SNP-array | | | |
| 2 | Liking $_{Gj}$ | Liking ratings for the four variants of Gj | | Labeled Affective Magnitude scale (0 = greatest imaginable dislike; 100 = greatest imaginable like) | [57] |
| | Liking $_{Cp}$ | Liking ratings for the four variants of Cp | | | |
| | Goldberg's Big Five Inventory questionnaire | Personality traits | 30 items | 7-point Likert scale (1 = does not apply to me at all, 7 = applies to me very well) | [58] |
| 3 | Intensity $_{Gj}$ | Intensity ratings for oral sensations from the four Variants of Gj | | generalized Labeled Magnitude Scale (0 = no sensation, 100 = the strongest imaginable sensation of any kind) | [59] |
| | Intensity $_{Cp}$ | Intensity ratings for oral sensations from the four variants of Cp | | | |
| | Screen for Child Anxiety Related Disorders questionnaire | Level of generalized, social, and school anxiety | 17 items | 3-point Likert scale (1 = almost never, 3 = often) | [60] |
| | Body Image Dimensional Assessment questionnaire | Level of body dissatisfaction | 4 items | Line scale from 1.8 to 5.2 | [61] |
| 4 | Intensity $_{PROP}$ | Intensity ratings from two PROP aqueous solutions | | generalized Labeled Magnitude Scale (0 = no sensation, 100 = the strongest imaginable sensation of any kind) | [59] |
| | Patient Health Questionnaire | Severity of depressive symptoms | 9 items | 4-point Likert scale (0 = never; 3 = nearly every day) | [46] |
| | Food-related and lifestyle habits | Monthly frequency of smoking cigarettes and e-cigs | 3 items | 7-point Likert scale (1 = never; 7 = 30 days) | |
| | | Weekly intake of beer, wine, liquors, and cocktails | 4 items | 7-point Likert scale (1 = never; 7 = more than once a day) | |
| | | Weekly intake of snacks and sweetened beverages | 14 items | 7-point Likert scale (1 = never; 7 = more than once a day) | |
| | | Daily use of social networks | 2 items | 5-point Likert scale (1 = less than 1 h; 5 = more than 4 h) | |

Part 1 (Table 2) began with the autonomous collection of a saliva sample and a tongue swab for metagenomic analyses. Participants then rated their familiarity with the same 57 food items assessed remotely and completed the Arnett Inventory of Sensation Seeking to evaluate their inclination for novel, varied, and intense experiences. An additional salivary sample was collected for genotyping before concluding Part 1 (Figure 1).

In Part 2 (Table 2), participants evaluated their liking of two independent sets of four variants each of two model foods (grapefruit juice, dark chocolate pudding) with varying sucrose concentrations. Sensory evaluations were interspersed with a 5 min break featuring logic games to maintain motivation. The following assessment of personality dimensions through the Goldberg's Big Five Inventory marked the end of this slot.

Part 3 (Table 2) focused on psychophysical responses to oral sensations elicited by the same model foods from Part 2. Participants then completed the Screen for Child Anxiety Related Disorders and the Body Image Dimensional Assessment to assess their generalized, social, and school anxiety and estimate their level of body dissatisfaction, respectively.

In Part 4 (Table 2), participants rated the perceived bitterness of two aqueous solutions containing PROP and reported the severity of depressive symptoms experienced in the past 2 weeks through the Patient Health Questionnaire. Next, they were asked about their smoking habits, alcohol, snacks, and sweetened beverages consumption, and habitual use of social networks before concluding the study.

Each task slot (Parts 1–4) was separated by a break (B1–B3) of at least 15 min (Figure 1), during which detailed verbal and practical instructions for the upcoming tasks were provided. Data were collected using the EyeQuestion software (version 5.5.0, Elst, The Netherlands), except for the EPIC-FFQ, which was completed in paper form. Data collection took place between November and December 2023, with lab sessions held from 9:00 AM to 12:00 PM. The following sections provide details on the measures used to achieve the aims of the current work.

### 2.3. *Sensory Tasks*

#### 2.3.1. Stimuli

We aimed to develop four variants of two food models that would elicit a range of oral sensations, including sweet, bitter, sour, astringent, and flavors (grapefruit and chocolate), with the potential to mimic the spectrum of intensities experienced in everyday foods. Given its strong acceptance at high intensities among adolescents [15], sweetness was selected as the target sensation for modulation. Each food model had to (a) be widely available on the Italian market and familiar to our target population; (b) evoke alarming oral sensations that could be masked by increasing sweetness levels while maintaining a congruent sensory profile; (c) be simple to prepare, store, portion, and consume at room temperature; (d) be accepted by omnivores, vegetarians, and vegans. As a result, commercially available 100% grapefruit juice (Gj) and dark chocolate pudding (Cp) base formulations [53] were deemed the most suitable options (Table 3).

For each food model, four different sucrose concentrations were initially chosen based on previous studies using Gj [41] and Cp [53], and expected to elicit an increasingly higher level of sweetness while progressively decreasing relevant alarming oral sensations. Given that these sucrose levels had been tested either in a different Gj base [41] or in adults (Cp) [53], an initial evaluation was conducted with a panel of 39 individuals (58.9% girls, mean age $\pm$ SD = 44.7 $\pm$ 14.5 y) experienced in sensory analysis (Pilot 1). Results (Figure S1a) indicated that slight adaptations to the sucrose span in both Gj and Cp were necessary to more effectively differentiate each series of products. Following these adjustments, a group of adolescents (Pilot 2, n = 16; 26.6% girls, mean age $\pm$ SD = 17.3 $\pm$ 0.4 y) rated the perceived intensities of oral sensations evoked by the optimized food models (Table 3) using the generalized Labeled Magnitude Scale (gLMS, 0 = no sensation, 100 = the strongest imaginable sensation of any kind), which revealed clearer sensory differences among variants of both Gj and Cp (Figure S1b). The suitability of both food models was later corroborated by our population (Part 3, Figure 1), as psychophysical responses (gLMS) to sweetness exhibited a systematic increase, while bitterness, sourness, and astringency decreased as sucrose concentration increased (Figure S2). For all testing, Gj was prepared by dissolving sucrose in the base juice, while the same methodology developed by Monteleone et al. [53] was used for preparation of the Cp variants.

Phenotypic responses to PROP (Part 4, Figure 1) were instead collected in duplicate from 10 mL aqueous solutions with 0.5447 g/L of PROP [53]. All stimuli were prepared the day before, stored at 4 °C overnight, and brought to ambient temperature 2 h prior to testing. Both food models and PROP solutions were served in 80 cc plastic glasses at ~20 °C, coded with a random 3-digit code, and tested under white warm light for liking and red light for intensity tasks (Figure 1).

**Table 3.** Food models and relative variants used in the present study. Ingredients (Brand), order of presentation of each variant within the liking and intensity tasks, sucrose concentrations added to each variant of Gj and Cp (g/kg), and the sensory ballot evaluated are listed. Target sensation is highlighted in bold.

| Product | Ingredients (Brand) | Variant | Sucrose (g/kg) | Order (Liking) | Order (Intensity) | Oral Sensations |
|---|---|---|---|---|---|---|
| Grapefruit juice (Gj) | Sucrose (Zucchero, Eridania S.p.A., Genoa, Italy) | P01 | 0 | 2 | 4 | **Sweet**, Sour, Bitter, Grapefruit |
| | | P02 | 40 | 3 | 1 | |
| | Grapefruit juice 100% (Puertosol, Eurospin Italia S.p.A, San Martino Buon Albergo, Italy) | P03 | 92 | 1 | 3 | |
| | | P04 | 160 | 4 | 2 | |
| Chocolate pudding (Cp) | Sucrose (Zucchero, Eridania S.p.A, Italy) | P01 | 0 | 2 | 4 | **Sweet**, Bitter, Astringent, Chocolate |
| | Chocolate pudding mix (Budino da zuccherare, Cameo S.p.A, Desenzano del Garda, Italy) | P02 | 60 | 3 | 1 | |
| | Cocoa powder (Cacao Amaro Perugina, Nestlé, Assago, Italy) | P03 | 138 | 1 | 3 | |
| | Water | P04 | 239 | 4 | 2 | |

### 2.3.2. Training

Each sensory assessment was preceded by detailed instructions on the psychophysical scaling methods to be used for the upcoming tasks. Before the liking task (Part 2, Figure 1), participants were trained to use the Labeled Affective Magnitude scale (LAM, 0 = greatest imaginable dislike, 100 = greatest imaginable like) according to common practices [57]. Subsequently, before the intensity tasks outlined in Parts 3 and 4 (Figure 1), special attention was given to minimizing potential artifacts in the use of the gLMS (Table 2) by (a) clarifying the meaning of the scale anchors and the sensory attributes being evaluated, providing simple descriptions (e.g., a dry mouth feeling led by unripe fruits for astringency) along with relevant food examples (e.g., lemon juice for sourness), (b) encouraging the use of the full scale to avoid categorical behaviors and to distinguish the intensity of a stimulus from its hedonic value, and (c) guiding participants to base the ratings on their daily sensory experiences across various modalities [40,59,62].

For individual calibration, participants provided psychophysical responses to 10 extraoral stimuli shortly after rating the intensity of oral sensations from each variant of both Gj and Cp series (n = 4 × 2) or PROP solution (n = 2). The items were presented in a fixed order, and stimuli representing various theoretical ranges on the gLMS were included in each set of products. Collectively, systematic differences were observed in parallel with the expected magnitude of the orientation stimuli (Figure S3). Furthermore, the effectiveness of the gLMS training was supported by both girls and boys rating all extraoral stimuli as equally intense, indicating consistent use and interpretation of the scale (Figure S4).

#### 2.3.3. Sensory Assessments

In the hedonic task (Part 2, Figure 1), ratings were obtained from two independent sets of four variants each of Gj and Cp (Table 3) using the LAM scale (Table 2). Participants were instructed to hold the whole sample of Gj (20 mL) or a full spoon of Cp (20 g) in their mouth for 5 s before swallowing and then rate their liking.

The same quantity of sample was used for the intensity assessment (Part 3, Figure 1). Unlike the preceding task, participants were asked to keep each variant of Gj and Cp in their mouth for 7 s, then swallow and wait 5 s before evaluating the perceived intensities using the gLMS scale (Table 2). The Gj series was always presented first, followed by the Cp series after a 5 min break during which participants engaged in logic games to maintain motivation (Figure 1). In contrast, the variants were presented in different fixed orders across the liking and intensity tasks (Table 3). This approach was designed to minimize excessive fluctuations in perceived intensities that could lead to inflated responses, prevent participants from associating the same presentation order with both tasks, and induce similar perceptual biases across individuals, thereby facilitating subsequent comparisons of sensory responses [63]. For the same purpose, the sensory ballot was presented with a fixed sequence (Table 3): the target sensation (sweetness) was rated first, followed by relevant attributes to Gj (sour, bitter) or Cp (bitter, astringency), with flavor evaluated last [63]. Before sensory evaluations, participants were asked to declare any allergies and/or intolerances to Gj and Cp ingredients (Table 3).

For PROP phenotyping (Part 4, Figure 1), participants retained each solution in their mouth for 10 s, then expectorated and waited 20 s before reporting the perceived bitterness (gLMS, Table 2). The average PROP values were used after confirming similar ratings among replicates ($W = 13{,}858$, $p = 0.295$).

A 60 s break was enforced after each tasting, during which participants were provided with mineral water and plain crackers to cleanse their palate.

### 2.4. *Questionnaires*

#### 2.4.1. The Patient Health Questionnaire (PHQ-9)

To assess depressive symptoms, we utilized the Italian version of the 9-item Patient Health Questionnaire [46], which has demonstrated psychometric soundness with diverse Italian adolescent populations, e.g., [64,65]. The PHQ-9 is widely recognized as the gold standard for diagnosing depressive disorders, exhibiting high sensitivity (89%) and specificity (88%) in primary care settings [66]. A 4-point Likert scale (0 = never, 3 = nearly every day) is employed to quantify the frequency of depressive symptoms experienced over the past 2 weeks. The sum of the nine items yields a total score ranging from 0 to 27, with higher values reflecting greater severity of depressive symptoms. In this study, the PHQ-9 demonstrated good internal consistency ($\alpha = 0.818$).

#### 2.4.2. The Screen for Child Anxiety Related Emotional Disorders (SCARED)

Anxiety levels were operationalized using a subset of the 38-item Screen for Child Anxiety Related Emotional Disorders adapted for the Italian adolescent population [60,67]. The SCARED employs a 3-point Likert scale (1 = almost never, 3 = often) to measure the frequency of anxiety-related symptoms across five factors: generalized anxiety (9 items), social anxiety (4 items), school anxiety (4 items), panic disorders (13 items), and separation anxiety (8 items). In the present study, only the generalized, school, and social anxiety subscales were collected. Each domain was scored by summing up the relevant items, with higher values indicating a greater inclination toward anxiety-related symptoms. The internal consistency of the subscales was acceptable to good, with Cronbach's $\alpha$s of 0.821, 0.747, and 0.684 for the generalized, social, and school anxiety, respectively.

#### 2.4.3. The Goldberg's Big Five Questionnaire (BIG-5)

The facets of personality were assessed using the 30-item Goldberg's Big Five questionnaire [68], which has been validated for use with Italian adolescents by Klimstra et al. [58]. Each personality dimension (agreeableness, conscientiousness, emotional stability, extraversion, and openness) was evaluated with six items on a 7-point Likert scale (1 = does not apply to me at all, 7 = applies to me very well). Composite subscale scores were then computed by averaging the ratings for each subscale, with the control items reversed. Each domain showed acceptable to good internal consistency, with Cronbach's $\alpha$s of 0.801 for agreeableness, 0.838 for conscientiousness, 0.778 for emotional stability, 0.845 for extraversion, and 0.671 for openness. For the purposes of this study, the emotional stability dimension was used and reversed to represent its opposite, neuroticism, which reflects the tendency to experience negative affective states [43].

#### 2.4.4. The Adult Picky Eating Questionnaire (APEQ)

As a food-related comorbidity of depressive symptoms [33], the Italian version of the 20-item Adult Picky Eating Questionnaire [55,69,70] was used to evaluate picky eating, defined as the reluctance to try familiar and novel foods [33]. Participants were asked to indicate how frequently they engage in picky eating behaviors on a 5-point Likert scale (1 = "Never"; 5 = "Always") across four dimensions: meal presentation (7 items), food variety (4 items), meal disengagement (3 items), and taste aversion (6 items). Scores were calculated as the average of all 20 items or within each subscale, with higher values indicating greater pickiness.

Due to the absence of validated measures for assessing picky eating among Italian adolescents and our prior validation being specific to adults over 18 [55], only the APEQ global score was used in this study. To preliminarily assess its psychometric properties, we tested its internal consistency ($\alpha = 0.808$), convergent and discriminant validity (Figure S5), and test-retest reliability, all yielding adequate results. For further details, please refer to Appendix A.

#### 2.4.5. The Body Image Dimensional Assessment (BIDA)

Lastly, participants completed the 4-item Body Image Dimensional Assessment [61], which measures the discrepancy between an individual's current body image and their idealized physique. Participants viewed four anonymized body shapes representing increasing weights along with a line scale from 1.8 to 5.2, before rating how closely their actual body shape aligned with their ideal figure (body dissatisfaction), the most attractive shape for the opposite sex (sexual body dissatisfaction), and the appearance of same-sex and -age peers (compared body dissatisfaction). Three sub-scores (ranging from $-100$

to 100) can be computed using the formulas provided by Sánchez-Miguel et al. [61], and a total BIDA score was derived as the mean of their absolute values, with higher scores reflecting greater body dissatisfaction [61].

Although validated only for Spanish adolescents aged 12–15 [61], the BIDA was still preferred over the recently validated 8-item BIBA [71] in the Italian context, as the latter was designed for younger children (ages 6–13). After an initial psychometric check, the BIDA total score showed good internal consistency ($\alpha = 0.763$), strong inter-item correlations (Table S2), and evidence of convergent and discriminant validity (Figure S6). Further details are provided in Appendix B.

2.4.6. Demographic and Lifestyle Variables

Alongside the remote (Table 1) and lab (Table 2) sessions, a range of demographic and lifestyle-related measures were collected. These included, but were not limited to, gender, age, and weight and height, with the latter used to calculate body mass index (BMI) in $kg/m^2$. Additionally, physical activity levels were estimated using the Italian short form of the International Physical Activity Questionnaire (IPAQ) [51].

2.4.7. The EPIC Food Frequency Questionnaire (EPIC-FFQ)

In the week preceding the lab session (Figure 1), participants filled out a paper-based version of the 163-item EPIC-FFQ, which was recently validated for Italian adolescents [45], to monitor their habitual diet over the past year. Participants were asked to report consumption frequencies for both individual foods (e.g., vegetables) and recipes (e.g., pasta, pizza), along with habitual cooking methods (e.g., baked, boiled, fried, grilled) across daily, weekly, monthly, and yearly intervals. To ensure accuracy, the questionnaire included images depicting portion sizes (small, medium, large) for various recipes (e.g., soups, stew) or food items within each food group (e.g., carrots for vegetables, cod for fish). Dietary data were treated according to Pala et al. [72] to estimate daily intake of energy (Kcal) and a list of macro- (e.g., carbohydrates, fats, proteins) and micronutrients (e.g., B vitamins, minerals) before further processing (Section 2.5).

*2.5. Data Analysis*

Given that the majority of variables did not meet the normality assumptions, gender differences in severity of depressive symptoms and their psychosocial correlates (generalized anxiety, neuroticism, picky eating, and body dissatisfaction) were first assessed using the Wilcoxon Rank Sum Test (W). To prepare for further analyses, the PHQ-9 scores, which exhibited right skewness ($\gamma = 1.421$), were transformed using the Yeo–Johnson method to approximate normality ($\gamma = 0.030$). Additionally, psychophysical responses to oral sensations evoked by each variant of Gj (sweet, sour, bitter) and Cp (sweet, bitter, astringent) were individually summed to derive six scores of global responsiveness ($\Sigma$), as outlined by Piochi et al. [73]. Importantly, hedonic and intensity ratings were available for 231 and 227 individuals, as one participant reported constraints to ingredients in Gj, and five in Cp.

Separate general linear models (GLMs) were then fitted to evaluate whether the associations between sensory perception and depressive symptoms differed by gender. In these models, the PHQ-9 scores served as the dependent variable, while both the main effects and the interactions between gender and each of the six taste global scores were included as predictors. To control for possible psychosocial confounders, each GLM was re-run with comorbid states or traits of depression included as covariates. According to the results, the global taste scores that exhibited significant interactions with gender as predictors of depressive symptoms, either with or without adjustments, were retained for further analysis. Data were then stratified by gender, and subsequent analyses were conducted separately for girls and boys.

Next, partial Spearman's rank correlation coefficients (ρ) were computed to examine gender-specific relationships between depressive symptoms and dietary habits, adjusting for age, BMI, and physical activity. Before analysis, nutrient data were screened for misreporting and implausible extreme values, as recommended by Welch et al. [74]. In brief, participants with more than 10 missing items (n = 8) due to misreporting or duplicate responses were excluded [74]. Next, basal metabolic rates specific to gender and age were estimated for each participant using the Schofield equation [75], and outliers (n = 6) were identified as those whose ratio of habitual energy intake to basal metabolic rate fell within the top or bottom 1% of the distribution [76]. As a result, data from 218 participants who had a demographic and lifestyle background similar to that of the original population (Table S3), were retained for analysis. Nutrient data were then individually adjusted for daily energy intake (Kcal) using the residual method [77] to account for variations in energy needs and to prevent inflation of the results.

Lastly, moderation analysis was applied to test whether varying vulnerabilities to correlates of depression led to variations in oral acuity that may prompt undesired dietary choices with long-term effects on mood. Moderation analysis evaluates how the effect of an independent variable (oral acuity) on a dependent variable (nutrient intake) varies with the level of a moderating variable (correlates of depression) [78]. Accordingly, each energy-adjusted nutrient was treated as a dependent variable, with both main effects and interaction terms between each global taste score (independent variable) and each comorbidity of depression (moderator) included as predictors. Again, these analyses were adjusted for age, BMI, and physical activity.

Bootstrapping with 10,000 iterations was used to robustly estimate 95% confidence intervals for all models, and continuous variables were scaled to unit variance prior to analysis to facilitate interpretability. Effect sizes are presented as bootstrapped standardized β coefficients, with squared semi-partial correlations ($sr^2$) included in moderation analysis to quantify the unique variance explained by each predictor [78]. The absence of severe multicollinearity and autocorrelation was confirmed by Variance Inflation Factor (VIF) values < 10 [79] and non-significant results from the Durbin–Watson test ($p > 0.05$), respectively. Lastly, data are summarized as median ± interquartile range (IQR) where applicable, with all tests being two-tailed and statistical significance set at $p < 0.05$. Data analysis was conducted using R version 4.3.1 [80].

## 3. Results

### 3.1. Rates of Depression and Gender Differences in Depressive Symptoms and Comorbid Traits

According to standard clinical cut-offs [46], our sample overall showed mild (PHQ-9 = 5–9) depressive symptoms (median ± IQR = 7 ± 5). Nonetheless, at least moderate depressive symptoms (PHQ-9 ≥ 10) were observed in 21.6% of participants. In detail, 1.7%, 27.6%, 49.1%, 13.4%, 6.0%, and 2.2% of adolescents reported experiencing no (PHQ-9 = 0), minimal (PHQ-9 = 1–4), mild (PHQ-9 = 5–9), moderate (PHQ-9 = 10–14), moderately severe (PHQ-9 = 15–19), and severe (PHQ-9 ≥ 20) depressive symptoms over the past two weeks, respectively. Notably, 36.1% of adolescent girls exhibited at least moderate depressive symptoms compared to 11.1% of boys.

As a result, significant ($p < 0.001$) gender differences in depression and related comorbidities emerged (Figure 2). Girls reported greater depressive symptoms, a higher inclination toward generalized anxiety, increased levels of neuroticism, and more engagement in picky eating behaviors compared to boys. In contrast, no significant differences were found between genders regarding overall body dissatisfaction (W = 6532.5, $p = 0.977$).

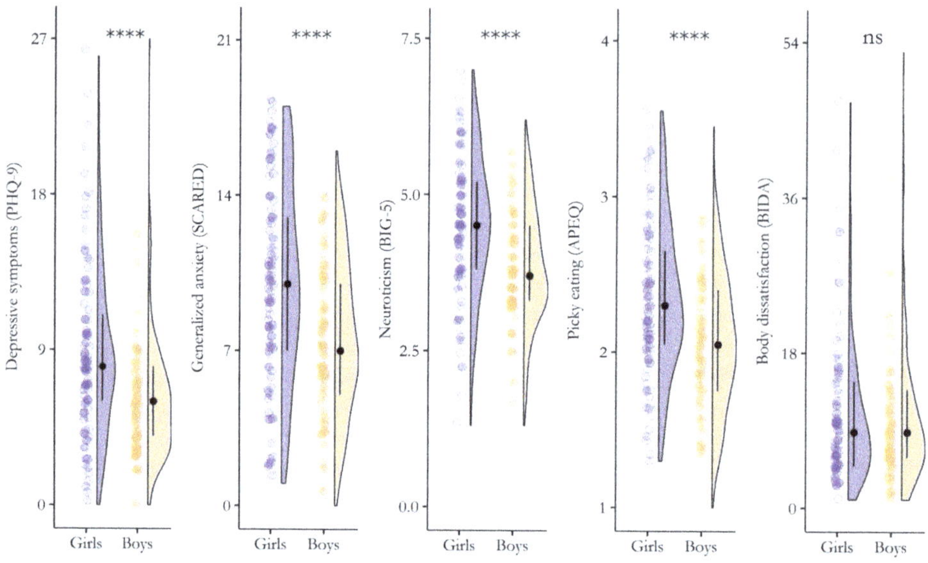


**Figure 2.** Differences (Wilcoxon Rank Sum Test) in the severity of depressive symptoms, generalized anxiety, neuroticism, picky eating behaviors, and body dissatisfaction between girls (slate blue) and boys (light tan). The plot shows raw data points (the "rain"), the kernel density estimate (the "cloud"), and the median (black filled circle) ± IQR (perpendicular black line). **** = $p < 0.0001$, ns = $p > 0.05$.

### *3.2. Associations Between Sensory Perception and Depression by Gender*

Next, we examined whether the associations between PHQ-9 scores and the six global scores of oral acuity varied by gender. Across all models, gender consistently showed a significant main effect on depressive symptoms ($p < 0.001$), whereas the opposite was true for all global taste scores (Table 4). This suggests that gender was a more influential predictor of PHQ-9 scores than orosensory responsiveness.

However, the relationship between sensory perception and depression was gender-specific. Notably, significant interaction effects were found in girls for global acuity for bitterness in Gj and astringency in Cp. In contrast, no significant or almost significant interactions were observed for sweetness in either food model, sourness in Gj ($p = 0.055$), or bitterness in Cp (Table 4). These findings indicate that greater responsiveness to alarming oral sensations predicted higher depressive symptoms in girls, but not in boys.

Similar patterns emerged after adjusting for psychosocial correlates of depression (Table 5). The main effects of global taste scores remained non-significant, while the predictive value of gender lost significance after adjustment. As expected, generalized anxiety and neuroticism were the strongest predictors of PHQ-9 scores across all models, with higher internalizing symptoms systematically linked to higher levels of depression.

Consistent with our previous findings, no significant interactions between gender and sweetness were observed in either food model. Similarly, the positive association between acuity for alarming oral sensations and PHQ-9 scores remained exclusive to girls, with significant interaction effects found for bitterness in Gj and for both bitterness and astringency in Cp (Table 5). Notably, both the crude (Table S4) and adjusted (Table S5) models showed no evidence of severe multicollinearity or autocorrelation (Durbin–Watson tests, $p > 0.05$), with VIFs well below the commonly used threshold of 10 [79]. Based on these results, only the global scores related to alarming oral sensations were retained for further analysis.

**Table 4.** Associations between global sensory responsiveness (Σ) and depressive symptoms as a function of gender. Bootstrapped β estimates, 95% confidence intervals, and *p*-values are provided. Statistically significant main and interaction effects are highlighted in bold. Gender [G]: girls, Gj: grapefruit juice, Cp: dark chocolate pudding.

| | Gj | | | | Cp | | |
|---|---|---|---|---|---|---|---|
| **Predictors** | **β** | **95% CI** | ***p* Value** | **Predictors** | **β** | **95% CI** | ***p* Value** |
| Σ sweet | 0.089 | −0.044–0.246 | 0.210 | Σ sweet | 0.089 | −0.088–0.263 | 0.298 |
| Gender [G] | 0.526 | 0.268–0.792 | **<0.001** | Gender [G] | 0.548 | 0.296–0.795 | **<0.001** |
| Σ sweet × Gender [G] | −0.161 | −0.453–0.156 | 0.310 | Σ sweet × Gender [G] | 0.021 | −0.248–0.270 | 0.897 |
| Σ sour | −0.024 | −0.126–0.093 | 0.670 | Σ bitter | −0.027 | −0.148–0.130 | 0.718 |
| Gender [G] | 0.538 | 0.288–0.786 | **<0.001** | Gender [G] | 0.550 | 0.287–0.808 | **<0.001** |
| Σ sour × Gender [G] | 0.260 | −0.005–0.519 | 0.055 | Σ bitter × Gender [G] | 0.221 | −0.104–0.555 | 0.186 |
| Σ bitter | −0.031 | −0.168–0.124 | 0.671 | Σ astringent | 0.036 | −0.07–0.153 | 0.506 |
| Gender [G] | 0.522 | 0.278–0.766 | **<0.001** | Gender [G] | 0.528 | 0.282–0.761 | **<0.001** |
| Σ bitter × Gender [G] | 0.260 | 0.000–0.503 | **0.050** | Σ astringent × Gender [G] | 0.310 | 0.004–0.520 | **0.048** |
| n | | 231 | | n | | 227 | |

**Table 5.** Associations between global responsiveness (Σ) to oral sensations and depressive symptoms by gender, controlling for psychosocial comorbidities of depression. Bootstrapped standardized β coefficients and 95% confidence intervals, as well as *p*-values, are listed. Statistically significant main and interaction effects are highlighted in bold. Gender [G]: girls, Gj: grapefruit juice Cp: dark chocolate pudding, SCARED: generalized anxiety, BIG-5 [N]: neuroticism, APEQ: picky eating, BIDA: body dissatisfaction.

| | Gj | | | | Cp | | |
|---|---|---|---|---|---|---|---|
| **Predictors** | **β** | **95% CI** | ***p* Value** | **Predictors** | **β** | **95% CI** | ***p* Value** |
| Σ sweet | 0.048 | −0.084–0.205 | 0.499 | Σ sweet | 0.044 | −0.108–0.203 | 0.546 |
| Gender [G] | 0.172 | −0.049–0.409 | 0.125 | Gender [G] | 0.183 | −0.051–0.417 | 0.117 |
| SCARED | 0.312 | 0.162–0.465 | **<0.001** | SCARED | 0.312 | 0.154–0.466 | **<0.001** |
| BIG-5 [N] | 0.204 | 0.045–0.350 | **0.012** | BIG-5 [N] | 0.206 | 0.050–0.356 | **0.007** |
| APEQ | 0.089 | −0.035–0.209 | 0.145 | APEQ | 0.079 | −0.042–0.201 | 0.216 |
| BIDA | 0.030 | −0.077–0.123 | 0.576 | BIDA | 0.029 | −0.083–0.127 | 0.631 |
| Σ sweet × Gender [G] | −0.137 | −0.375–0.092 | 0.229 | Σ sweet × Gender [G] | −0.008 | −0.242–0.203 | 0.920 |
| Σ sour | −0.028 | −0.133–0.101 | 0.645 | Σ bitter | −0.063 | −0.171–0.077 | 0.353 |
| Gender [G] | 0.193 | −0.031–0.427 | 0.091 | Gender [G] | 0.189 | −0.045–0.433 | 0.114 |
| SCARED | 0.299 | 0.148–0.452 | **<0.001** | SCARED | 0.315 | 0.164–0.466 | **<0.001** |
| BIG-5 [N] | 0.210 | 0.057–0.354 | **0.008** | BIG-5 [N] | 0.205 | 0.050–0.349 | **0.007** |
| APEQ | 0.082 | −0.040–0.204 | 0.178 | APEQ | 0.079 | −0.037–0.205 | 0.189 |
| BIDA | 0.034 | −0.067–0.125 | 0.519 | BIDA | 0.050 | −0.067–0.139 | 0.440 |
| Σ sour × Gender [G] | 0.155 | −0.069–0.367 | 0.172 | Σ bitter × Gender [G] | 0.244 | 0.004–0.483 | **0.047** |
| Σ bitter | −0.079 | −0.202–0.059 | 0.242 | Σ astringent | −0.040 | −0.152–0.084 | 0.511 |
| Gender [G] | 0.186 | −0.034–0.418 | 0.094 | Gender [G] | 0.186 | −0.045–0.413 | 0.110 |
| SCARED | 0.294 | 0.145–0.445 | **0.001** | SCARED | 0.306 | 0.155–0.459 | **<0.001** |
| BIG-5 [N] | 0.221 | 0.069–0.362 | **0.005** | BIG-5 [N] | 0.193 | 0.036–0.343 | **0.013** |
| APEQ | 0.074 | −0.046–0.190 | 0.222 | APEQ | 0.079 | −0.043–0.202 | 0.209 |
| BIDA | 0.044 | −0.054–0.133 | 0.382 | BIDA | 0.047 | −0.063–0.143 | 0.403 |
| Σ bitter × Gender [G] | 0.238 | 0.019–0.442 | **0.035** | Σ astringent × Gender [G] | 0.269 | 0.034–0.458 | **0.029** |
| n | | 231 | | n | | 227 | |

### 3.3. *Associations Between Dietary Habits and Depressive Symptoms by Gender*

After stratifying by gender, we examined the correlations between depressive symptoms and dietary habits. Several commonalities emerged between girls and boys, though depressive symptoms were associated with a greater number of dietary outcomes in girls (Table 6).

**Table 6.** Partial correlation between the severity of depressive symptoms and nutrient intake by gender. Spearman's ρ coefficients, adjusted for age, BMI, and physical activity, along with *p*-values, are listed. Statistically significant values ($p < 0.05$) are indicated in bold.

| Nutrients | | Girls | | Boys | |
|---|---|---|---|---|---|
| | | ρ | *p* Value | ρ | *p* Value |
| Carbohydrates | Carbohydrates | 0.171 | **0.004** | 0.053 | 0.303 |
| | Simple sugars | 0.041 | 0.498 | 0.060 | 0.245 |
| | Fibers | −0.161 | **0.007** | −0.049 | 0.344 |
| Fats | Fats | −0.188 | **0.002** | −0.053 | 0.303 |
| | Animal fats | −0.087 | 0.148 | 0.027 | 0.602 |
| | Vegetable fats | −0.192 | **0.001** | −0.120 | **0.021** |
| | Saturated fats | −0.109 | 0.071 | 0.051 | 0.326 |
| | Monounsaturated fats | −0.199 | **0.001** | −0.097 | 0.061 |
| | Polyunsaturated fats | −0.143 | **0.017** | −0.080 | 0.120 |
| | Linoleic acid | −0.137 | **0.023** | −0.066 | 0.202 |
| | Linolenic acid | −0.170 | **0.004** | 0.014 | 0.782 |
| | Oleic acid | −0.205 | **0.001** | −0.106 | **0.041** |
| Proteins | Proteins | −0.166 | **0.006** | −0.103 | **0.047** |
| | Animal proteins | −0.163 | **0.006** | −0.087 | 0.094 |
| | Vegetable proteins | 0.053 | 0.378 | −0.111 | **0.032** |
| Alcohol | Alcohol | 0.176 | **0.003** | 0.148 | **0.004** |
| Minerals | Calcium | −0.064 | 0.284 | 0.031 | 0.544 |
| | Iron | −0.286 | **0.000** | −0.058 | 0.263 |
| | Phosphorus | −0.135 | **0.025** | −0.046 | 0.372 |
| | Potassium | −0.264 | **0.000** | 0.078 | 0.133 |
| | Sodium | −0.026 | 0.672 | −0.093 | 0.071 |
| | Zinc | −0.177 | **0.003** | −0.063 | 0.226 |
| Vitamins | Vitamin A | −0.040 | 0.506 | 0.050 | 0.337 |
| | Vitamin B1 | −0.275 | **0.000** | −0.080 | 0.122 |
| | Vitamin B2 | −0.107 | 0.076 | −0.013 | 0.806 |
| | Vitamin B3 | −0.159 | **0.008** | −0.022 | 0.677 |
| | Vitamin B6 | −0.311 | **0.000** | −0.069 | 0.184 |
| | Vitamin B9 | −0.127 | **0.034** | −0.044 | 0.399 |
| | Vitamin C | −0.110 | 0.068 | 0.069 | 0.184 |
| | Vitamin D | −0.201 | **0.001** | 0.048 | 0.357 |
| | Vitamin E | −0.192 | **0.001** | −0.133 | **0.010** |
| | β-carotene | −0.131 | **0.029** | 0.001 | 0.981 |
| n | | 93 | | 125 | |

In both genders, PHQ-9 scores were positively correlated ($p < 0.05$) with alcohol intake and exhibited inverse associations with daily consumption of proteins, vegetable fats, oleic acid, and vitamin E (Table 6).

It is noteworthy that, in addition to a positive correlation with carbohydrate intake, depressive symptoms were inversely correlated ($p < 0.05$) with the intake of a lengthy list of beneficial nutrients exclusively in girls. These included, but were not limited to, fibers, polyunsaturated fats, vitamin B6, vitamin D, and zinc. In contrast, habitual intake of vegetable proteins showed an inverse correlation with PHQ-9 scores in boys, but not in girls (Table 6).

### 3.4. Moderation Analysis

Lastly, we performed several moderation analyses to examine whether varying vulnerabilities to comorbidities of depression influenced the link between global responsiveness to alarming oral sensations and diet. In general, psychosocial correlates of depression primarily moderated the associations between dietary habits and responsiveness to alarming oral sensations in girls (Figure 3). For girls, the moderation effect from all significant models (F $p$-value < 0.05) accounted for additional variance ($sr^2$) ranging from 1.8% to 7.8% for generalized anxiety, 0.0% to 7.7% for neuroticism, 1.0% to 6.8% for picky eating, and 4.7% to 9.8% for body dissatisfaction. In contrast, for boys, the incremental variance attributed to these variables was generally lower and often non-significant ($p > 0.05$), ranging from 0.0% to 5.5% for generalized anxiety, 0.5% to 4.1% for neuroticism, 0.0% to 3.1% for picky eating, and 0.0% to 0.9% for body dissatisfaction.

Specifically, comorbidities of depression moderated ($p < 0.05$) the association between responsiveness to bitterness, astringency, and (to a lesser extent) sourness and the consumption of various healthful nutrients (Figure 3). For instance, generalized anxiety ($sr^2 = 0.078$), neuroticism ($sr^2 = 0.057$), and picky eating ($sr^2 = 0.045$) moderated the negative association between bitter taste acuity and habitual fiber intake. Similarly, neuroticism exerted a moderating role on the inverse association between bitterness perception and habitual intake of proteins ($sr^2 = 0.040$), calcium ($sr^2 = 0.077$), and phosphorus ($sr^2 = 0.040$). Additionally, body dissatisfaction was found to moderate the negative association between the intake of vitamins B1 ($sr^2 = 0.050$), B2 ($sr^2 = 0.050$), B3 ($sr^2 = 0.062$), and B6 ($sr^2 = 0.092$) and responsiveness to bitterness (in Cp), astringency, bitterness (in Gj), and sourness, respectively (Figure 3). In other words, adolescent girls with higher levels of comorbid traits of depression showed stronger negative associations between responsiveness to alarming oral sensations and the consumption of these nutrients.

Importantly, all models were free from multicollinearity (VIF < 10) [79] and autocorrelation issues, with VIFs ranging from 1.005 to 1.837 for girls and 1.014 to 1.259 for boys. Additionally, the results from the Durbin–Watson test were non-significant ($p > 0.05$) across all analyses.

To facilitate comprehension of the moderating effects, Figure 4 depicts examples of how varying levels of correlates of depression influence the strength of associations between sensory perception and dietary intake by gender. Full details on bootstrapped standardized $\beta$ coefficients and 95% confidence intervals, as well as $p$-values, $sr^2$, and adjusted $R^2$ for all significant models (F $p$-value < 0.05) highlighting a moderating effect of generalized anxiety, neuroticism, picky eating, and body dissatisfaction on the relationship between acuity for alarming oral sensations and nutrient intake by gender are provided in Tables S6–S9.

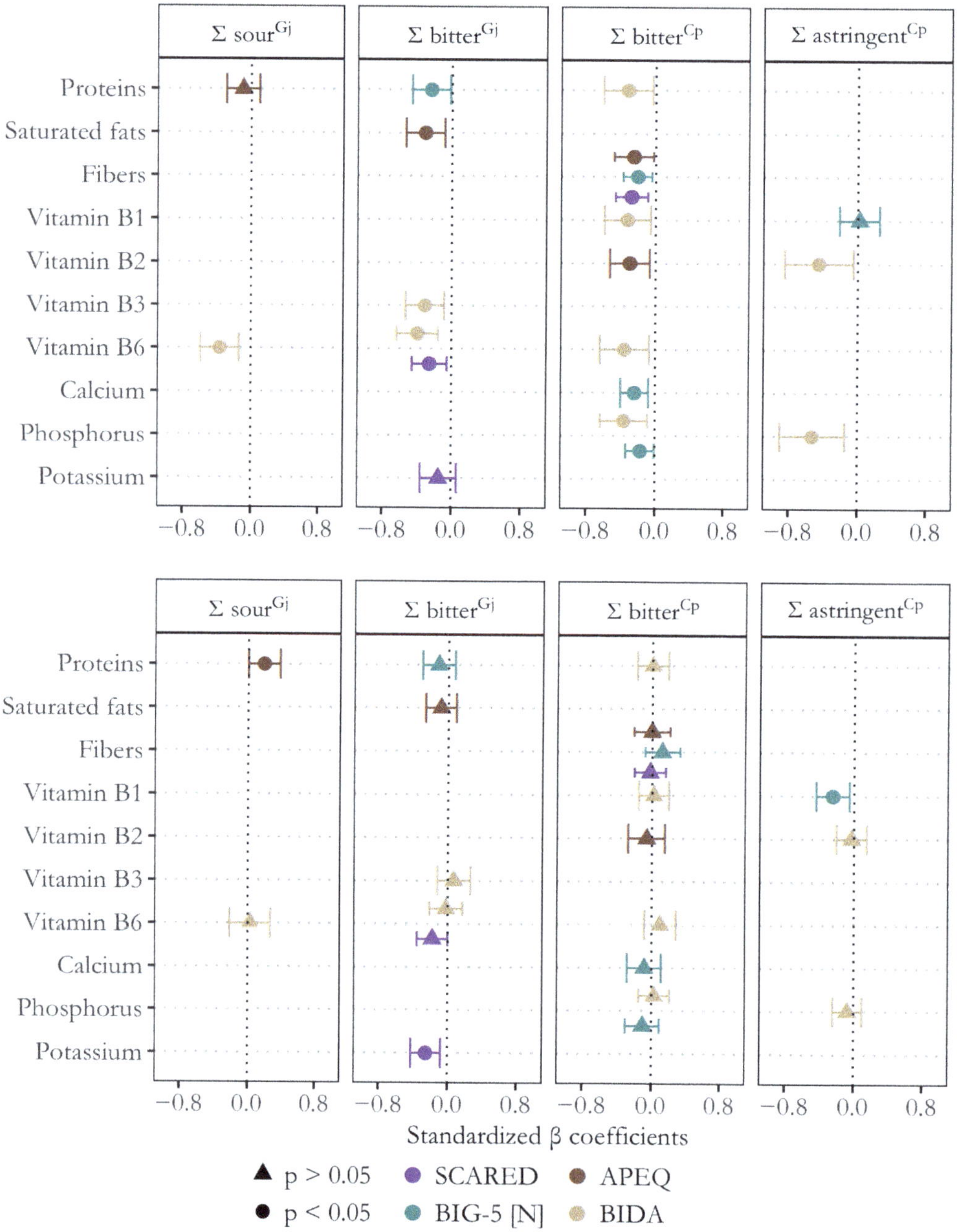


**Figure 3.** Significant moderating effects of depression comorbidities on the links between global responsiveness (Σ) to alarming oral sensations and nutrient intake in girls (**top**) and boys (**bottom**). The plot shows bootstrapped standardized β coefficients and 95% confidence intervals, along with *p*-values (circles: $p < 0.05$, triangles: $p > 0.05$). SCARED (generalized anxiety), BIG-5 [N] (neuroticism), APEQ (picky eating), and BIDA (body dissatisfaction).

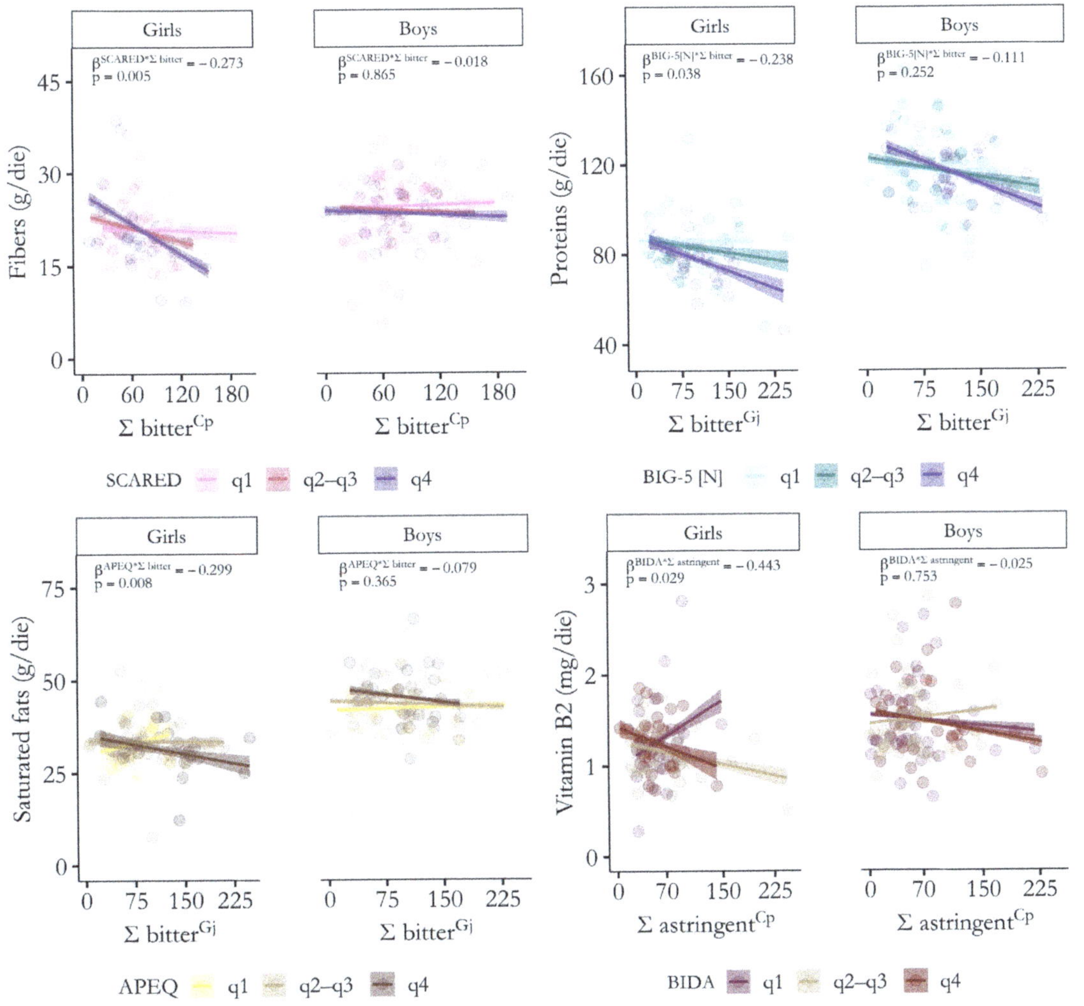


**Figure 4.** Significant moderating effects of depression comorbidities on the relationship between global responsiveness (Σ) to alarming oral sensations and dietary habits by gender. Simple slopes illustrate how internalizing symptom severity (q1: lowest quartile, q2–q3: second and third quartiles, q4: highest quartile) may exacerbate the association between sensory perception and nutrient intake (fibers: top left, proteins: top right, saturated fats: bottom left, vitamin B2: bottom right). Bootstrapped standardized β coefficients and *p*-values for the interaction term (Σ taste * correlates of depression) from moderation analyses are shown. SCARED (generalized anxiety), BIG-5 [N] (neuroticism), APEQ (picky eating), and BIDA (body dissatisfaction).

## 4. Discussion

In this study, we investigated how gender influences the associations between sensory perception, depression, and diet during adolescence. Given the differing vulnerabilities between girls and boys, e.g., [7], we further examined whether common comorbidities of depression moderated the impact of sensory perception on dietary choices. Our findings first confirmed that girls are more susceptible to mood disturbances and revealed that acuity for bitterness and astringency is positively associated with depressive symptoms exclusively in this group. Additionally, we corroborated known associations between depression and poor dietary habits and found that higher levels of internalizing symptoms in girls exacerbated the inverse relationship between heightened responsiveness to alarming oral sensations and the intake of several nutrients.

### *4.1. Adolescent Girls Exhibit Higher Levels of Internalizing Symptoms than Boys*

The concerning global trends in the prevalence of depression among adolescents, e.g., [3,7], were also evident in our sample. Using a PHQ-9 score $\geq$ 10 [46] as a sensitive (85%) and specific (89%) indicator of early major depressive disorders [66], 21.6% of adolescents exceeded this threshold. This finding is consistent with recent data from similar regions, where 16% of European [3] and 20.8% of Italian adolescents [81] are currently at risk of developing clinical depression. Moreover, we corroborated established gender differences in susceptibility to internalizing symptoms. In our sample, girls reported depressive symptoms at a rate three times higher than boys (36.1% vs. 11.1%), thus slightly surpassing the 2017–2018 estimates (28.7% vs. 10.4%) from a larger Italian sample [81] and underscoring the growing incidence of mental health difficulties among adolescents [3,7].

It was therefore unsurprising to observe girls exhibiting higher levels of common comorbid traits of depression, such as generalized anxiety, neuroticism, and picky eating [7,33,42,43]. However, interpreting the latter finding is less straightforward. The literature on variations in the engagement of picky eating behaviors by gender is sparse and inconsistent, often focusing on younger children and varying in assessment tools and definitions of picky eating itself [82]. For instance, Marchi and Cohen [83] found that pickiness was more common in girls (n = 326) than in boys (n = 333) across the ages of 1–10, 9–18, and 11–21 over a 10-year follow-up. In contrast, no significant gender differences were found in other cross-sectional [84] or prospective [85] studies.

Despite this mixed evidence, it is reasonable to speculate that our findings reflect underlying gender differences in domains of anxiety, which have been reported to prompt selective eating behaviors, e.g., [33]. This is further supported by systematic gender disparities observed in the links between pickiness and domains of anxiety within our sample, with stronger associations in girls for generalized (girls: $\rho = 0.334$, $p < 0.001$; boys: $\rho = 0.205$, $p = 0.017$), social (girls: $\rho = 0.313$, $p = 0.002$; boys: $\rho = 0.193$, $p = 0.025$), and school anxiety (girls: $\rho = 0.371$, $p < 0.001$; boys: $\rho = 0.205$, $p = 0.017$). However, further research is needed to confirm or refute this hypothesis within diverse nonclinical adolescent populations.

Surprisingly, our data also revealed no gender variations in overall body dissatisfaction. This may tentatively be attributed to the multifaceted nature of the BIDA total score, which assesses various aspects of body dissatisfaction. Indeed, girls felt their actual body shape was more rounded compared to their ideal figure, whereas boys perceived their physique to be leaner than what they viewed as most attractive to the opposite sex (Figure S7). Hence, while our findings contrast with those of Sánchez-Miguel et al. [61], they align with the current understanding that weight concerns and the pursuit of muscularity are key contributors of body dissatisfaction for girls and boys, respectively (for review, see [86]).

### *4.2. Greater Responsiveness to Bitterness and Astringency Is Associated with Higher Severity of Depressive Symptoms in Adolescent Girls, but Not in Boys*

We found no main effects of sensory responsiveness on depressive symptoms, thus suggesting that taste *per se* is not a relevant predictor of depressive symptoms in healthy adolescents. While comparisons with the existing literature should be made cautiously due to differing methods for assessing taste function that may not be correlated [40], our result aligns with prior research involving nonclinical adults showing no associations between depression and thresholds for sweet, sour, bitter, or salty tastes [35], intensity ratings of chocolate and vanilla milks [35], or recalled saltiness intensities for 10 food items [36]. However, it contrasts with studies observing either positive [31,34] or negative [37] associations between depressive symptoms and intensity ratings for sweet and bitter-tasting aqueous solutions.

Nonetheless, we provide evidence that sensory perception may differentially predict mood disturbances by gender. Specifically, responsiveness to bitterness and astringency was associated with the severity of depressive symptoms exclusively in girls, an effect observed in both the crude and adjusted models. This finding is somewhat consistent with a Japanese work involving young adults (n = 70, mean age ± SD = 20.7 ± 0.3 y) that explored gender- and menstrual-cycle-related associations between internalizing symptoms (depression, anxiety) and sweet taste recognition thresholds [87]. In that study, depression was positively correlated with sweet taste sensitivity (i.e., decreased thresholds) uniquely in women during the luteal phase, with no correlations detected in men or in women during the follicular phase.

However, a similarity with the findings of Nagai et al. [87] exists in the gender-specific association between a high responsiveness to oral stimuli and the severity of internalizing symptoms. Intriguingly, the luteal phase group was the sole group to show a positive correlation between sweet sensitivity and anxiety, which was independent of baseline differences in anxiety between genders or menstrual cycle groups [87]. While the authors hypothesized that this pattern might be attributed to higher levels of luteinizing hormones [87], we can further speculate that the predictive value of sensory perception regarding depressive symptoms might become more evident when comparing individuals experiencing distinct physiological states and/or possessing varying vulnerabilities to psychological constructs that predispose them to heightened affective responses [88,89].

Although the menstrual cycle was not considered in our study, we still lend support to this notion through three key points. First, adolescent girls exhibited higher anxiety-related traits relative to boys. Second, putative physiological influences on findings can be reasonably ruled out, as no gender differences were observed in either PROP acuity (Figure S8) or global responsiveness to oral sensations (Figure S9). Lastly, external cues are unlikely to fully account for the observed differences, as both genders had similar demographic and lifestyle backgrounds (Table S1), were equally familiar (Food Familiarity Questionnaire, Figure 1) with grapefruit juice (W = 6158, $p = 0.423$) and dark chocolate (W = 6051, $p = 0.294$), and reported similar hedonic ratings for all variants of Gj and Cp, except for the highest sucrose concentration in both food models (Table S1).

### *4.3. Comorbidities of Depression Moderate the Links Between Sensory Perception and Dietary Habits with Long-Term Adverse Effects on Mood in Adolescent Girls*

Consistent with the existing literature, e.g., [10–13], we noted that various dietary outcomes were inversely correlated with PHQ-9 scores, with this relationship being more pronounced in girls. Specifically, alcohol use combined with low intake of proteins, vegetable fats, oleic acid, and vitamin E was associated with depressive symptoms in both genders. Similar patterns, though not statistically significant in boys, were observed with low daily consumption of animal proteins, mono- and polyunsaturated fats, linoleic acid, B vitamins (B1, B6, B9), iron, phosphorus, and zinc.

Several mechanisms have been proposed to explain how inadequate intake of these nutrients might exacerbate depression, and vice versa. Depression may prompt unhealthy eating habits, and nutrient deficiencies may impair brain function by disrupting the synthesis and activity of monoamines (serotonin, norepinephrine, dopamine) or by increasing oxidative stress, both of which are thought to be involved in the pathophysiology of depression (for review, see [14]). For example, amino acids from both animal and plant proteins are key for monoamine production, while omega-3 fatty acids (e.g., linolenic acid) enhance serotonin function and are believed to counteract oxidative stress through their potent anti-inflammatory effects. Similarly, B-group vitamins and iron support neurotransmitter metabolism, and zinc is essential for the optimal functioning of brain regions involved in mood regulation and cognitive functions [10]. Consequently, it is not surprising that dietary supplementation with nutrients involved in the homeostasis of brain functions has frequently led to significant improvements in depressive symptoms across diverse adolescent populations [10]. Our findings thus reinforce the substantial evidence supporting a mutual link between adequate nutrition and mental health during adolescence.

In this context, we observed that comorbidities of depression exacerbated the negative relationships between acuity for alarming oral stimuli and the intake of various nutrients associated with higher PHQ-9 scores, particularly among girls. Importantly, the overall direction of these effects was consistent across genders and aligned with current knowledge, e.g., [15,90], with greater responsiveness generally serving as a barrier to consuming beneficial nutrients that also support mood regulation (Tables S6–S9). However, in models accounting for the moderating effects of correlates of depression, significant associations emerged almost exclusively in girls. This suggests that other biological, non-biological, or psychosocial variables may better explain how variations in perception of alarming oral sensations affect dietary habits in adolescent boys, warranting further investigation.

Conversely, we can tentatively discuss why depression-related states or traits might strengthen the association between sensory perception and suboptimal dietary choices in girls. Alarming oral sensations, which signal potentially aversive post-ingestive effects, naturally elicit negative arousal. Even minor stimulation can trigger a cascade of psychobiological responses, which individuals may cope with differently before deciding whether to accept or reject the stimulus. Beyond intensity, the complexity and novelty of sensory stimuli also contribute to arousal, with responses shaped by expectations about the upcoming experience. A discrepancy between expected and perceived sensations may therefore lead to exaggerated attention on specific product cues, thereby intensifying negative emotional states [39]. This might explain why the associations between bitterness in Cp and diet were more pronounced than those observed in Gj. It is possible that participants had well-formed expectations about the taste qualities of both food models, with bitterness being more anticipated in Gj than in Cp. Therefore, a mismatch between expected and actual taste might have evoked negative arousal, which was possibly further promoted by participants' greater familiarity with chocolate puddings ($W = 838$, $p < 0.001$). This, in turn, would have magnified the discrepancy between anxiety-related traits and highlighted stronger moderating effects. However, further studies are needed to confirm this hypothesis.

Alternatively, it is noteworthy that a higher vulnerability to internalizing symptoms often reflects a generalized hyperresponsiveness to environmental inputs and stressors, such as noise, tactile stimuli [91], pain [92] or danger-signaling smells [93]. While this trend is observed in both girls and boys, notable gender differences in coping strategies for stress exist [89]. Generally speaking, girls are more likely to notice and report even subtle emotional changes, while boys tend to exhibit more passive behaviors that shift toward a more feminine coping style only once signs of distress become evident [89,94]. Such heightened reactivity to stressors might thus translate into intensified attentional bias and negative affective states in response to negatively arousing stimuli, thereby reinforcing avoidant behaviors when alarming oral sensations are experienced in food. Conversely, a similar pattern might only emerge in boys when depression-related states or traits and extremely intense arousing sensations occur concurrently. To either confirm or refute this hypothesis, future studies with adolescent populations exhibiting similar levels of psychosocial correlates of depression among genders and exposed to food stimuli that combine multiple sources of arousal are necessary.

### 4.4. Strengths and Limitations

To our knowledge, this study is the first to examine the associations between depressive symptoms, taste perception, and dietary habits in a nonclinical adolescent population. Its key strengths include the use of one of the largest samples ever employed in this line of research, a thorough data collection protocol, and the use of real food models to assess a range of oral sensations at intensities that closely mirror real-life experiences. In response to calls for novel research on the role of taste in adolescent food choices [24], we also extend the existing literature by testing previously overlooked oral sensations (sourness, astringency). Furthermore, this work presents the first empirical evidence linking astringency perception with depressive symptoms in healthy adolescent girls. Lastly, we provide preliminary support for the psychometric validity of the APEQ and the BIDA for use with Italian adolescents, which have the potential to offer deeper insights into the determinants of eating habits in this age group, pending their comprehensive psychometric validation.

Nevertheless, our results should also be interpreted in light of several limitations. Firstly, while we employed ecologically valid food stimuli and sensory assessments, we did not test other relevant tastes (salty, umami) or chemesthetic sensations (e.g., pungency) that may influence adolescents' dietary habits. Additionally, our focus was on a limited set of food models, which elicited a narrow range of sensory qualities typically encountered in daily experiences. Whether the results can be replicated with a broader variety of foods differing in sensory, chemical, and physical properties should thus be probed in future investigations.

Secondly, the dietary data were based on self-reports, which can be subject to motivational and recall biases, as well as under- or over-reporting practices [95]. Despite meticulous data preprocessing to minimize these biases and a strong alignment with existing knowledge, the potential for inaccuracies remains. In order to enhance the reliability of data, future research should incorporate more precise and less burdensome approaches, such as technological aids that facilitate memory and portion estimation, ideally within repeated measures frameworks. In this vein, the integration of 24-hour recalls with digital tools (e.g., mobile apps, image-based assessments) holds promise in mitigating prevalent sources of error in adolescent populations.

Thirdly, the gender-based sensitivity analysis may have lacked sufficient statistical power to detect subtle yet meaningful effects, particularly given the relatively modest variance explained by the moderation models (Tables S6–S9). Besides suggesting that other factors, including genetic, non-genetic (e.g., oral microbiome), and psychosocial variables not included in this study, might further elucidate how taste perception influences dietary outcomes and potentially affects mood in the long term, this prompts the need to devote future efforts to include larger and adequately powered samples.

Fourthly, the internalizing symptoms investigated herein are inherently interrelated and might have acted as confounding or overlapping variables in the observed associations. Case–control studies or novel research involving non-clinically gender-stratified groups that differ in key psychosocial traits will be instrumental in further validating and extending our findings. Lastly, due to the study's design, causality can not be inferred. Novel longitudinal studies are thus needed to determine whether taste perception indirectly exacerbates depressive symptoms through inadequate food choices or whether the opposite is true.

### 4.5. Conclusions

In addition to confirming a greater susceptibility to mood disturbances in girls compared to boys, along with associations between internalizing symptoms and habitual nutrient intake, this study methodologically contributes to expanding knowledge on picky eating and body image in Italian adolescents by supporting the initial psychometric validity of tools currently used across different age groups (APEQ) or geographic contexts (BIDA). More importantly, we present the first evidence linking heightened responsiveness to bitterness and astringency, evoked by variants of food models designed to mimic a range of intensities encountered in daily food experiences, with greater severity of depressive symptoms exclusively in girls. While our findings require replication across a broader range of food stimuli, oral sensations, adolescent populations, and longitudinal frameworks, they remain highly relevant in supporting future gender-tailored, sensory-based strategies aimed at promoting adherence to dietary patterns that may confer long-term benefits for mental health. In conclusion, this study advances our understanding of gender-specific vulnerabilities to mood disturbances during adolescence and suggests that the links between depressive symptoms, taste perception, and dietary habits are gender-specific, possibly shaped by differing coping strategies for internalizing symptoms.

**Supplementary Materials:** The following supporting information can be downloaded at: https://www.mdpi.com/article/10.3390/nu17101653/s1, Figure S1: Sensory profiles of variants for each food model (Gj, Cp) evaluated in Pilot 1 and Pilot 2; Figure S2: Variations in psychophysical responses (gLMS) to oral sensations evoked by Gj and Cp variants across increasing sucrose concentrations; Figure S3: Differences in recalled intensities between the extraoral stimuli employed during the gLMS training; Figure S4: Girls and boys rated the ten extraoral stimuli employed during the gLMS training as equally intense; Figure S5: Correlations between the APEQ total score and both theoretically similar and dissimilar constructs; Figure S6: Correlations between the BIDA total and subscale scores and theoretically similar and dissimilar factors; Figure S7: Differences in body dissatisfaction and sexual body dissatisfaction between girls and boys; Figure S8: Variations in PROP responsiveness between genders; Figure S9: Differences in global sensory acuity by gender; Figure S10: Differences between genders in liking for all variants of grapefruit juice and dark chocolate pudding; Table S1: Demographic, lifestyle, and diet-related characteristics of participants; Table S2: Inter-item correlations between subscale scores of the Body Image Dimensional Assessment (BIDA); Table S3: Demographic, lifestyle, and diet-related characteristics of participants whose FFQ data passed the screening for misreporting and outliers; Table S4: Variance Inflation Factor and results from the Durbin-Watson test for the unadjusted general linear models assessing whether the relationship between global taste scores and depressive symptoms varies by gender; Table S5: Variance Inflation Factor and results from the Durbin-Watson test for the adjusted general linear models assessing whether the relationship between global scores of sensory responsiveness and depressive symptoms varies by gender; Table S6: Moderating effects of generalized anxiety on the link between global responsiveness to alarming oral sensations and habitual nutrient intake, as a function of gender; Table S7: Moderating effects of neuroticism on the link between global responsiveness to alarming oral sensations and habitual nutrient intake, as a function of gender; Table S8: Moderating effects of picky eating on the link between global responsiveness to alarming oral sensations and habitual nutrient intake, as a function of gender; Table S9: Moderating effects of body dissatisfaction on the link between global responsiveness to alarming oral sensations and habitual nutrient intake, as a function of gender.

**Author Contributions:** Conceptualization, L.M., L.F., S.C., I.E., M.P.C., P.G. and F.G.; Data curation, L.M.; Formal analysis, L.M.; Funding acquisition, L.M., P.G. and F.G.; Investigation, L.M., L.F., S.C., I.E. and F.G.; Methodology, L.M., L.F., I.E. and F.G.; Project administration, F.G.; Software, L.M.; Supervision, F.G.; Validation, L.M.; Visualization, L.M.; Writing—original draft, L.M.; Writing—review and editing, L.M., L.F., S.C., I.E., M.P.C., P.G. and F.G. All authors have read and agreed to the published version of the manuscript.

**Funding:** We acknowledge the support of the MUR PNRR project iNEST—Interconnected Nord-Est Innovation Ecosystem (ECS00000043), funded by the European Union under NextGenerationEU. In addition, we recognize the funding provided by the PROmedLIFE project—Novel food products for the promotion of a Mediterranean lifestyle and healthy diet—grant agreement nr. 2132 (PRIMA programme supported by the European Union) for the support given to Lara Fontana.

**Institutional Review Board Statement:** Ethical approval for the involvement of human subjects in this study was granted by the Research Ethics Committee of the University of Trento, Reference number 2023-047, 28 September 2023.

**Informed Consent Statement:** Written informed consent was obtained from parents or legal guardians of participants, while written assent was secured from all adolescents involved in this study.

**Data Availability Statement:** Data are available from the corresponding author upon reasonable requests.

**Acknowledgments:** We extend our sincere thanks to all the volunteers and their families who participated in this study. Special thanks go to Luana Budano and the teachers from Istituto Tecnico Agrario, San Michele all'Adige (Trento, Italy), as well as to Caterina Bonapace and Federico Andermarcher from Istituto Martino Martini, Mezzolombardo (Trento, Italy), for their invaluable assistance in optimizing logistics throughout the project phases. We also wish to express our gratitude to Martina Moretton and Jessica Zambanini (Edmund Mach Foundation, San Michele all'Adige, Trento, Italy) for their significant contribution during data collection. Furthermore, we greatly appreciate the support of Sabina Sieri and Chiara Roncallo (Istituto Nazionale dei Tumori, Milan) for their indispensable help in processing the EPIC-FFQ.

**Conflicts of Interest:** The authors declare no conflicts of interest.

**Disclaimer:** During the preparation of this work the authors used ChatGPT (version 4o mini) in order to improve grammar and readability of the manuscript. After using this tool, the authors reviewed and edited the content as needed and take full responsibility for the content of the publication.

## Appendix A

In this work, we only used the APEQ total score due to our previous validation being specific to adults over 18 [55] and the lack of validated measures for picky eating in Italian adolescents. Hence, to preliminarily assess its suitability for this age group, we tested the APEQ score for internal reliability (Cronbach's $\alpha$), convergent and discriminant validity (Spearman's rho coefficients), and for test-retest reliability (Intra-Class Coefficients, ICC). Overall, we found good internal consistency ($\alpha = 0.808$), adequate evidence of convergent or discriminant validity (Figure S5), and test-retest reliability. In line with previous reports, e.g., [33,55,69,70], the APEQ total score was positively correlated ($p < 0.001$) with theoretically similar states or traits such as depressive symptoms, domains of anxiety, food neophobia, and neuroticism, while showing no significant correlations with either BMI ($\rho = 0.012$, $p = 0.861$) or age ($\rho = 0.014$, $p = 0.829$). Additionally, several inverse correlations between picky eating and familiarity with plant-based foods (Figure S5) further supported the discriminant validity of the APEQ. Finally, the APEQ total score exhibited good test-retest reliability (ICC = 0.751; 95% CI: 0.669–0.813, $p < 0.001$) when reassessed in a subset of our sample (n = 190, 34.2% girls, mean age $\pm$ SD = 14.4 $\pm$ 0.64 y, mean BMI $\pm$ SD = 21.1 $\pm$ 3.2 kg/m$^2$) 30 days after data collection.

## Appendix B

Although the BIDA has been validated only for Spanish adolescents (ages 12–15) [61], it was still favored over the recently validated 8-item BIBA in the Italian context [71], as the latter was designed for children and early adolescents (ages 6–13). Despite this, the BIDA showed preliminary psychometric soundness, as evidenced by (a) a good internal consistency ($\alpha = 0.763$); (b) an acceptable 30-day test-retest reliability (n = 190, ICC = 0.557, 95% CI: 0.410–0.668, $p < 0.001$); (c) strong inter-item correlations ($\rho > 0.500$, $p < 0.001$) (Table S2); and (d) an initial proof of its convergent and divergent validity (Figure S6). Particularly, convergent validity was supported by positive correlations between the BIDA composite score and its subscales with BMI [61] and anxiety domains, while divergent validity was corroborated by a negative association between the BIDA total score and engagement in physical activity [96] (Figure S6).

## References


1. World Health Organization. Depressive Disorder (Depression). Available online: https://www.who.int/news-room/fact-sheets/detail/depression (accessed on 19 February 2025).
2. World Health Organization. Mental Health of Adolescents. Available online: https://www.who.int/news-room/fact-sheets/detail/adolescent-mental-health (accessed on 24 July 2024).
3. Shorey, S.; Ng, E.D.; Wong, C.H.J. Global Prevalence of Depression and Elevated Depressive Symptoms among Adolescents: A Systematic Review and Meta-analysis. *Br. J. Clin. Psychol.* **2022**, *61*, 287–305. [CrossRef] [PubMed]
4. Racine, N.; McArthur, B.A.; Cooke, J.E.; Eirich, R.; Zhu, J.; Madigan, S. Global Prevalence of Depressive and Anxiety Symptoms in Children and Adolescents During COVID-19. *JAMA Pediatr.* **2021**, *175*, 1142. [CrossRef] [PubMed]
5. Essau, C.A.; Lewinsohn, P.M.; Seeley, J.R.; Sasagawa, S. Gender Differences in the Developmental Course of Depression. *J. Affect. Disord.* **2010**, *127*, 185–190. [CrossRef] [PubMed]
6. Hankin, B.L.; Abramson, L.Y.; Moffitt, T.E.; Silva, P.A.; McGee, R.; Angell, K.E. Development of Depression from Preadolescence to Young Adulthood: Emerging Gender Differences in a 10-Year Longitudinal Study. *J. Abnorm. Psychol.* **1998**, *107*, 128–140. [CrossRef]
7. Keyes, K.M.; Platt, J.M. Annual Research Review: Sex, Gender, and Internalizing Conditions among Adolescents in the 21st Century—Trends, Causes, Consequences. *J. Child Psychol. Psychiatry* **2024**, *65*, 384–407. [CrossRef]
8. Caspi, A.; Houts, R.M.; Ambler, A.; Danese, A.; Elliott, M.L.; Hariri, A.; Harrington, H.L.; Hogan, S.; Poulton, R.; Ramrakha, S.; et al. Longitudinal Assessment of Mental Health Disorders and Comorbidities Across 4 Decades Among Participants in the Dunedin Birth Cohort Study. *JAMA Netw. Open* **2020**, *3*, e203221. [CrossRef]
9. Ghafari, M.; Nadi, T.; Bahadivand-Chegini, S.; Doosti-Irani, A. Global Prevalence of Unmet Need for Mental Health Care among Adolescents: A Systematic Review and Meta-Analysis. *Arch. Psychiatr. Nurs.* **2022**, *36*, 1–6. [CrossRef]
10. Chopra, C.; Mandalika, S.; Kinger, N. Does Diet Play a Role in the Prevention and Management of Depression among Adolescents? A Narrative Review. *Nutr. Health* **2021**, *27*, 243–263. [CrossRef]
11. Oddy, W.H.; Hickling, S.; Smith, M.A.; O'Sullivan, T.A.; Robinson, M.; de Klerk, N.H.; Beilin, L.J.; Mori, T.A.; Syrette, J.; Zubrick, S.R.; et al. Dietary Intake of Omega-3 Fatty Acids and Risk of Depressive Symptoms in Adolescents. *Depress. Anxiety* **2011**, *28*, 582–588. [CrossRef]
12. Sinclair, R.; Millar, L.; Allender, S.; Snowdon, W.; Waqa, G.; Jacka, F.; Moodie, M.; Petersen, S.; Swinburn, B. The Cross-Sectional Association between Diet Quality and Depressive Symptomology amongst Fijian Adolescents. *PLoS ONE* **2016**, *11*, e0161709. [CrossRef]
13. Huang, P.; O'Keeffe, M.; Elia, C.; Karamanos, A.; Goff, L.M.; Maynard, M.; Cruickshank, J.K.; Harding, S. Fruit and Vegetable Consumption and Mental Health across Adolescence: Evidence from a Diverse Urban British Cohort Study. *Int. J. Behav. Nutr. Phys. Act.* **2019**, *16*, 19. [CrossRef] [PubMed]
14. Pano, O.; Martínez-Lapiscina, E.H.; Sayón-Orea, C.; Martinez-Gonzalez, M.A.; Martinez, J.A.; Sanchez-Villegas, A. Healthy Diet, Depression and Quality of Life: A Narrative Review of Biological Mechanisms and Primary Prevention Opportunities. *World J. Psychiatry* **2021**, *11*, 997–1016. [CrossRef] [PubMed]
15. Daly, A.N.; O'Sullivan, E.J.; Kearney, J.M. Considerations for Health and Food Choice in Adolescents. *Proc. Nutr. Soc.* **2022**, *81*, 75–86. [CrossRef]
16. Drewnowski, A.; Mennella, J.A.; Johnson, S.L.; Bellisle, F. Sweetness and Food Preference. *J. Nutr.* **2012**, *142*, 1142S–1148S. [CrossRef] [PubMed]

17. Köster, E.P. Diversity in the Determinants of Food Choice: A Psychological Perspective. *Food Qual. Prefer.* **2009**, *20*, 70–82. [CrossRef]
18. Diószegi, J.; Llanaj, E.; Ádány, R. Genetic Background of Taste Perception, Taste Preferences, and Its Nutritional Implications: A Systematic Review. *Front. Genet.* **2019**, *10*, 1272. [CrossRef]
19. Bartoshuk, L.M.; Duffy, V.B.; Miller, I.J. PTC/PROP Tasting: Anatomy, Psychophysics, and Sex Effects. *Physiol. Behav.* **1994**, *56*, 1165–1171. [CrossRef]
20. Feeney, E.L.; O'Brien, S.A.; Scannell, A.G.M.; Markey, A.; Gibney, E.R. Genetic and Environmental Influences on Liking and Reported Intakes of Vegetables in Irish Children. *Food Qual. Prefer.* **2014**, *32*, 253–263. [CrossRef]
21. Borazon, E.Q.; Villarino, B.J.; Magbuhat, R.M.T.; Sabandal, M.L. Relationship of PROP (6-n-Propylthiouracil) Taster Status with Body Mass Index, Food Preferences, and Consumption of Filipino Adolescents. *Food Res. Int.* **2012**, *47*, 229–235. [CrossRef]
22. Joseph, P.V.; Reed, D.R.; Mennella, J.A. Individual Differences Among Children in Sucrose Detection Thresholds. *Nurs. Res.* **2016**, *65*, 3–12. [CrossRef]
23. Baranowski, T.; Baranowski, J.C.; Watson, K.B.; Jago, R.; Islam, N.; Beltran, A.; Martin, S.J.; Nguyen, N.; Tepper, B.J. 6-n-Propylthiouracil Taster Status Not Related to Reported Cruciferous Vegetable Intake among Ethnically Diverse Children. *Nutr. Res.* **2011**, *31*, 594–600. [CrossRef] [PubMed]
24. Bawajeeh, A.O.; Albar, S.A.; Zhang, H.; Zulyniak, M.A.; Evans, C.E.L.; Cade, J.E. Impact of Taste on Food Choices in Adolescence—Systematic Review and Meta-Analysis. *Nutrients* **2020**, *12*, 1985. [CrossRef] [PubMed]
25. Barragán, R.; Coltell, O.; Portolés, O.; Asensio, E.M.; Sorlí, J.V.; Ortega-Azorín, C.; González, J.I.; Sáiz, C.; Fernández-Carrión, R.; Ordovas, J.M.; et al. Bitter, Sweet, Salty, Sour and Umami Taste Perception Decreases with Age: Sex-Specific Analysis, Modulation by Genetic Variants and Taste-Preference Associations in 18 to 80 Year-Old Subjects. *Nutrients* **2018**, *10*, 1539. [CrossRef]
26. Overberg, J.; Hummel, T.; Krude, H.; Wiegand, S. Differences in Taste Sensitivity between Obese and Non-Obese Children and Adolescents. *Arch. Dis. Child.* **2012**, *97*, 1048–1052. [CrossRef] [PubMed]
27. Spence, C. Do Men and Women Really Live in Different Taste Worlds? *Food Qual. Prefer.* **2019**, *73*, 38–45. [CrossRef]
28. Spence, C. What Is the Link between Personality and Food Behavior? *Curr. Res. Food Sci.* **2022**, *5*, 19–27. [CrossRef]
29. Ileri-Gurel, E.; Pehlivanoglu, B.; Dogan, M. Effect of Acute Stress on Taste Perception: In Relation with Baseline Anxiety Level and Body Weight. *Chem. Senses* **2013**, *38*, 27–34. [CrossRef]
30. Dess, N.K.; Edelheit, D. The Bitter with the Sweet: The Taste/Stress/Temperament Nexus. *Biol. Psychol.* **1998**, *48*, 103–119. [CrossRef]
31. Platte, P.; Herbert, C.; Pauli, P.; Breslin, P.A.S. Oral Perceptions of Fat and Taste Stimuli Are Modulated by Affect and Mood Induction. *PLoS ONE* **2013**, *8*, e65006. [CrossRef]
32. Spinelli, S.; Pierguidi, L.; Gavazzi, G.; Dinnella, C.; De Toffoli, A.; Prescott, J.; Monteleone, E. Skin Conductance Responses to Oral Stimuli: The Role of Taste Quality and Intensity, and Personality Traits. *Food Qual. Prefer.* **2023**, *109*, 104917. [CrossRef]
33. Kauer, J.; Pelchat, M.L.; Rozin, P.; Zickgraf, H.F. Adult Picky Eating. Phenomenology, Taste Sensitivity, and Psychological Correlates. *Appetite* **2015**, *90*, 219–228. [CrossRef] [PubMed]
34. Dess, N.K.; Chapman, C.D. Individual Differences in Taste, Body Weight, and Depression in the "Helplessness" Rat Model and in Humans. *Brain Res. Bull.* **1990**, *24*, 669–676. [CrossRef] [PubMed]
35. Scinska, A.; Sienkiewicz-Jarosz, H.; Kuran, W.; Ryglewicz, D.; Rogowski, A.; Wrobel, E.; Korkosz, A.; Kukwa, A.; Kostowski, W.; Bienkowski, P. Depressive Symptoms and Taste Reactivity in Humans. *Physiol. Behav.* **2004**, *82*, 899–904. [CrossRef]
36. Ferraris, C.; Scarlett, C.J.; Bucher, T.; Beckett, E.L. Liking of Salt Is Associated with Depression, Anxiety, and Stress. *Chem. Senses* **2023**, *48*, 1–13. [CrossRef]
37. Al'absi, M.; Nakajima, M.; Hooker, S.; Wittmers, L.; Cragin, T. Exposure to Acute Stress Is Associated with Attenuated Sweet Taste. *Psychophysiology* **2012**, *49*, 96–103. [CrossRef]
38. Berlyne, D.E. Arousal and Reinforcement. In *Nebraska Symposium on Motivation*; University of Nebraska Press: Lincoln, NE, USA, 1967.
39. Prescott, J.; Spinelli, S. Arousal and the Modulation of Sensory Experience: Evidence from Food-Related Emotions. *Philos. Trans. R. Soc. B Biol. Sci.* **2024**, *379*, 20230255. [CrossRef]
40. Webb, J.; Bolhuis, D.P.; Cicerale, S.; Hayes, J.E.; Keast, R. The Relationships Between Common Measurements of Taste Function. *Chemosens. Percept.* **2015**, *8*, 11–18. [CrossRef]
41. Ervina, E.; Almli, V.L.; Berget, I.; Spinelli, S.; Sick, J.; Dinnella, C. Does Responsiveness to Basic Tastes Influence Preadolescents' Food Liking? Investigating Taste Responsiveness Segment on Bitter-Sour-Sweet and Salty-Umami Model Food Samples. *Nutrients* **2021**, *13*, 2721. [CrossRef]
42. Cummings, C.M.; Caporino, N.E.; Kendall, P.C. Comorbidity of Anxiety and Depression in Children and Adolescents: 20 Years After. *Psychol. Bull.* **2014**, *140*, 816. [CrossRef]

43. Kuyken, W.; Watkins, E.; Holden, E.; Cook, W. Rumination in Adolescents at Risk for Depression. *J. Affect. Disord.* **2006**, *96*, 39–47. [CrossRef]
44. Goldfield, G.S.; Moore, C.; Henderson, K.; Buchholz, A.; Obeid, N.; Flament, M.F. Body Dissatisfaction, Dietary Restraint, Depression, and Weight Status in Adolescents. *J. Sch. Health* **2010**, *80*, 186–192. [CrossRef] [PubMed]
45. Huybrechts, I.; Miglio, R.; Mistura, L.; Grioni, S.; Pozzebon, I.; Odorifero, C.; Borea, R.; Gitto, A.; Terrafino, M.; Scipioni, M. Relative Validity of an Italian EPIC Food Frequency Questionnaire for Dietary Factors in Children and Adolescents. A Rizzoli Orthopedic Institute Study. *Nutrients* **2021**, *13*, 1245. [CrossRef] [PubMed]
46. Kroenke, K.; Spitzer, R.L.; Williams, J.B.W. The PHQ-9: Validity of a brief depression severity measure. *J. Gen. Intern. Med.* **2001**, *16*, 606–613. [CrossRef] [PubMed]
47. Menghi, L.; Cliceri, D.; Fava, F.; Pindo, M.; Gaudioso, G.; Giacalone, D.; Gasperi, F. Salivary Microbial Profiles Associate with Responsiveness to Warning Oral Sensations and Dietary Intakes. *Food Res. Int.* **2023**, *171*, 113072. [CrossRef]
48. Champely, S. *pwr: Basic Functions for Power Analysis*; R package Version 1.3-0; Stephane Champely: Lyon, France, 2020; Volume 1.
49. *ISO 8589*; Sensory analysis—General Guidance for the Design of Test Rooms. International Organization for Standardization: Genève, Switzerland, 2007.
50. De Backer, C.J.S.; Hudders, L. Meat Morals: Relationship between Meat Consumption Consumer Attitudes towards Human and Animal Welfare and Moral Behavior. *Meat Sci.* **2015**, *99*, 68–74. [CrossRef]
51. Mannocci, A.; Di Thiene, D.; Del Cimmuto, A.; Masala, D.; Boccia, A.; De Vito, E. International Physical Activity Questionnaire: Validation and Assessment in an Italian Sample. *Ital. J. Public Health* **2010**, *7*, 369–376. [CrossRef]
52. Buunk-Werkhoven, Y.A.B.; Dijkstra, A.; Van Der Schans, C.P. Determinants of Oral Hygiene Behavior: A Study Based on the Theory of Planned Behavior. *Community Dent. Oral Epidemiol.* **2011**, *39*, 250–259. [CrossRef]
53. Monteleone, E.; Spinelli, S.; Dinnella, C.; Endrizzi, I.; Laureati, M.; Pagliarini, E.; Sinesio, F.; Gasperi, F.; Torri, L.; Aprea, E.; et al. Exploring Influences on Food Choice in a Large Population Sample: The Italian Taste Project. *Food Qual. Prefer.* **2017**, *59*, 123–140. [CrossRef]
54. Laureati, M.; Spinelli, S.; Monteleone, E.; Dinnella, C.; Prescott, J.; Cattaneo, C.; Proserpio, C.; De Toffoli, A.; Gasperi, F.; Endrizzi, I.; et al. Associations between Food Neophobia and Responsiveness to "Warning" Chemosensory Sensations in Food Products in a Large Population Sample. *Food Qual. Prefer.* **2018**, *68*, 113–124. [CrossRef]
55. Menghi, L.; Endrizzi, I.; Cliceri, D.; Zampini, M.; Giacalone, D.; Gasperi, F. Validating the Italian Version of the Adult Picky Eating Questionnaire. *Food Qual. Prefer.* **2022**, *101*, 104647. [CrossRef]
56. Smorti, M.; Guarnieri, S. A Study on the Validity of the Arnett Inventory of Sensation Seeking (AISS) in an Italian Adolescent Sample. *Int. J. Adv. Psychol.* **2013**, *2*, 10–18.
57. Schutz, H.G.; Cardello, A.V. A Labeled Affective Magnitude (LAM) Scale for Assessing Food Liking/Disliking. *J. Sens. Stud.* **2001**, *16*, 117–159. [CrossRef]
58. Klimstra, T.A.; Crocetti, E.; Hale, W.W., III; Fermani, A.; Meeus, W.H.J. Big Five Personality Dimensions in Italian and Dutch Adolescents: A Cross-Cultural Comparison of Mean-Levels, Sex Differences, and Associations with Internalizing Symptoms. *J. Res. Pers.* **2011**, *45*, 285–296. [CrossRef]
59. Bartoshuk, L.M.; Duffy, V.B.; Green, B.G.; Hoffman, H.J.; Ko, C.W.; Lucchina, L.A.; Marks, L.E.; Snyder, D.J.; Weiffenbach, J.M. Valid Across-Group Comparisons with Labeled Scales: The GLMS versus Magnitude Matching. *Physiol. Behav.* **2004**, *82*, 109–114. [CrossRef]
60. Crocetti, E.; Hale, W.W.; Fermani, A.; Raaijmakers, Q.; Meeus, W. Psychometric Properties of the Screen for Child Anxiety Related Emotional Disorders (SCARED) in the General Italian Adolescent Population: A Validation and a Comparison between Italy and The Netherlands. *J. Anxiety Disord.* **2009**, *23*, 824–829. [CrossRef]
61. Sánchez-Miguel, P.A.; Vaquero-Solís, M.; Sánchez-Oliva, D.; Pulido-González, J.J.; Segura-García, C.; Tapia-Serrano, M.A. Validation of the Body Image Dimensional Assessment in Adolescents from Spanish High School. *Eat. Weight. Disord.* **2021**, *26*, 1749–1756. [CrossRef]
62. Hayes, J.E.; Allen, A.L.; Bennett, S.M. Direct Comparison of the Generalized Visual Analog Scale (GVAS) and General Labeled Magnitude Scale (GLMS). *Food Qual. Prefer.* **2013**, *28*, 36–44. [CrossRef]
63. Menghi, L.; Cliceri, D.; Fava, F.; Pindo, M.; Gaudioso, G.; Stefani, E.; Giacalone, D.; Gasperi, F. Variations in Oral Responsiveness Associate with Specific Signatures in the Gut Microbiota and Modulate Dietary Habits. *Food Qual. Prefer.* **2023**, *106*, 104790. [CrossRef]
64. Graziano, S.; Spanò, B.; Majo, F.; Righelli, D.; Vincenzina, L.; Quittner, A.; Tabarini, P. Rates of Depression and Anxiety in Italian Patients with Cystic Fibrosis and Parent Caregivers: Implementation of the Mental Health Guidelines. *Respir. Med.* **2020**, *172*, 106147. [CrossRef]

65. Correale, C.; Falamesca, C.; Tondo, I.; Borgi, M.; Cirulli, F.; Truglio, M.; Papa, O.; Vagnoli, L.; Arzilli, C.; Venturino, C.; et al. Depressive Anxiety Symptoms in Hospitalized Children with Chronic Illness during the First Italian COVID-19 Lockdown. *Children* **2022**, *9*, 1156. [CrossRef]
66. Manea, L.; Gilbody, S.; McMillan, D. Optimal Cut-off Score for Diagnosing Depression with the Patient Health Questionnaire (PHQ-9): A Meta-Analysis. *Can. Med. Assoc. J.* **2012**, *184*, E191–E196. [CrossRef] [PubMed]
67. Birmaher, B.; Khetarpal, S.; Brent, D.; Cully, M.; Balach, L.; Kaufman, J.; Neer, S.M.K. The Screen for Child Anxiety Related Emotional Disorders (SCARED): Scale Construction and Psychometric Characteristics. *J. Am. Acad. Child Adolesc. Psychiatry* **1997**, *36*, 545–553. [CrossRef] [PubMed]
68. Goldberg, L.R. The Development of Markers for the Big-Five Factor Structure. *Psychol. Assess.* **1992**, *4*, 26. [CrossRef]
69. Ellis, J.M.; Galloway, A.T.; Mary Webb, R.; Martz, D.M. Measuring Adult Picky Eating: The Development of a Multidimensional Self-Report Instrument. *Psychol. Assess.* **2017**, *29*, 955–966. [CrossRef]
70. He, J.; Ellis, J.M.; Zickgraf, H.F.; Fan, X. Translating, Modifying, and Validating the Adult Picky Eating Questionnaire for Use in China. *Eat. Behav.* **2019**, *33*, 78–84. [CrossRef]
71. Segura-Garcia, C.; Aloi, M.; Carbone, E.A.; Staltari, F.A.; Rania, M.; Papaianni, M.C.; Vaquero-Solís, M.; Tapia-Serrano, M.Á.; Sánchez-Miguel, P.A.; Liuzza, M.T. Development, Validation, and Measurement Invariance of the Body Image Bidimensional Assessment (BIBA) in Italian and Spanish Children and Early Adolescent Samples. *Int. J. Environ. Res. Public Health* **2023**, *20*, 3626. [CrossRef]
72. Pala, V.; Sieri, S.; Palli, D.; Salvini, S.; Berrino, F.; Bellegotti, M.; Frasca, G.; Tumino, R.; Sacerdote, C.; Fiorini, L.; et al. Diet in the Italian Epic Cohorts: Presentation of Data and Methodological Issues. *Tumori J.* **2003**, *89*, 594–607. [CrossRef]
73. Piochi, M.; Dinnella, C.; Spinelli, S.; Monteleone, E.; Torri, L. Individual Differences in Responsiveness to Oral Sensations and Odours with Chemesthetic Activity: Relationships between Sensory Modalities and Impact on the Hedonic Response. *Food Qual. Prefer.* **2021**, *88*, 104112. [CrossRef]
74. Welch, A.A.; Luben, R.; Khaw, K.T.; Bingham, S.A. The CAFE Computer Program for Nutritional Analysis of the EPIC-Norfolk Food Frequency Questionnaire and Identification of Extreme Nutrient Values. *J. Hum. Nutr. Diet.* **2005**, *18*, 99–116. [CrossRef]
75. Schofield, W.N. Predicting Basal Metabolic Rate, New Standards and Review of Previous Work. *Hum. Nutr. Clin. Nutr.* **1985**, *39*, 5–41.
76. Mazzeo, T.; Roncoroni, L.; Lombardo, V.; Tomba, C.; Elli, L.; Sieri, S.; Grioni, S.; Bardella, M.T.; Agostoni, C.; Doneda, L.; et al. Evaluation of a Modified Italian European Prospective Investigation into Cancer and Nutrition Food Frequency Questionnaire for Individuals with Celiac Disease. *J. Acad. Nutr. Diet.* **2016**, *116*, 1810–1816. [CrossRef] [PubMed]
77. Willett, W.; Howe, G.; Kushi, L. Adjustment for Total Energy Intake in Epidemiologic Studies. *Am. J. Clin. Nutr.* **1997**, *65*, 1220S–1228S. [CrossRef] [PubMed]
78. Hayes, A.F. *Introduction to Mediation, Moderation, and Conditional Process Analysis: A Regression-Based Approach*; Guilford publications: New York, NY, USA, 2017; ISBN 146253466X.
79. O'Brien, R.M. A Caution Regarding Rules of Thumb for Variance Inflation Factors. *Qual. Quant.* **2007**, *41*, 673–690. [CrossRef]
80. R Core Team. *R: A Language and Environment for Statistical Computing*; R Foundation for Statistical Computing: Vienna, Austria, 2023.
81. Donato, F.; Triassi, M.; Loperto, I.; Maccaro, A.; Mentasti, S.; Crivillaro, F.; Elvetico, A.; Croce, E.; Raffetti, E. Symptoms of Mental Health Problems among Italian Adolescents in 2017–2018 School Year: A Multicenter Cross-Sectional Study. *Environ. Health Prev. Med.* **2021**, *26*, 67. [CrossRef]
82. Taylor, C.M.; Wernimont, S.M.; Northstone, K.; Emmett, P.M. Picky/Fussy Eating in Children: Review of Definitions, Assessment, Prevalence and Dietary Intakes. *Appetite* **2015**, *95*, 349–359. [CrossRef]
83. Marchi, M.; Cohen, P. Early Childhood Eating Behaviors and Adolescent Eating Disorders. *J. Am. Acad. Child Adolesc. Psychiatry* **1990**, *29*, 112–117. [CrossRef]
84. Jacobi, C.; Schmitz, G.; Stewart Agras, W. Is Picky Eating an Eating Disorder? *Int. J. Eat. Disord.* **2008**, *41*, 626–634. [CrossRef]
85. Mascola, A.J.; Bryson, S.W.; Agras, W.S. Picky Eating during Childhood: A Longitudinal Study to Age 11 years. *Eat. Behav.* **2010**, *11*, 253–257. [CrossRef]
86. Karazsia, B.T.; Murnen, S.K.; Tylka, T.L. Is Body Dissatisfaction Changing across Time? A Cross-Temporal Meta-Analysis. *Psychol. Bull.* **2017**, *143*, 293–320. [CrossRef]
87. Nagai, M.; Matsumoto, S.; Endo, J.; Sakamoto, R.; Wada, M. Sweet Taste Threshold for Sucrose Inversely Correlates with Depression Symptoms in Female College Students in the Luteal Phase. *Physiol. Behav.* **2015**, *141*, 92–96. [CrossRef]
88. Handy, A.B.; Greenfield, S.F.; Yonkers, K.A.; Payne, L.A. Psychiatric Symptoms Across the Menstrual Cycle in Adult Women: A Comprehensive Review. *Harv. Rev. Psychiatry* **2022**, *30*, 100–117. [CrossRef] [PubMed]

89. Hankin, B.L.; Mermelstein, R.; Roesch, L. Sex Differences in Adolescent Depression: Stress Exposure and Reactivity Models. *Child Dev.* **2007**, *78*, 279–295. [CrossRef] [PubMed]
90. Dinnella, C.; Morizet, D.; Masi, C.; Cliceri, D.; Depezay, L.; Appleton, K.M.; Giboreau, A.; Perez-Cueto, F.J.A.; Hartwell, H.; Monteleone, E. Sensory Determinants of Stated Liking for Vegetable Names and Actual Liking for Canned Vegetables: A Cross-Country Study among European Adolescents. *Appetite* **2016**, *107*, 339–347. [CrossRef]
91. Ben-Sasson, A.; Carter, A.S.; Briggs-Gowan, M.J. Sensory Over-Responsivity in Elementary School: Prevalence and Social-Emotional Correlates. *J. Abnorm. Child Psychol.* **2009**, *37*, 705–716. [CrossRef]
92. Wilner, J.G.; Vranceanu, A.-M.; Blashill, A.J. Neuroticism Prospectively Predicts Pain among Adolescents: Results from a Nationally Representative Sample. *J. Psychosom. Res.* **2014**, *77*, 474–476. [CrossRef]
93. Cortese, B.M.; Uhde, T.W.; Schumann, A.Y.; McTeague, L.M.; Sege, C.T.; Calhoun, C.D.; Danielson, C.K. Anxiety-Related Shifts in Smell Function in Children and Adolescents. *Chem. Senses* **2021**, *46*, bjab051. [CrossRef]
94. Shi, P.; Yang, A.; Zhao, Q.; Chen, Z.; Ren, X.; Dai, Q. A Hypothesis of Gender Differences in Self-Reporting Symptom of Depression: Implications to Solve Under-Diagnosis and Under-Treatment of Depression in Males. *Front. Psychiatry* **2021**, *12*, 589687. [CrossRef]
95. Thompson, F.E.; Subar, A.F. Dietary Assessment Methodology. In *Nutrition in the Prevention and Treatment of Disease*; Elsevier: Amsterdam, The Netherlands, 2017; pp. 5–48.
96. Añez, E.; Fornieles-Deu, A.; Fauquet-Ars, J.; López-Guimerà, G.; Puntí-Vidal, J.; Sánchez-Carracedo, D. Body Image Dissatisfaction, Physical Activity and Screen-Time in Spanish Adolescents. *J. Health Psychol.* **2018**, *23*, 36–47. [CrossRef]

nutrients

MDPI

*Article*

# Development of a Trauma-Informed, Culturally Sensitive Eating-Disorder-Specific Nutrition-Focused Physical Examination Tool: A Modified Delphi Study

**Donna Gallagher, Alainn Bailey, Laura Byham-Gray, Diane Rigassio-Radler and Jane Ziegler ***

Department of Clinical and Preventive Nutrition Sciences, School of Health Professions, Rutgers Health, Newark, NJ 07107, USA; eatpeacefully@gmail.com (D.G.); aab297@shp.rutgers.edu (A.B.); byhamgld@shp.rutgers.edu (L.B.-G.); rigassdl@shp.rutgers.edu (D.R.-R.)

* Correspondence: ziegleja@shp.rutgers.edu; Tel.: +1-973-972-3274

**Abstract: Objective:** This study aimed to identify components of a trauma-informed, culturally sensitive eating-disorder-specific nutrition-focused physical examination (ED-NFPE) tool. **Method:** The modified Delphi methodology was used to develop a consensus on the clinical relevance of nine evidence-informed ED-specific nutrition examination domains and 46 components within those domains. Purposive sampling was used to recruit a panel of registered dietitian nutritionist (RDN) experts in the ED field. The panelists responded to survey statements on a five-point Likert scale. The panelists also provided qualitative feedback on domain and component additions, exclusions, modifications, and trauma-informed culturally sensitive examination practice techniques. **Results:** Twenty-two RDN expert panelists completed Round One of the study, and eighteen panelists completed Round Two (82% retention). Twenty-one were female. Fifteen panelists had ten or more years of experience in ED dietetics. Fifty percent held an advanced practice credential from the International Association of Eating Disorders Professionals (IAEDP). After the two survey rounds, the nine ED-NFPE domains and 46 of the 48 components achieved a consensus for clinical relevance. The panelists proposed two new examination components and provided qualitative feedback for trauma-informed culturally sensitive practice techniques in all nine domains. **Conclusions:** This modified Delphi study design was chosen to reach a consensus on developing an ED-NFPE tool, as there are few current evidence-based guidelines for nutrition examinations in ED care. An NFPE tool specifically designed to detect the nutrition-related findings of individuals with EDs would strengthen the overall nutrition assessment. RDNs at every level of care and with all degrees of experience could use an ED-NFPE to inform patient treatment goals.

**Keywords:** eating disorders; trauma-informed; culturally sensitive; nutrition assessment; nutrition-focused physical examination; body image; weight-inclusive; registered dietitian nutritionist; Delphi study

## 1. Introduction

Eating disorders (EDs) are complex biopsychosocial illnesses [1]. Individuals of every age, weight, shape, and size can develop an ED [2]. They often affect marginalized and vulnerable populations such as people of color, certain religious groups, people that are lesbian, gay, bisexual, transgender, queer, and others (LGBTQ+), and those with neurodivergence and physical disabilities [3–7]. Individuals with EDs may have co-occurring mental or medical illnesses and a history of trauma [8–13]. Nine percent, or 28.8 million Americans, will experience an ED in their lifetime [14]. The global lifetime ED prevalence is nearly eight percent [15].

All ED subtypes are associated with nutrition-related medical complications, and al-though each individual with an ED may have a unique presentation, complications occur primarily due to ED behaviors such as restrictive eating, bingeing, purging, enema use, laxative, diuretic, and other substance abuse, overhydration, dehydration, excessive exercise, and others [2,16–28]. Complications of ED behaviors can be physical, psychological, and nutritional and include cardiorespiratory complications, bone, muscle, and fat changes, hydration and electrolyte imbalances, impaired cognition and psychosocial functioning, intraoral and extraoral issues, dermatologic abnormalities, parotid gland enlargement, gastrointestinal symptoms, and other complications affecting one or more organ systems (Figure 1) [13,16–26,29–53]. Awareness of unique ED-related complications, timely ED-specific nutritional, psychiatric, psychological, and medical assessments, and treatment are critical factors in ED recovery [54–56]. Most complications associated with EDs can be reversed through effective treatment, including the cessation of ED behaviors, proper nutrition rehabilitation, and weight restoration (if needed) [18,37,57].

**Figure 1.** Nutrition-related signs and symptoms of eating disorder behaviors [13,16–26,29–53].

In addition to a comprehensive assessment of an individual's eating and exercise patterns, height and weight history, and family history of EDs or other psychiatric illnesses, the American Psychiatric Association (APA) recommends that multidisciplinary initial physical assessments of individuals with EDs be performed [58]. These include the evaluation of body temperature, heart rate, resting and orthostatic blood pressure, pulse, height, weight, body mass index (BMI), and physical signs of malnutrition and purging behaviors [58]. The APA also recommends a wide-ranging biochemical assessment, an electrocardiogram, and a complete review of body systems [58]. Lastly, the APA emphasizes that individuals with EDs should receive a culturally appropriate, person-centered treatment plan that includes psychological, psychiatric, nutritional, and medical experts [58].

Although national and international organizations recommend comprehensive assessments, there are few current ED-specific nutrition assessment tools, clinical guidelines, or standards for examining or assessing this population for nutrition-related complications [3,27,58–70]. Table 1 outlines the primary ED diagnoses, several ED behavior types, and co-occurring considerations for an ED nutrition assessment.

**Table 1.** Eating disorder diagnoses, behavior types, and co-occurring considerations.

| Eating Disorder Diagnoses [28] | Eating Disorder Behavior Types [27,28,45,71–73] | Co-occurring Considerations [2,4,12,13,28,74–82] |
|---|---|---|
| Anorexia nervosa restricting type | Restrictive eating [28] | Anxiety disorders [12] |
| Anorexia nervosa binge/purge type | Bingeing [28] | Bipolar disorder [12] |
| Avoidant/restrictive food intake disorder | Purging [28] | Borderline personality disorder [12] |
| Binge eating disorder | Laxative/enema abuse [27] | Major depressive disorder [12] |
| Bulimia nervosa | Diuretic abuse [27] | Obsessive-compulsive disorder [12] |
| Other specified feeding or eating disorder | Overhydration | Autism spectrum disorder [5,7] |
| Pica | Dehydration [20] | Attention-deficit disorder [5] |
| Rumination disorder | Chewing and spitting [27] | Intellectual developmental disorder [5] |
| Unspecified feeding or eating disorder | Ipecac abuse [20] | Substance use disorders [12] |
| | Regurgitating and re-chewing food [28] | Trauma history [10] |
| | Over exercise [27] | Associated medical conditions [13] |
| | Night eating [83] | Socioeconomic status/food insecurity [84] |
| | Insulin manipulation (type 1 diabetes) [85] | Age [66,77] |
| | Eating non-food items [28] | Biological sex [2] |
| | Supplement misuse [27] | Gender identity [4] |
| | Alcohol or drug abuse [86] | Sexual orientation [4,80] |
| | | Body weight and size [87,88] |
| | | Race [74] |
| | | Ethnicity [74] |
| | | Religion [74] |
| | | Physical disability [74] |
| | | Athlete [75] |
| | | Extreme concern over weight, shape [89] |

Registered dietitian nutritionists (RDNs) play a central role in ED care, as nutritional rehabilitation is one of the cornerstones of short- and long-term ED treatment [3,27]. The Academy of Nutrition and Dietetics (AND) standards of practice for RDNs include performing a comprehensive nutrition assessment, establishing a nutrition diagnosis(es), developing a nutrition intervention plan, monitoring and evaluating this plan, and conducting patient discharge or modifying the level of care provided [3,59,90,91].

The nutrition assessment domains of the nutrition care process include (1) client history, (2) anthropometric measurements, (3) biochemical, medical, and diagnostic tests, (4) food and nutrition history, (5) nutrition-focused physical examination (NFPE), and (6) ongoing reassessment(s), monitoring, and evaluation [3,42,59,63,64,70,90,92–95].

According to the APA, in individuals with EDs, guilt, embarrassment, and shame about ED behaviors and symptoms may prevent them from providing a complete and accurate food- and nutrition-related history [28]. Additionally, weight and body mass index (BMI) are not always diagnostic features of individuals with diagnoses other than anorexia nervosa, such as bulimia nervosa, binge eating disorder, "unspecified feeding or eating disorder", and "other specified feeding or eating disorder". These individuals may present in a "larger" body or with a BMI in a "healthy" range [28,63,66,96–98]. Lastly, individuals with EDs often present with normal biochemical and medical diagnostic test findings despite illness severity [3,27].

Because food and nutrition history, weight and BMI, and biochemical and medical diagnostic tests may not capture the physical signs and symptoms of malnutrition or the degree or severity of an ED, a systematic head-to-toe assessment tool (NFPE) is a crucial component of the nutrition assessment process [23,27,54,66,74,90,99].

The NFPE assessment tool evolved from the Subjective Global Assessment (SGA) tool and the American Society of Parenteral and Enteral Nutrition (ASPEN)/AND criteria for malnutrition [69,100]. The SGA is a validated nutrition assessment tool developed in the 1980s [100]. Based on patient history and physical examination findings, the SGA has been used by RDNs and other medical professionals globally to help determine the presence and severity of malnutrition [100].

In 2009, etiology-based diagnostic criteria for malnutrition were defined by the European Society for Clinical Nutrition and Metabolism (ESPEN) and, similarly, by a working group of the American Society of Parenteral and Enteral Nutrition (ASPEN) and the Academy of Nutrition and Dietetics (AND) [69,101]. The AND/ASPEN criteria were developed from the SGA and then expanded to include a minimum of two among the following items: insufficient energy intake, weight loss, loss of muscle mass and subcutaneous fat stores, fluid retention, and reduction in functional status [69,100]. Malnutrition was also defined in the context of acute illness or injury, chronic illness, or social or environmental circumstances [69].

The NFPE is utilized by RDNs in various clinical settings worldwide to assess micronutrient and macronutrient deficiencies, excesses, and imbalances and support a malnutrition diagnosis [28,64,92,93,102–105]. Muscle and fat stores, fluid retention (hydration status), functional status (grip strength), and other signs and symptoms of nutrition-related complications can be assessed through the NFPE [42,92,93,102–104]. The AND consistently updates the standards of practice and professional performance for RDNs (novice through to expert) and includes an NFPE to assess for malnutrition and overall nutrition status [3,59,70,92,102–104,106].

The examination techniques of inspection, palpation, percussion, and auscultation are used in the traditional NFPE to identify nutrition-related signs and symptoms arising from disease pathology [42,70,102–104,106]. The traditional NFPE domains include (1) general appearance, (2) vital signs, (3) skin, (4) nails, (5) head and hair, (6) eyes and nose, (7) mouth (intra and extraoral), (8) neck and chest, (9) abdomen, and (10) musculoskeletal system [92,93,102,104].

There are no evidence-based guidelines or standards for the NFPE of individuals with EDs [3,27,59,99]. The traditional NFPE lacks the specificity to thoroughly capture the unique nutritional complications commonly seen in individuals with EDs [93,99]. The conventional NFPE does not incorporate the principles of trauma-informed care, patient-centeredness, or culturally sensitive guidelines for its use in clinical practice, which are key considerations for providing the highest-quality healthcare (Table 2) [3,27,90,93,99,107–110].

According to the Substance Abuse and Mental Health Services Administration (SAMHSA), a trauma-informed approach in healthcare contributes to a positive treatment outcome. Trauma-informed care prioritizes the physical and emotional safety of survivors [109,110].

**Table 2.** Principles of trauma-informed care, patient-centeredness, cultural sensitivity.

| Trauma-Informed Care [109,110] | Patient Centeredness [107,108] | Cultural Sensitivity [74,107] | Unique Considerations for TIC and Cultural Sensitivity [4,6,64,74,78,79,87,88,111–113] |
| --- | --- | --- | --- |
| Create a physically and emotionally safe environment | Regard the patient as a person, not a diagnosis | Understand the patient's values, beliefs, and treatment preferences | LGBTQ+ [4,64,111,113,114]—medical trauma is common; stigma in health care is often experienced; gender-affirming communication is essential. |
| Establish trust, clarity, and appropriate boundaries | Share responsibility and power with the patient | Build relationship trust and boundaries | Individuals in higher-weight bodies [6,87,88]—independent of weight, body image concerns can impact self-esteem and lead to ED behaviors; weight stigma can reduce quality of care; may have severe symptoms despite body weight or size. |
| Assure patients they have choice and control, and emphasize patient empowerment | Build an appropriate provider–patient relationship | Bear in mind one's own biases and beliefs | Religious groups [74,78,79,112]—religious rules may play a role in food choices and fasting, causing short-term deficiencies unrelated to ED behaviors; some religions have guidelines around modesty, upholding no physical contact, while thoroughly covering the head and/or body with headwear and specific clothing. These guidelines necessitate the RDN's exhibition of respect. |
| Build a collaborative relationship while sharing power | Express unconditional positive regard and respect toward the patient | Possess knowledge and understanding of various cultures and minority groups | Racial and ethnic differences [74,79]—there are potential differences in the presentation, signs, and symptoms associated with EDs in BIPOC; dark-skinned individuals are also more prone to skin conditions like xerosis and acanthosis nigricans and vitamin D deficiency than light-skinned individuals. |
| Encourage the support of others who have a history of trauma | | | |
| Provide services that are inclusive of individuals with various backgrounds while avoiding stereotypes and biases | | | |

TIC—trauma-informed care; BIPOC—black, indigenous, and people of color.

Although not all individuals with EDs have a trauma history, trauma safeguards must be instituted in ED assessments and treatment [115]. Individuals with a trauma history may have a heightened fear of threat, high emotional reactivity, and sensitivity to criticism [9]. A trauma history may also predict ED treatment withdrawal and poor recovery outcomes [9,115]. Alternatively, a patient's perception of a safe and supportive treatment environment can help to remove obstacles to effective treatment [115].

Person-centered, culturally sensitive guidelines ensure that the treatment provider treats the patient as a unique person, not a diagnosis [107,108]. The treatment provider should exhibit awareness of their personal and professional biases, bear knowledge of various cultures, races, ethnicities, and religions, and understand the health disparities of marginalized groups [74,107].

A trauma-informed ED-specific nutrition-focused physical examination (ED-NFPE) tool, conducted with cultural sensitivity, could more thoroughly and adeptly identify nutrition-related complications in individuals with EDs and improve patients' experiences and clinical outcomes [3,27,74,93,107–110]. The results of this ED-NFPE, in combination with other nutrition assessment findings from food and nutrition history, medical and psychosocial history, biochemical data, and medical and diagnostic tests and procedures, could help to inform goals for nutrition rehabilitation [3,27,74,90,99].

This study aimed to identify the domains and components of an ED-NFPE tool for RDNs to utilize at various levels of ED care to assess nutrition-related complications in individuals with EDs. The study also aimed to identify trauma-informed, culturally sensitive practice tools and techniques based on the principles of trauma-informed care in behavioral health science [110,116]. To meet these objectives, the study's first aim was to conduct a literature review to identify evidence to inform the development of the ED-NFPE tool. The second aim was to develop a web-based survey and recruit ED RDN expert panelists [117–120]. The second aim used the modified Delphi methodology to establish a consensus among the expert panelists on the clinical relevance of the proposed ED-NFPE domains and components and to obtain feedback on a trauma-informed, culturally sensitive examination approach for inclusion in the ED-NFPE tool [121]. The research team employed the grounded theory research strategy to analyze the qualitative data and support the study objectives [122].

## 2. Materials and Methods

### 2.1. Study Design

#### 2.1.1. Delphi Approach Introduction

The Delphi and modified Delphi methods are widely used in mental health and ED research to develop clinical guidelines and improve mental health training, treatment, and cultural competence [60,65,123–125]. The modified Delphi method was chosen due to the complexities in the ED population and the need for expert opinions on ED care [121]. A modified Delphi approach, with iterative survey rounds and anonymous in-depth "discussion" and "communication" among expert panelists, will likely generate a strong consensus among the expert panel and reduce the impact of potential groupthink [121]. A literature review of the most commonly seen nutrition-related complications in EDs was conducted, and an analysis of the traditional NFPE domains and common nutrition-related complications in other clinical patient populations was performed. From the literature review, a pre-selected set of survey statements was developed. The modified method permitted a more focused study that could be conducted in six months with two to three rounds [121]. This type of study would also potentially maximize panelist participation and retention, as the time commitment was limited [121]. The researchers used the post-positivism research paradigm framework to obtain values-based, holistic, qualitative panelist feedback for trauma-informed, culturally sensitive examination techniques [126]. Three feedback rounds were determined a priori; however, only two were conducted to develop sufficient feedback and "communication" among panelists to achieve a consensus [121].

#### 2.1.2. Literature Review

An electronic literature review was conducted using the PubMed, CINAHL, and Scopus databases to identify primary research articles, review articles, and case studies on medical, dental, and nutrition assessment techniques and the current assessment methods used in ED care and to identify the common nutrition-related signs and symptoms resulting from the ED behaviors of all ED subtypes in children, adolescents, and adults.

The keywords and MeSH terms used in numerous groupings included "feeding and eating disorders" (MeSH), "eating disorders", "anorexia nervosa", "bulimia or bulimic or bulimia nervosa", "avoidant restrictive food intake disorder or ARFID", "atypical anorexia nervosa", "pica", "eating disorders not otherwise specified", "physical assessment", "nutrition assessment", "nutrition-focused physical examination", "skin conditions", "skin signs", "dermatologic signs", "Russell's sign", "micronutrient deficiencies", "macronutrient deficiencies", "malnutrition", "dehydration", "overhydration", "hydration", "mouth symptoms", "oral symptoms", "vital signs", "cognition", "fat loss", "muscle loss", "hair", "nails", "eyes", "neurological consequences", "abdominal issues", "parotid gland", "cardiovascular", "eating disorder treatment guidelines", "gaps in eating disorders treatment", and "eating disorder nutrition treatment guidelines".

Several articles were hand-selected from the reference lists of primary research, review articles, and PubMed-suggested articles. Duplicate articles were eliminated. Published guidelines from international organizations such as the Academy of Nutrition and Dietetics, Academy for Eating Disorders, Australian and New Zealand Academy for Eating Disorders, American Psychological Association, American Psychiatric Association, and National Institute for Health and Care Excellence were retrieved from searches on Google Scholar, UpToDate, and reference lists from the chosen articles. Articles were also reviewed to clarify potential circumstances common in patients with eating disorders that should be addressed before and during the examination. These include patients in larger bodies, patients with a trauma history, racial, ethnic, and religious diversity, neurodiverse patients, LGBTQ+ patients, and patients with gastrointestinal diseases or other medical or mental health comorbidities.

The search was limited to articles written in English. The other inclusion criteria were years of publication from 1987 to 2025, adults, adolescents, and children, inpatient and outpatient settings, all genders, and all subtypes of eating disorders. Articles written before 2000 were included for historical value as relevant to the development of the NFPE.

The ED-specific nutrition-related complications identified through the literature search were cross-referenced with the traditional NFPE domains and common clinical findings in other clinical patient populations and organized into examination domains that were more ED-specific and could be conducted with the primary techniques of discussion and observation. The literature search identified nine ED-specific examination domains, including (1) anthropometrics, (2) general survey, cognition, and neuropsychiatric symptoms, (3) vital signs, (4) bone loss and injury, fat and muscle stores, (5) hydration status, (6) skin, hands, and nails, (7) hair, eyelashes, eyebrows, and eyes, (8) orofacial (intra and extraoral) and neck, and (9) abdomen (gastrointestinal) [92,93,102,104].

As an alternative to the traditional NFPE examination techniques of inspection, palpation, percussion, and auscultation, the examination techniques for the ED-NFPE tool included (1) discussion and (2) observation, (3) and measuring blood pressure, temperature, radial pulse, capillary refill rate, and grip strength (with explicit patient permission) [93,102]. Discussion and observation examination techniques (with explicit patient permission) were selected by the authors to accurately assess individuals for nutrition-related complications in most domains while adhering to the principles of trauma-informed care, patient-centered care, and cultural sensitivity (i.e., ensuring patient safety and maintaining appropriate boundaries) [3,27,74]. Measuring blood pressure, temperature, radial pulse, capillary refill rate, and grip strength requires physical contact to be performed and was, therefore, excluded from the observation and discussion technique categories and organized into a separate technique. The identified examination domains and common ED findings that informed all survey statements for an ED-NFPE tool are outlined in Table 3: Eating-Disorder-Specific Nutrition-Focused Physical Examination Tool.

**Table 3.** Eating-Disorder-Specific Nutrition-Focused Physical Examination Tool.

| Examination Domain and Components | Examination Technique | Normal Findings | Common ED-Specific Examination Findings | Possible Nutritional Causes | Comments/ Non-Nutritional Causes |
| --- | --- | --- | --- | --- | --- |
| Anthropometrics | Discussion of weight, weight history, and weight-related experiences Measure height and weight (with explicit patient/guardian permission) when weight restoration is a treatment goal, and measure against previous growth charts (if child or adolescent) [63] | Normal weight for height, weight, height, and stature for age and developmental stage, following previous growth curve trajectory (if child or adolescent) [93,99] | Weight loss, weight gain, variation from prior growth curve trajectory (if child or adolescent), and history of weight bullying and/or weight stigma [18,63,127] | Malnutrition, excessive exercise, binge eating, weight-related eating disorder antecedents, current weight-related challenges, and non-nutrition-related weight-related circumstances [23,44,63] | Medications and non-nutrition related illnesses [27] |
| General Survey Neuropsychiatric, Cognition, and Mood | Discussion and observation of engagement and alertness | Engagement and alertness [93,99] | Lack of engagement, weakness, sleep disturbances, and fatigue [44,50,63] | Malnutrition, dehydration, excessive exercise, and night eating syndrome [18–20,44,63,71,81] | Neurodiversity, autism spectrum disorders, ADHD, or other learning disabilities, and medication type and dose [5,27] |
| | Discussion and observation of cognition, mood, and speech | Appropriate answers to questions, age-appropriate behavior, and cognition [93,99] | Depressed mood, feels cold, memory loss, cognitive impairment, flat affect, anxiety, dizziness, slowed speech, tingling of hands and/or feet, skeletal muscle cramps or paralysis, and nerve palsy [18,44,45,50,63,81] | Thiamine, pyridoxine, vitamin B12 deficiencies, macronutrient deficiencies, calorie deficiency, dehydration, medication thyroid medication misuse, hypothyroidism, and excessive exercise [18–20,44,63,71,81] | Neurodiversity, autism spectrum disorders, ADHD, or other learning disabilities, and medication type and dose [5,27] |
| Vital signs and temperature | Discussion and obtain radial pulse (heart rate), blood pressure (BP) (supine to sitting or sitting to standing), capillary refill rate, and temperature (with explicit patient/guardian permission) | Pulse: 60–100 pulses/min BP: <130 mmHg/85 mmHg Capillary refill time: <2 s Temperature: 96.4–99.1 °F [93,99] | Bradycardia (<60 pulses/min in adults, <50 pulses/min in adolescents and children), tachycardia > 100 pulses/min in adults, arrhythmia, orthostatic hypotension (decrease in systolic BP of 25 mmHg from supine to sitting or sitting to standing in adults, decrease in systolic BP of >20 mmHg, and a decrease in diastolic BP of >10 mm/Hg (for children and adolescents), hypertension, hypothermia, cool extremities, lightheadedness, chest pain, fatigue, hot flashes, dyspnea, and capillary refill time $\geq$ 2 s [20,33,63,128–130] | Malnutrition, weight loss, dehydration, purging, excessive exercise, laxative or diuretic abuse, thyroid medication misuse, electrolyte imbalances (hypokalemia, hypomagnesemia, hypophosphatemia), refeeding syndrome, and hypothyroidism (euthyroid sick syndrome) [20,21,44, 63,129,131,132] | Antipsychotic, antidepressant, and mood stabilizer medications [128] |

**Table 3.** *Cont.*

| Examination Domain and Components | Examination Technique | Normal Findings | Common ED-Specific Examination Findings | Possible Nutritional Causes | Comments/ Non-Nutritional Causes |
|---|---|---|---|---|---|
| Bone Loss and Injury<br>Fat Stores<br>Muscle Stores | Discussion of potential bone loss, fracture, or injury | No bone loss or age-related bone loss and no fracture or bone-related injury | Osteopenia, osteoporosis, fracture, and over-exercise-associated stress fracture or injury [49,133] | Micronutrient and macronutrient deficiency, low estrogen levels, hypothyroidism, and over-exercise [134] | Age-related osteoporosis or athletic injury [135,136] |
| | Discussion and observation (with explicit patient/guardian permission) of the orbital region | Minimal fat wasting and slight bulge in the orbital fat pad [93,99] | Reduction in fat stores in all regions, including the face, hollowing, dark color, loose skin under eyes, and delayed pubertal development [41,63] | Macronutrient and calorie deficiencies and excessive exercise [23,44,63] | Hormone replacement therapy in gender-affirming treatment can alter body composition [137] |
| | Discussion and observation (with explicit patient/guardian permission) of interosseous hand muscles, temple region, and grip strength measurement | Firm, well-defined muscles of normal size and shape bilaterally [93,99] | Changes in body composition, decreased muscle mass, muscle weakness, reduced grip strength, muscle pain, hollow, flattened muscle, concave temple region, cramping, rhabdomyolysis, and delayed pubertal development [20,41,63,138–140] | Protein and calorie malnutrition, excessive exercise, dehydration, and laxative abuse [23,44,63] | Hormone replacement therapy in gender-affirming treatment can alter body composition [137] |
| Hydration Status | Discussion and observation (with explicit patient/guardian permission) of hands, feet, ankles, abdomen, and eyelids | Normal mucus membranes and no edema [93,99] | Dehydration—sunken eyes, dark area under eyes, loss of skin turgor, dry, cracked lips, headache, dizziness, concentrated urine, and overhydration—puffy eyes, edema in abdominal area, peripheral edema, and frequent urination [20,21,38,128,141] | Lack of appropriate fluid intake, excessive fluid intake (water loading), excessive caffeine intake, protein deficiency, thiamin deficiency, purging, laxative or diuretic abuse, ipecac abuse, insulin misuse, excessive exercise, pseudo-Bartter's syndrome, rebound or refeeding edema, and refeeding syndrome [20,21,44,52,128] | Include a detailed discussion of fluid, caffeine, diuretic, and laxative intake in food/nutrition history to determine if the patient uses fluid and caffeine to mask hunger |
| Skin<br>Hands and Nails<br>Hair, Eyelashes, and Eyebrows<br>Eyes | Discussion and observation (with explicit patient/guardian permission) of skin | Uniform color, texture, moisture, and temperature [93,99,102] | Carotenoderma on palms of hands, poor wound healing, pallor, xerosis, acanthosis nigricans, hirsutism, dermatitis, acne, loss of turgor, cool temperature, scars on backs of hands from purging, general bruising, bruising over bony prominences, striae distensae, cyanosis, signs of self-harm (burns, cuts—especially on arms, legs, and abdomen), and dermatitis artefacta [20,29,63,142] | Iron, folate, vitamin B12 deficiency, zinc deficiency, macronutrient deficiencies or excesses, poor fluid status, fat deficiency, vitamin A deficiency, excessive intake of beta-carotene-containing foods, essential fatty acid deficiency, calorie deficiency, purging, peripheral vasoconstriction, excessive exercise, and binge eating [29,44,63,102] | Increased facial hair from hormone replacement therapy (transgender patients) [137] and self-harm |

**Table 3.** *Cont.*

| Examination Domain and Components | Examination Technique | Normal Findings | Common ED-Specific Examination Findings | Possible Nutritional Causes | Comments/ Non-Nutritional Causes |
| --- | --- | --- | --- | --- | --- |
| | Discussion and observation (with explicit patient/guardian permission) of hands and nails | Smooth, standard color and shape, and less than two seconds capillary refill time (CRT) [93,99,102] | Ridges, koilonychia, dry, peeling, or short nails, pale nail bed, bleeding cuticles, interosseous muscle loss, poor circulation, Russell's sign, and slow CRT [29,63,102] | Iron, protein, zinc, folate, magnesium, selenium deficiency, macronutrient, micronutrient, and calorie deficiency, dehydration, and purging hypovolemia [29,102] | Nail and cuticle biting |
| | Discussion and observation (with explicit patient/guardian permission) of hair, eyes, eyelashes, eyebrows, and fat pads below the eyes | Uniform color, texture, and amount of hair, eyelashes, and eyebrows [93,99] | Alopecia, easily pluckable, brittle, dry hair, loss of eyebrows and eyelashes, and lanugo on face and torso [29,63,102] | Macronutrient, iron, zinc, essential fatty acid deficiency, hypothyroidism, severe malnutrition, and weight loss [102] | Trichotillomania, antidepressants [102], and hormone replacement therapy (transgender patients) [137] |
| | | Clear conjunctivae, moist, and pink membranes [93,99] | Pale conjunctivae, sunken, dry appearance, and subconjunctival hemorrhage [20,63,143] | Dehydration, vitamin A deficiency, iron deficiency, fat loss, and purging [143] | |
| Mouth<br>Neck | Discussion and observation (with explicit patient/guardian permission) of teeth, lips, gums, mucosa, tongue, breath, and voice | A normal amount of healthy teeth, pink, smooth lips with no sores, pink gums, red, moist tongue with papillae, and a rough appearance [93,99,102,104] | Tooth erosion, enamel erosion, tooth sensitivity, tooth breakage, missing teeth, dental caries, angular cheilitis, red, cracked lips, dark red (magenta) tongue, excessive or minimal saliva production, intra and extraoral mouth sores, palatal scratches or ulcer, gingival recession, swollen bleeding gums, redness in back of mouth/throat, oral bleeding, oropharyngeal dysphasia, halitosis, hyperactive or hypoactive gag reflex, and hoarse voice [18,20,22,30,32,39,43, 47,63,102,104,143,144] | Macronutrient deficiency, bingeing, purging, biting, chewing on hard foods (ice and hard candy), using hard implements to purge (i.e., toothbrush), eating non-food items, chewing and spitting, consuming large amounts of food in a short period of time (binge eating), regurgitation of stomach acid, iron, riboflavin, niacin, pyridoxine, vitamin B12, folate, deficiency, and ketoacidosis [20,47,102,104,145] | |
| | Discussion and observation (with explicit patient/guardian permission) of the neck | Fatty, triangular shape, and unilobular [93,146] | Parotid and submandibular gland pain and enlargement [63,147] | Bingeing and purging [63,147] | |

**Table 3.** *Cont.*

| Examination Domain and Components | Examination Technique | Normal Findings | Common ED-Specific Examination Findings | Possible Nutritional Causes | Comments/ Non-Nutritional Causes |
|---|---|---|---|---|---|
| Abdomen | Discussion and observation (with explicit patient/guardian permission) of the abdomen | Flat, round, or scaphoid appearance, normal bowel sounds, and minimal self-reported discomfort [93,99] | Distension, bloating, nausea, fluid accumulation, flatulence, epigastric discomfort, extreme scaphoid abdomen, constipation, blood in stool, rectal fissure, reflux, hard abdomen, early satiety, gastric dilatation, constipation, diarrhea, cathartic colon, hemorrhoids, and rectal prolapse [18,63,81,144,148] | Gastroparesis, irritable bowel syndrome, celiac disease, Crohn's disease, muscle loss and fat loss from malnutrition, excessive exercise, superior mesenteric artery syndrome, laxative abuse, pelvic floor dysfunction, binge eating, and straining from constipation [11,39,44,53,63,81,143,149] | Antidepressant medications, anti-anxiety medications, antipsychotic medications, and mood stabilizers |

The search also identified the six principles of trauma-informed care and culturally sensitive guidelines [58,74,107–110,116]. These principles and guidelines provided background to evaluate the panelists' qualitative open-ended feedback. After validation through evidentiary support, the panelists' responses will also inform the future development of trauma-informed, culturally sensitive instructional guidelines for ED-NFPE tools [3–6,8,12,13,23,50,63,72,74–80].

### 2.2. Survey Development

The literature search results informed the construction of the Qualtrics survey (Qualtrics, Provo, Utah), the "Eating Disorders Nutrition Assessment Tool" Survey (survey is available in Supplementary Information) [150]. The survey consisted of panelist consent, instructions, demographic and professional information, and a series of closed- and open-ended statements organized by examination domain. An introduction and description of each domain preceded the closed-ended statements. Each statement was designed so the panelists could rate their agreement with the clinical relevance of the domains and components on a 5-point Likert scale (strongly disagree, disagree, neither agree nor disagree, agree, or strongly agree) [123,151].

An open-ended statement requesting panelist feedback for domain and component addition, omission, or modification followed the set of closed-ended statements. Lastly, the survey included an open-ended statement for each domain, stating, "If you feel the (anthropometrics) domain is clinically relevant, please provide feedback on any particular practice tools or techniques you may currently employ or would recommend promoting a trauma-informed, culturally specific assessment. Leave blank if you prefer not to answer or have no additional feedback".

### 2.3. Survey Distribution and Panelist Recruitment

The study was conducted from May through to August 2024. Purposive sampling was used to recruit a panel of RDN experts in ED practice. Several ED organizations were contacted to request their members' email addresses for study participation or to post the study invitation on their professional discussion boards or professional interest group online forums. The organizations included in recruitment were the Academy for Eating Disorders (AED), the Academy of Nutrition and Dietetics (AND) Behavioral Health Nutrition Practice Group, Eating Disorder Registered Dietitians and Professionals (EDRD Pro), and the International Federation of Eating Disorder Dietitians (IFEDD) [117–120,152].

Organization leaders posted an invitation to participate in an initial screening survey on professional online forums or an invitation was directly emailed to nutrition professionals from these professional interest groups and ED organizations. The invitation included a brief introduction to the research study, research team, Delphi methodology, and study purpose. The screening survey included a brief set of questions to assess panelist eligibility.

If potential expert panelists met the criteria from the screening survey and chose to participate, they were sent a second personalized email including a link to the survey. Participants were informed of the time commitment of their participation. To maintain interest and reduce the risk of attrition, participants were offered the opportunity to participate in a raffle for a USD 50.00 Amazon e-gift card for their participation [153].

The inclusion criteria required that the expert panelists were RDNs with five or more years of experience in ED treatment or held an advanced practice credential through the International Association of Eating Disorders Professionals (IAEDP), with a minimum of 2500 supervised hours of work in the ED field [152]. Experts were not required to have advanced training in trauma-informed care. Although panelist group size can vary in Delphi studies, the study sought to recruit 20 or more panelists [123].

### 2.4. Ethics

The Rutgers University Institutional Review Board approved the study protocol and consent for panelists' screening and participation (Pro2023002454). The authors followed the recommended Conducting and Reporting of Delphi Studies (CREDES) guidelines outlined by Jünger and colleagues and the Reporting guidelines for Delphi techniques in health sciences: A methodological review by Spranger et al. [154,155]. The panelists were informed that their demographic information and survey responses would be de-identified. Their anonymous responses would be included in future rounds for other panelists to review [156].

### 2.5. Survey Process

Round One included study background information, panelist consent, instructions, and demographic and professional information. The panelists were instructed to respond to statements about the clinical relevance of the nine ED-NFPE domains and 46 components within these domains by rating their agreement with the statements on a 5-point Likert scale [123,151]. The panelists were also given the option to provide feedback for domain and component addition, omission, modification, and practice tools or techniques they may currently employ or recommend to promote a trauma-informed, culturally specific ED-NFPE.

The panelists were required to complete and return their Round One survey via email within three weeks. After Round One data collection and analysis, the panelists were provided with a summary of their responses, mean group responses, and all other panelists' de-identified responses to open-ended statements to permit the panelists to compare their responses to other panelists' responses and revise them if desired. The panelists were given another week to review and revise their responses if desired. The round then closed, and a final data analysis was conducted to determine if consensus was achieved on any domains and components. Open-ended qualitative responses were reviewed for potential additions, omissions, modifications, and approaches. Survey items from Round One that the panelists added were verified through an additional literature review and added based on the results of a review of evidence. Feedback on a trauma-informed, culturally sensitive approach was analyzed for shared content. These comments will be evaluated for evidentiary support in future studies.

Round Two followed the same sequence as Round One. Items from Round One that achieved consensus were visible to the panelists in Round Two but were closed for further comments. Round Two contained statements that did not reach consensus in Round One and statements for the two additional components recommended by the panelists. As in Round One, the panelists were provided with a results summary and the opportunity to review and revise their responses if desired. Of the 48 final components, 46 reached a consensus after Round Two. There was no upward scoring trend for the additional two components in Round Two. It was, therefore, determined by the research team that a third round was not indicated. The results of Round Two determined a final consensus on domains, components, and approaches.

*2.6. Data Analysis*

IBM SPSS (IBM Corp. Released 2020. IBM SPSS Statistics for Windows, Version 27.0. Armonk, NY, USA: IBM Corp.) was used for data analysis [157]. Descriptive statistics (frequency and percentage) were used for panelist demographic information. The results for the clinical relevance of the ED-NFPE domains and components are reported as means and standard deviations (SDs). All research team members systematically read, hand-coded, and performed content analysis on the data from the qualitative responses to identify patterns in certain concepts or themes [122]. Based on previous Delphi studies on NFPE development, mental health, and ED practice, consensus was defined a priori as ≥75% of panelists answering "neither agree nor disagree", "agree", or "strongly agree" using the 5-point Likert scale, or as a rank of 4 ± 1 SD for each domain or component [65,151,158–160].

## 3. Results

*3.1. Panelists*

In total, 22 of the 44 panelists who participated in the eligibility screening survey met the eligibility criteria, consented to participate, and completed Round One (Table 4). Ninety-five percent identified as female. All panelists were 30 years old or older. Two-thirds had ten or more years of experience in ED dietetics. Fourteen worked thirty or more hours per week. Most panelists worked in an outpatient clinic or private practice. Three panelists were employed in a partial hospitalization program (PHP), intensive outpatient program (IOP), or "other" setting. All panelists were RDNs, with 50% holding an advanced practice credential from IAEDP. In the demographics portion of the survey, the panelists were not asked which country they lived in. They were, however, recruited from several international professional organizations, including IAEDP and IFEDD. Eighteen panelists completed Round Two (82% retention).

**Table 4.** Demographics and professional characteristics of eating disorder dietitian expert panelists.

| **Variable (N = 22)** | *n* | % |
|---|---|---|
| **Gender** | | |
| Female | 21 | 95.5 |
| **Age** | | |
| 30–39 | 7 | 31.8 |
| 40–49 | 5 | 22.7 |
| 50+ | 9 | 40.9 |
| **Years Employed in Eating Disorders Dietetics** | | |
| 5–9 years | 7 | 31.8 |
| 10–14 years | 3 | 13.6 |
| 15 years or more | 12 | 54.5 |

**Table 4.** *Cont.*

| Variable (N = 22) | *n* | % |
|---|---|---|
| **Professional Practice Status** | | |
| Full-time (≥30 h/week) | 14 | 63.6 |
| Part-time (≤29 h/week) | 6 | 27.3 |
| Other | 2 | 9.1 |
| **Eating Disorders Treatment Setting Where Employed** | | |
| Outpatient Clinic or Private Practice | 19 | 86.4 |
| Other | 3 | 13.6 |
| **Nutrition Profession** | | |
| Registered Dietitian Nutritionist | 22 | 100.0 |
| **International Association of Eating Disorders Professionals (IAEDP) Accreditation** | | |
| Yes | 11 | 50.0 |
| No | 11 | 50.0 |
| **IAEDP Advanced Practice Credential** | | |
| Certified Eating Disorder Specialist (CEDS) | 6 | 27.3 |
| Certified Eating Disorder Specialist-Consultant (CEDS-C) | 5 | 22.7 |

### *3.2. Round One*

All nine domains (100%) and 44 of 46 components (95.6%) achieved consensus in Round One (Table 5). Two components, "BMI" and "measuring weight when weight restoration is not a treatment goal", did not reach consensus.

**Table 5.** Round One domains' and components' clinical relevance, consensus determination.

| Domains and Components of Examination | Clinical Relevance Group Mean (N = 22) | Standard Deviation | Frequency (%) Rating 3 to 5 | Consensus |
|---|---|---|---|---|
| **Domain 1 Anthropometrics** | 4.55 | 0.51 | 100.0 | yes |
| **Components A1–A6 Anthropometrics** | | | | |
| A1 BMI | 2.59 | 1.30 | 59.0 | no |
| A2 Measuring height and growth trajectory | 4.95 | 0.21 | 100.0 | yes |
| A3 Measuring weight when weight restoration is a goal | 4.68 | 0.57 | 100.0 | yes |
| A4 Discussing weight changes, body image, and weight experiences when weight restoration is a goal | 4.68 | 0.48 | 100.0 | yes |
| A5 Measuring weight when weight restoration is not a goal | 2.00 | 0.93 | 22.7 | no |
| A6 Discussing weight changes, body image, and weight experiences when weight restoration is not a goal | 3.86 | 1.39 | 77.3 | yes |
| **Domain 2 General Survey, Cognition, and Neuropsychiatric Symptoms** | 4.95 | 0.21 | 100.0 | yes |
| **Components G1–G3 General Survey, Cognition, and Neuropsychiatric Symptoms** | | | | |
| G1 General Survey Assessment | 4.95 | 0.21 | 100.0 | yes |
| G2 Cognitive Assessment | 4.82 | 0.40 | 100.0 | yes |
| G3 Neuropsychiatric Symptoms Assessment | 4.86 | 0.35 | 100.0 | yes |
| **Domain 3 Vital Signs** | 4.86 | 0.35 | 100.0 | yes |
| **Components V1–V5 Vital Signs** | | | | |
| V1 Discussing vital signs and symptoms | 4.91 | 0.29 | 100.0 | yes |
| V2 Measuring blood pressure | 4.60 | 0.73 | 100.0 | yes |
| V3 Measuring orthostatic blood pressure | 4.82 | 0.50 | 100.0 | yes |
| V4 Measuring heart rate using radial pulse | 4.73 | 0.55 | 100.0 | yes |
| V5 Measuring temperature | 4.09 | 1.15 | 86.3 | yes |

**Table 5.** *Cont.*

| Domains and Components of Examination | Clinical Relevance Group Mean (N = 22) | Standard Deviation | Frequency (%) Rating 3 to 5 | Consensus |
|---|---|---|---|---|
| **Domain 4 Bone Loss and Injury, Body Fat, and Muscle Stores** | 4.62 (n = 21) | 0.59 | 100.0 (n = 21) | yes |
| **Components B1–B5 Bone Loss and Injury, Body Fat, and Muscle Stores** | | | | |
| B1 Discussing bone health | 4.77 | 0.43 | 100.0 | yes |
| B2 Discussing body fat changes | 3.64 | 1.22 | 77.3 | yes |
| B3 Observing body fat changes | 3.77 | 1.11 | 86.4 | yes |
| B4 Discussing body muscle changes | 4.05 | 0.95 | 90.9 | yes |
| B5 Observing body muscle changes | 4.14 | 0.89 | 90.9 | yes |
| **Domain 5 Hydration Status** | 4.81 (n = 21) | 0.40 | 95.2 (n = 21) | yes |
| **Components H1–H8 Hydration Status** | | | | |
| H1 Discussing dehydration signs and symptoms | 4.68 | 0.65 | 100.0 | yes |
| H2 Observing dehydration signs and symptoms | 4.86 | 0.35 | 100.0 | yes |
| H3 Discussing overhydration signs and symptoms | 4.59 | 0.67 | 100.0 | yes |
| H4 Observing overhydration signs and symptoms | 4.73 | 0.46 | 100.0 | yes |
| H5 Discussing abdominal edema signs and symptoms | 4.18 | 0.91 | 100.0 | yes |
| H6 Observing abdominal edema signs and symptoms | 4.41 | 0.80 | 100.0 | yes |
| H7 Discussing peripheral edema signs and symptoms | 4.36 | 0.79 | 100.0 | yes |
| H7 Observing peripheral edema signs and symptoms | 4.55 | 0.67 | 100.0 | yes |
| **Domain 6 Skin, Hands, and Nails** | 4.41 | 0.67 | 100.0 | yes |
| **Components S1–S4 Skin, Hands, and Nails** | | | | |
| S1 Discussing skin and hands | 4.27 | 0.70 | 100.0 | yes |
| S2 Observing skin and hands | 4.36 | 0.79 | 100.0 | yes |
| S3 Discussing nails | 4.18 | 0.73 | 100.0 | yes |
| S4 Observing nails | 4.32 | 0.78 | 100.0 | yes |
| **Domain 7 Hair, Eyelashes, Eyebrows, and Eyes** | 4.23 | 0.87 | 95.5 | yes |
| **Components E1–E6 Hair, Eyelashes, Eyebrows, and Eyes** | | | | |
| E1 Discussing hair | 4.27 | 0.63 | 100.0 | yes |
| E2 Observing hair | 4.27 | 0.83 | 95.2 | yes |
| E3 Discussing eyebrows and eyelashes | 3.59 | 1.01 | 86.4 | yes |
| E4 Observing eyebrows and eyelashes | 3.82 | 1.14 | 86.4 | yes |
| E5 Discussing eyes | 3.95 | 0.90 | 100.0 | yes |
| E6 Observing eyes | 4.18 | 0.96 | 95.5 | yes |
| **Domain 8 Intraoral, Extraoral, and Neck** | 4.14 | 1.08 | 86.4 | yes |
| **Components O1–O6 Intraoral, Extraoral, and Neck** | | | | |
| O1 Discussing intraoral | 4.05 | 0.90 | 95.5 | yes |
| O2 Observing intraoral | 3.91 | 1.11 | 86.4 | yes |
| O3 Discussing extraoral | 3.95 | 1.00 | 90.9 | yes |
| O4 Observing extraoral | 4.00 | 1.20 | 81.8 | yes |
| O5 Discussing neck | 3.95 | 0.79 | 100.0 | yes |
| O6 Observing neck | 4.14 | 0.89 | 95.5 | yes |
| **Domain 9 Abdomen (Gastrointestinal)** | 4.86 | 0.35 | 100.0 | yes |

**Table 5.** *Cont.*

| Domains and Components of Examination | Clinical Relevance Group Mean (N = 22) | Standard Deviation | Frequency (%) Rating 3 to 5 | Consensus |
|---|---|---|---|---|
| **Components G1–G3 Abdomen (Gastrointestinal)** | | | | |
| G1 Discussing gastrointestinal signs and symptoms | 4.82 | 0.40 | 100.0 | yes |
| G2 Discussing abdomen | 4.50 | 0.67 | 100.0 | yes |
| G3 Observing abdomen | 4.09 | 1.11 | 86.4 | yes |

BMI—body mass index.

Several panelists recommended two new components. These components, "capillary refill time" and "grip strength", were reviewed for evidentiary support through a literature search and added to the Round Two survey in the "vital signs" and "bone, body fat, and muscle" domains, respectively.

From the open-ended statements, the panelists provided feedback for revisions, stating that the discussion and observation of components should be case-dependent and carefully considered and timed because of the topic sensitivity and to prevent the reinforcement of eating disorder cognitions. Additionally, the panelists suggested that other treatment providers should perform some measurements, observations, and discussions, especially if the RDN performs telehealth treatment. Several panelists felt that they needed additional training to assess and interpret some component findings.

The panelists were also asked to provide feedback, based on their extensive experience, on practice tools to promote a trauma-informed, culturally specific approach to examination components. Through hand-coding, frequently mentioned responses emerged (Table 6).

**Table 6.** Rounds One and Two feedback for trauma-informed, culturally sensitive techniques.

| Domain | Feedback * |
|---|---|
| Anthropometrics | The RDN should ask the patient for permission before weighing them, assess their fears about being weighed, and discuss with the patient whether they (the patient) should be informed of their weight or if a blind weight (where the patient does not know the number) should be performed.<br><br>When exact numbers may be triggering, the RDN should use phrasing like "up, down, the same", "moving in the right direction", or "on target".<br><br>The RDN should convey to the patient that weight is only one component of the overall assessment and not the sole determinant of their health.<br><br>The RDN should adopt a weight-inclusive approach when discussing concerns about body image.<br><br>Individuals, especially those with larger bodies and those who may have a history of weight-related emotional stress, weight bias, and weight shaming or bullying, should be permitted to provide consent for any discussion about weight, shape, size, weight history, and body image. |

**Table 6.** *Cont.*

| Domain | Feedback * |
| --- | --- |
| General Survey, Cognition, and Neuropsychiatric Symptoms | Neuropsychiatric symptoms could lead to feelings of being misunderstood and potentially be harmful to some patients, particularly individuals who are neurodivergent. Therefore, this domain would need to be neuro-affirming.<br><br>While asking the patient questions about these symptoms, the RDN should observe the patient's ability to process and respond to questions or statements.<br><br>The RDN should discuss social connection, the patient's willingness and ability to engage in everyday social activities (after-school activities, friend gatherings, church, family gatherings, etc.), and the frequency of "meltdown/shutdown" experiences in clients with autism spectrum disorders.<br><br>All components in this domain should consider the involvement of environmental factors like trauma. By considering non-nutrition contributors, the effect of the patient's nutritional status on cognitive and neuropsychiatric symptoms and changes may be more accurately determined. |
| Vital Signs | The RDN should inform the patient of each component and explain how and why it will be performed. The patient should provide consent before any vital sign tests are conducted.<br><br>The RDN should assess the patient's comfort level and reassure them of their autonomy and safety.<br><br>Blood pressure cuffs should be sized appropriately to accommodate individuals of all sizes and prevent potential traumatization.<br><br>The RDN may ask the patient to perform some of these tests independently if the patient prefers to do so.<br><br>RDNs who prefer not to perform these components or choose not to touch their patients (radial pulse and blood pressure) can ask the patient to have their medical provider perform them and share the results with the RDN. |
| Bone Loss and Injury, Body Fat, and Muscle Stores | The RDN should only conduct discussions with patients in this domain if relevant to ED recovery.<br><br>Discussions should honor the patient's experience, autonomy, and safety, and each topic should be approached with consent and at a pace that meets the person where they can engage in the conversation.<br><br>Body fat discussions can be re-traumatizing for some individuals with EDs. The clinical relevance of discussions on body fat and muscle should outweigh the patient's distress.<br><br>Assessment results should be considered in the context of family frame size and body type.<br><br>When weight restoration and nutrition rehabilitation are necessary, body fat and muscle changes should be discussed before changes occur to help the patient understand what to expect during the nutritional rehabilitation process.<br><br>Because the topics of body fat and body muscle are sensitive topics for individuals with EDs, discussions should be case-dependent and primarily used when the loss of body muscle and fat is a concern.<br><br>Discussion may be more relevant than observation (with specific exceptions such as orbital fat stores)<br><br>The RDN could provide body fat and body muscle information as a component of exposure therapy, but must be mindful not to reinforce ED cognitions.<br><br>Due to anti-fat bias, conversations regarding fat stores may be more sensitive than discussions on muscle or bone health.<br><br>The RDN should limit feedback on the patient's body appearance but provide general education about the importance of adequate body stores. |

**Table 6.** *Cont.*

| Domain | Feedback * |
| --- | --- |
| Hydration Status | The RDN should screen for body dysmorphia before inquiring about swelling in the abdomen and ankles.<br><br>Because "water loading" is common before a patient is weighed, the RDN should ask the patient to use the restroom before the weigh-in. |
| Skin, Hands, and Nails | Due to skin and nail quality variations among individuals of various races and ethnicities, the RDN should not make assumptions about a patient's "normal" skin and nail quality. The RDN should initiate a thoughtful discussion of nail and skin changes to assure cultural competence.<br><br>The RDN should be clear and thoughtful before observing components in this domain so that the patient feels safe and autonomous. The RDN should also discuss the context for the assessment and ask for explicit permission to observe the patient's skin or hands before the evaluation. |
| Hair, Eyelashes, Eyebrows, and Eyes | To assess TGNC patients, the RDN should be educated on the principles of gender-affirming care and the potential physical changes associated with gender transition.<br><br>For patients who have difficulty making eye contact (neurodiversity or trauma), the RDN should not require the patient to make eye contact during an eye assessment. |
| Intraoral, Extraoral, and Neck | The RDN must be highly skilled and keenly aware of the possible adverse emotional response to this type of assessment because intraoral and extraoral areas are potentially personal and private areas of the body, especially in patients with a trauma history.<br><br>Skin color can impact mucous membrane and gingiva color; therefore, to ensure cultural competence, the RDN should not make assumptions about a patient's "normal" mucous membrane and gingiva color but should ask them what is "normal" for them. |
| Abdomen (Gastrointestinal) | The RDN should include open-ended questions that permit discussions of race-specific or family-specific gastrointestinal conditions or predispositions.<br><br>The RDN should listen sensitively to patients' concerns about their abdominal or gastrointestinal issues, even if the problem is not physiologically based.<br><br>The RDN should be aware that functional gastrointestinal issues are common in patients with EDs, and some patients with functional gastrointestinal disorders may have a history of sexual trauma. |

ED—eating disorder, RDN—registered dietitian nutritionist, TGNC—transgender and gender non-conforming, * Feedback was extracted and paraphrased (not verbatim).

Each panelist was emailed a feedback form and could revise their ratings and responses by requesting a new survey link. Two panelists (9%) requested a new survey link and revised their responses.

### 3.3. *Round Two*

The statements that did not achieve a consensus on clinical relevance in Round One, body mass index (BMI) and "measuring weight when weight restoration is not a treatment goal" did not achieve consensus in Round Two (Table 7). The responses (N = 18) for these components remained low, with a frequency rating of ≤55.6%. The two newly recommended components, "assessing capillary refill rate" and "measuring grip strength", achieved consensus, with frequency ratings of 88.9% and 100.0%, respectively. No additional additions, omissions, or revisions were suggested. Several new comments for a trauma-informed, culturally sensitive approach were offered, while various comments from Round One were re-stated in Round Two (Table 6).

**Table 7.** Round Two domains' and components' clinical relevance, consensus determination.

| Domains and Components of Examination | Clinical Relevance Group Mean (N = 18) | Standard Deviation | Frequency (%) Rating 3 to 5 | Consensus |
|---|---|---|---|---|
| **Domain 1 Anthropometrics** | | | | |
| **Components A1, A5 Anthropometrics** | | | | |
| A1 BMI | 2.78 | 1.57 | 55.5 | no |
| A5 Measuring weight when weight restoration is not a goal | 2.83 | 1.04 | 55.6 | no |
| **Domain 3 Vital Signs** | | | | |
| **Component V6 Vital Signs** | | | | |
| V6 New Component: Assessing capillary refill rate | 3.67 | 1.14 | 88.9 | yes |
| **Domain 4 Bone Loss and Injury, Body Fat, and Muscle Stores** | | | | |
| **Component B6 Vital Signs** | | | | |
| B6 New Component: Measuring grip strength | 3.50 | 0.99 | 100.0 | yes |

BMI—body mass index.

As in Round One, the panelists were emailed personalized feedback forms and offered the opportunity to revise their responses. No panelists requested a new survey link for response revision.

At the end of Round Two, all nine assessment domains (100.0%) and 46 of the 48 (95.8%) final components achieved clinical relevance consensus with no ascendant response change in Round Two for the two remaining components, indicating no need for a third survey round. The research team evaluated the results and closed the study.

### 3.4. *Domain 1—Anthropometrics*

All panelists determined that Domain 1 was clinically relevant in an ED-NFPE (100%, n = 22). The panelists endorsed measuring height and growth trajectory using growth charts in children and adolescents and using BMI when plotting it against their growth curve, considering pubertal stage, family heritage, family eating patterns, and other medical and psychological considerations. Clarification was provided in Round Two that measurements in this domain and others could either be performed by the RDN or obtained by another clinician, shared with the RDN, and discussed with the patient in the nutrition assessment. The panelists proposed assessing medical conditions or medications that could affect weight and nutrient needs for all patients.

#### Panelist Feedback

In the open-ended statement where panelists could provide practice techniques promoting a trauma-informed, culturally specific assessment, multiple panelists recommended asking the patient for permission before weighing them, assessing their fears about being weighed, and discussing with the patient whether they (the patient) should be informed of their weight or if blind weighing (where the patient does not know the number) should be performed. When exact numbers may be triggering, the panelists recommended that the RDN use phrasing like "up, down, the same", "moving in the right direction", or "on target". The panelists also suggested that the RDN convey to the patient that weight is only one component of the overall assessment and not the sole determinant of their health. Individuals, especially those with larger bodies and those who may have a history of weight-related emotional stress, weight bias, and weight shaming or bullying, should be permitted to provide consent for any discussion about weight, shape, size, weight history, and body image.

### 3.5. Domain 2—General Survey, Cognition, and Neuropsychiatric Symptoms

Domain 2 and its three components were unanimously rated as clinically relevant (100%, n = 22). The panelists recommended having access to other clinicians to provide additional details for the general survey, cognition, and neuropsychiatric symptom assessment. Some panelists suggested that other qualified professionals conduct this domain and that the results be shared with the RDN. The panelists indicated that in addition to asking the patient if they are taking medication regularly and as prescribed, they should ask if they have been prescribed any medication that they may not be taking as prescribed, as this might impact neuropsychiatric symptoms.

#### Panelist Feedback

The panelists stated that neuropsychiatric symptoms could lead to feelings of being misunderstood and potentially be harmful to some patients, particularly individuals who are neurodivergent. Therefore, this domain would need to be neuro-affirming. Additionally, the panelists suggested that the RDN should observe the patient's ability to process and respond to questions or statements while asking the patient questions about these symptoms. To be accurately assessed, multiple panelists believed that all components in this domain should consider the involvement of environmental factors like trauma. By considering non-nutrition contributors, the effect of the patient's nutritional status on cognitive and neuropsychiatric symptoms and changes may be more accurately determined.

### 3.6. Domain 3—Vital Signs

Vital signs achieved consensus (n = 22), with a 100% frequency rating. Four of the five components were unanimously rated as clinically relevant and achieved a consensus (100%, n = 22). Assessing capillary refill rate was added to the Round Two survey based on the panelist recommendations and additional literature review, and consensus (88.9%) was reached for clinical relevance. In addition to obtaining vital signs in a nutrition examination, several panelists recommended acquiring baseline, historical, or "normal" values from other treatment providers.

#### Panelist Feedback

The panelists suggested that the patient be informed of each component, how and why it will be performed, and that consent be provided before conducting any vital sign tests. The panelists stressed the importance of assessing the patient's comfort level and giving reassurance about their autonomy and safety. According to the panelist feedback, blood pressure cuffs should be sized appropriately to accommodate individuals of all sizes and prevent potential traumatization. If preferred, the panelists proposed that the RDN may ask the patient to perform some of these tests independently or have them performed by a medical clinician and shared with the RDN. The panelists mentioned that RDNs who choose not to touch their patients can also ask a medical provider to measure blood pressure and heart rate.

### 3.7. Domain 4—Bone Loss and Injury, Body Fat, and Muscle Stores

This domain achieved consensus among all panelists (100%, n = 21). All five components achieved consensus, with frequency ratings between 77.3% and 100%. One panelist recommended an additional component, measuring grip strength. This component was added to Round Two and achieved consensus, with a frequency rating of 100% (n = 18).

The panelists recommended that the RDN observe and discuss any concerning changes during each session (e.g., muscle wasting, facial fat loss, etc.), not solely during the examination. The panelists affirmed that grip strength is an easily measured, non-invasive tool to assess malnutrition and can demonstrate muscle loss without emphasizing weight, shape, or body fat.

Panelist Feedback

The panelists recommended only conducting discussions with patients in this domain if relevant to ED recovery. According to the panelists, discussions should honor patients' experience, autonomy, and safety, approaching each topic with consent and at a pace that meets the person where they can engage in the conversation. Body fat discussions can be re-traumatizing for some individuals with EDs. The clinical relevance of discussions on body fat and muscle should outweigh a patient's distress. The panelists suggested that the RDN limits feedback on the patient's body appearance but provides general education about the importance of adequate body stores.

### *3.8. Domain 5—Hydration Status*

Hydration status achieved consensus (n = 21), with a 95.2% frequency rating. The discussion and observation of signs and symptoms of dehydration, overhydration, and abdominal and peripheral edema achieved unanimous consensus (n = 22), with a 100% frequency rating. There was some concern about the panelists' inexperience in performing edema assessments and when providing telehealth; therefore, the panelists suggested that a medical provider perform the edema assessment. To assess objectivity, the panelists recommended screening for body dysmorphia before inquiring about edema.

Panelist Feedback

The panelists recommended screening for body dysmorphia before inquiring about swelling in the abdomen and ankles. Because "water loading" is common before a patient is weighed, the panelists suggested asking the patient to use the bathroom to urinate just before being weighed.

### *3.9. Domain 6—Skin, Hands, and Nails*

This domain and its four components achieved unanimous consensus (100%, n = 22). Several panelists stated that RDNs could make clinically relevant observations in this domain, but may need additional training to detect signs and symptoms specific to EDs.

Panelist Feedback

Due to skin and nail quality variations among individuals of various races and ethnicities, the panelists suggested that the RDN should not make assumptions about a patient's "normal" skin and nail quality. According to the panelists, the RDN should initiate a thoughtful discussion of nail and skin changes to ensure cultural competence. To be trauma-informed, the panelists stated that the RDN should be very clear and thoughtful before observing these components so that the patient feels safe and autonomous. The RDN should also discuss the context for the assessment and ask for explicit permission to observe the patient's skin or hands before the evaluation.

### *3.10. Domain 7—Hair, Eyelashes, Eyebrows, and Eyes*

This domain achieved consensus, with a 95.5% frequency rating. The six components achieved clinical relevance consensus, with frequency ratings between 86.4% and 100%. In addition to the RDN's assessment, the panelists suggested that the patient's perception of hair changes (including eyelashes and eyebrows) may also be clinically relevant. The panelists recommended that the RDN ask the patient if they take medications that might impact eye health.

#### Panelist Feedback

To assess transgender and gender non-conforming (TGNC) patients, the panelists recommended that the RDN be educated on the principles of gender-affirming care and the potential physical changes associated with gender transition. According to the panelist recommendations, for patients who have difficulty making eye contact (neurodiversity or trauma), the RDN should not require the patient to make eye contact during an eye assessment.

### *3.11. Domain 8—Intraoral, Extraoral, and Neck*

The intraoral, extraoral, and neck domain and its six components achieved consensus, with frequency ratings between 86.4% and 100%. The panelists agreed that several critical micronutrient deficiencies are easily visible on the tongue and gingiva and suggested assessing chewing and swallowing issues.

The panelists were more comfortable assessing extraoral versus intraoral changes and more willing to discuss them without observing them. Several panelists stated that medical and dental providers should perform these assessments and share the results with the RDN. RDNs should, however, observe the parotid glands.

#### Panelist Feedback

The panelists stated that intraoral and extraoral areas are potentially personal and private areas of the body, especially in patients with a trauma history. Therefore, the RDN must be highly skilled and keenly aware of a possible adverse emotional response to this type of assessment. Skin color can impact mucous membrane and gingiva color; therefore, according to the panelists, the RDN should not make assumptions about a patient's "normal" mucous membrane and gingiva color. The RDN should also ask the patient questions to ensure cultural competence.

### *3.12. Domain 9—Abdomen (Gastrointestinal)*

This domain achieved unanimous consensus (100%, n = 22). The three components within this domain also achieved consensus (86.4–100%). Several panelists stated they would feel comfortable discussing abdominal and gastrointestinal symptoms, but would not observe the abdomen. They would refer the patient to a medical provider and request that they share the results.

#### Panelist Feedback

The panelists recommended including open-ended questions that permit discussions of race-specific or familial-specific gastrointestinal conditions or predispositions. They also suggested listening sensitively to patients' concerns about their abdominal or gastrointestinal issues, even if the problem is not physiologically based. According to the panelists, RDNs should be aware that functional gastrointestinal issues are common in patients with

EDs, and some patients with functional gastrointestinal disorders may have a history of sexual trauma.

## 4. Discussion

Individuals with EDs can experience unique, often life-threatening nutrition-related complications [16,20,21,39,57]. The early detection of these complications and a swift nutrition care plan implemented by an RDN are essential components of treatment [27]. Because the traditional NFPE may not thoroughly capture ED-specific nutrition-related complications, an ED-NFPE tool designed to detect these findings could improve the overall nutrition assessment and help to inform treatment [99].

The results of two similar studies indicate the need for disease-specific and population-specific NFPE tools for new and experienced RDNs in various specialties [159,160]. A modified Delphi study was conducted in 2022 by Pike et al. to establish a consensus on the components of the traditional NFPE that would be clinically relevant for detecting the nutrition-related conditions seen explicitly in athletes [159]. In 2023, a modified Delphi study was conducted by Bathgate et al. to examine the need for a modified pediatric NFPE tool specific for infants and children with bronchopulmonary dysplasia (BPD) [160].

In this study, 22 RDN panelists in Round One and 18 in Round Two found the nine ED-NFPE examination domains and 46 of 48 domain components to be clinically relevant for incorporation into an ED-NFPE. Primary ED-NFPE techniques of discussion and observation (with explicit patient permission) were proposed in the place of inspection, palpation, percussion, and auscultation, the traditional NFPE techniques. By conducting a less "invasive" examination, the RDN can assess nutrition-related complications adeptly while ensuring patient autonomy and safety and respecting religious and cultural rules and preferences. The results indicated that the expert panelists predominantly supported these examination techniques.

A unique feature of this study was the qualitative panelist feedback provided for trauma-informed, culturally sensitive practice techniques. The RDN panelists recognized potential trauma and its impact on an individual's treatment experience. They provided qualitative responses to minimize the risk of inadvertent re-traumatization and optimize treatment adherence and outcomes [9,110,116].

The panelists also provided numerous suggestions for providing a culturally sensitive NFPE, including maintaining a weight-inclusive treatment approach, assessing a patient's comfort level around weight and body image, and conveying that weight is not the sole determinant of health (state vs. weight) [161].

Although the NFPE is an assessment tool used by RDNs in various clinical settings and performed on numerous patient populations, a common theme in the panelists' comments was a lack of comfort in performing some examination techniques in several domains. Some panelists stated they would feel comfortable discussing but not observing (i.e., abdominal and intraoral) specific symptoms and observing but not discussing (body fat and muscle) others. They would prefer that a medical provider perform some examination components and share the results with the RDN. The panelists also felt that they needed more training to perform some examination components. Lastly, the panelists who performed telehealth expressed limitations regarding what examination components they could perform.

### 4.1. Strengths and Limitations

Although the selection of domains and components for the ED-NFPE was based on an extensive literature search, a limitation of this study is that the researchers may have unintentionally introduced bias in statement wording and the choice of ED-specific domains and examination techniques. RDN panelists unfamiliar or uncomfortable with the domains and components may have introduced bias into their survey responses, and their answers may have been influenced by their personal or professional bias [125]. Although all panelists had wide-ranging experience in ED care, it was unclear if they had extensive experience in trauma-informed care or exhibited cultural, racial, or other forms of diversity, possibly introducing bias into their responses. Detailed narratives preceded survey statements. However, the panelists may not have completely understood the statements, and an opportunity for statement clarification was only made available in the following round (Round Two) [65].

Another limitation was that consensus cutoffs included neutral ratings. "Neither agree nor disagree" responses may not have indicated that the panelists supported the statement. Lastly, while panelist group size varies between 10 and 100 in most published Delphi studies, a limitation of this study is the panelist group size of 22 [123].

A significant strength of this study is that consensus was achieved in all nine domains and 46 of the final 48 components. Another strength and core objective of the study is the extensive qualitative feedback that the panelists provided in all domains. The retention rate was high, with 22 panelists in Round One and 18 panelists in Round Two (81.8% retention).

### 4.2. Implications of Findings

A thorough trauma-informed, culturally sensitive ED-NFPE tool will close the gap in ED patient care by enabling RDNs to skillfully conduct detailed, comprehensive nutrition assessments of individuals in this population. An ED-NFPE could be utilized at the onset and throughout treatment to evaluate changes in nutritional status during the re-nourishment and recovery process. RDNs could use this tool at the individual outpatient level of care through to the acute inpatient treatment level.

RDNs in various clinical and non-clinical settings and those with limited ED expertise could use an ED-NFPE tool as a resource to help identify potential ED-related nutritional findings, provide appropriate treatment referrals, and improve patient care.

Several factors, including a lack of experience or knowledge of the NFPE, time constraints, and a lack of confidence, autonomy, and training, may limit RDNs' use of the NFPE [162,163]. Formal NFPE training for general practice RDNs, RDNs in ED care, and students enrolled in accredited dietetics programs could provide ample knowledge and hands-on experience to remove potential barriers to conducting comprehensive disease-specific or population-specific nutrition examinations, such as an ED-NFPE [164].

## 5. Conclusions

In-depth ED-specific assessments and multidisciplinary treatment are vital to reduce or eliminate ED behaviors and minimize, eliminate, or reverse potential nutrition-based medical complications [3]. The RDN is a crucial treatment team member at every level of care [3,27]. To provide optimal ED nutrition care, evidence-based standards and guidelines must be developed for RDNs. The results of this study show that a trauma-informed, culturally competent ED-NFPE tool can not only improve the nutrition assessment of individuals with EDs, but can potentially improve the overall patient experience.

Future studies can build on this study's results to develop and validate ED-NFPE tool components and evidence-based guidelines for their use, explore the tool's impact on patient outcomes, and obtain patient feedback on the examination experience. The further development and validation of an ED-NFPE tool and professional NFPE training for RDNs will help to direct nutrition interventions and improve patient outcomes.

**Supplementary Materials:** The following supporting information can be downloaded at: https://www.mdpi.com/article/10.3390/nu17091449/s1, the "Eating Disorders Nutrition Assessment Tool" Survey.

**Author Contributions:** D.G., A.B., L.B.-G., D.R.-R. and J.Z. equally contributed to the conceptualization and methodology of the study; D.G. contributed to the curation and analysis of the data, contributed to the interpretation of the data, and drafted the manuscript. All authors have read and agreed to the published version of the manuscript.

**Funding:** This research received no external funding.

**Institutional Review Board Statement:** The Rutgers University Institutional Review Board approved the study protocol and consent for panelists' screening and participation (Pro2023002454, 27 March 2024). The study was conducted by the Declaration of Helsinki.

**Informed Consent Statement:** Informed consent was obtained from all subjects involved in the study.

**Data Availability Statement:** The data supporting the conclusions of this article will be made available by the corresponding author on request due to privacy reasons.

**Conflicts of Interest:** The authors declare no conflicts of interest.

## Abbreviations

The following abbreviations are used in this manuscript:

| | |
|---|---|
| ED | Eating Disorder |
| NFPE | Nutrition Focused Physical Examination |
| RDN | Registered Dietitian Nutritionist |
| IAEDP | International Association of Eating Disorder Professionals |
| LGBTQ+ | Lesbian, Gay, Bisexual, Transgender, Queer, and Others |
| APA | American Psychiatric Association |
| BMI | Body Mass Index |
| SAMHSA | Substance Abuse and Mental Health Services Administration |
| TGD | Transgender and Gender Diverse |
| AED | Academy for Eating Disorders |
| AND | Academy of Nutrition and Dietetics |
| EDRD Pro | Eating Disorder Registered Dietitians and Professionals |
| IFEDD | International Federation of Eating Disorder Dietitians |
| CREDES | Conducting and Reporting of Delphi Studies |
| SD | Standard Deviation |
| PHP | Partial Hospitalization Program |
| IOP | Intensive Outpatient Program |
| BIPOC | Black, Indigenous, and People of Color |
| BPD | Bronchopulmonary Dysplasia |

## References


1. Maine, M.; Bunnell, D.W. Chapter 1—A Perfect Biopsychosocial Storm: Gender, Culture, and Eating Disorders. In *Treatment of Eating Disorders*; Maine, M., McGilley, B.H., Bunnell, D.W., Eds.; Academic Press: San Diego, CA, USA, 2010; pp. 3–16.
2. National Association of Anorexia Nervosa and Associated Disorders. Eating Disorder Statistics. Available online: https://anad.org/eating-disorder-statistic/ (accessed on 8 June 2024).
3. Hackert, A.N.; Kniskern, M.A.; Beasley, T.M. Academy of Nutrition and Dietetics: Revised 2020 Standards of Practice and Standards of Professional Performance for Registered Dietitian Nutritionists (Competent, Proficient, and Expert) in Eating Disorders. *J. Acad. Nutr. Diet.* **2020**, *120*, 1902–1919. [CrossRef] [PubMed]
4. Diemer, E.W.; Grant, J.D.; Munn-Chernoff, M.A.; Patterson, D.A.; Duncan, A.E. Gender Identity, Sexual Orientation, and Eating-Related Pathology in a National Sample of College Students. *J. Adolesc. Health* **2015**, *57*, 144–149. [CrossRef] [PubMed]

5. Li, Z.; Halls, D.; Byford, S.; Tchanturia, K. Autistic characteristics in eating disorders: Treatment adaptations and impact on clinical outcomes. *Eur. Eat. Disord. Rev.* **2022**, *30*, 671–690. [CrossRef] [PubMed]
6. Ralph, A.F.; Brennan, L.; Byrne, S.; Caldwell, B.; Farmer, J.; Hart, L.M.; Heruc, G.A.; Maguire, S.; Piya, M.K.; Quin, J.; et al. Management of eating disorders for people with higher weight: Clinical practice guideline. *J. Eat. Disord.* **2022**, *10*, 121. [CrossRef]
7. Solmi, F.; Bentivegna, F.; Bould, H.; Mandy, W.; Kothari, R.; Rai, D.; Skuse, D.; Lewis, G. Trajectories of autistic social traits in childhood and adolescence and disordered eating behaviours at age 14 years: A UK general population cohort study. *J. Child Psychol. Psychiatry* **2021**, *62*, 75–85. [CrossRef]
8. Convertino, A.D.; Morland, L.A.; Blashill, A.J. Trauma exposure and eating disorders: Results from a United States nationally representative sample. *Int. J. Eat. Disord.* **2022**, *55*, 1079–1089. [CrossRef]
9. Brewerton, T.D. An Overview of Trauma-Informed Care and Practice for Eating Disorders. *J. Aggress. Maltreatment Trauma* **2019**, *28*, 445–462. [CrossRef]
10. Backholm, K.; Isomaa, R.; Birgegård, A. The prevalence and impact of trauma history in eating disorder patients. *Eur. J. Psychotraumatol.* **2013**, *4*, 22482. [CrossRef]
11. Mascolo, M.; Geer, B.; Feuerstein, J.; Mehler, P.S. Gastrointestinal comorbidities which complicate the treatment of anorexia nervosa. *Eat. Disord.* **2017**, *25*, 122–133. [CrossRef]
12. Udo, T.; Grilo, C.M. Psychiatric and medical correlates of DSM-5 eating disorders in a nationally representative sample of adults in the United States. *Int. J. Eat. Disord.* **2019**, *52*, 42–50. [CrossRef]
13. Riehl, M.E.; Scarlata, K. Understanding disordered eating risks in patients with gastrointestinal conditions. *J. Acad. Nutr. Diet.* **2022**, *122*, 491–499. [CrossRef] [PubMed]
14. Harvard T.H. Chan—School of Public Health. Strategic Training Initiative for the Prevention of Eating Disorders. Report: Economic Costs of Eating Disorders. Available online: https://www.hsph.harvard.edu/striped/report-economic-costs-of-eating-disorders/ (accessed on 16 July 2024).
15. Galmiche, M.; Déchelotte, P.; Lambert, G.; Tavolacci, M.P. Prevalence of eating disorders over the 2000–2018 period: A systematic literature review. *Am. J. Clin. Nutr.* **2019**, *109*, 1402–1413. [CrossRef] [PubMed]
16. Mehler, P.S.; Blalock, D.V.; Walden, K.; Kaur, S.; McBride, J.; Walsh, K.; Watts, J. Medical findings in 1,026 consecutive adult inpatient–residential eating disordered patients. *Int. J. Eat. Disord.* **2018**, *51*, 305–313. [CrossRef]
17. Hanachi, M.; Dicembre, M.; Rives-Lange, C.; Ropers, J.; Bemer, P.; Zazzo, J.F.; Poupon, J.; Dauvergne, A.; Melchior, J.C. Micronutrients Deficiencies in 374 Severely Malnourished Anorexia Nervosa Inpatients. *Nutrients* **2019**, *11*, 792. [CrossRef] [PubMed]
18. Westmoreland, P.; Krantz, M.J.; Mehler, P.S. Medical Complications of Anorexia Nervosa and Bulimia. *Am. J. Med.* **2016**, *129*, 30–37. [CrossRef]
19. Wassenaar, E.; Friedman, J.; Mehler, P.S. Medical complications of binge eating disorder. *Psychiatr. Clin.* **2019**, *42*, 275–286. [CrossRef]
20. Forney, K.J.; Buchman-Schmitt, J.M.; Keel, P.K.; Frank, G.K.W. The medical complications associated with purging. *Int. J. Eat. Disord.* **2016**, *49*, 249–259. [CrossRef]
21. Gaudiani, J.L.; Brinton, J.T.; Sabel, A.L.; Rylander, M.; Catanach, B.; Mehler, P.S. Medical outcomes for adults hospitalized with severe anorexia nervosa: An analysis by age group. *Int. J. Eat. Disord.* **2016**, *49*, 378–385. [CrossRef]
22. Lin, J.A.; Woods, E.R.; Bern, E.M. Common and Emergent Oral and Gastrointestinal Manifestations of Eating Disorders. *Gastroenterol. Hepatol.* **2021**, *17*, 157–167.
23. Peebles, R.; Sieke, E.H. Medical complications of eating disorders in youth. *Child Adolesc. Psychiatr. Clin.* **2019**, *28*, 593–615. [CrossRef]
24. Strumia, R. Eating disorders and the skin. *Clin. Dermatol.* **2013**, *31*, 80–85. [CrossRef]
25. Lourenço, M.; Azevedo, Á.; Brandão, I.; Gomes, P.S. Orofacial manifestations in outpatients with anorexia nervosa and bulimia nervosa focusing on the vomiting behavior. *Clin. Oral Investig.* **2018**, *22*, 1915–1922. [CrossRef] [PubMed]
26. Gibson, D.; Workman, C.; Mehler, P.S. Medical complications of anorexia nervosa and bulimia nervosa. *Psychiatr. Clin.* **2019**, *42*, 263–274. [CrossRef]
27. Academy for Eating Disorders Nutrition Working Group. Guidebook for Nutrition Treatment of Eating Disorders. Available online: https://higherlogicdownload.s3.amazonaws.com/AEDWEB/27a3b69a-8aae-45b2-a04c-2a078d02145d/UploadedImages/Publications_Slider/FINAL_AED_Purple_Nutrition_Book.pdf (accessed on 14 July 2024).
28. American Psychiatric Association. Feeding and Eating Disorders. In *Diagnostic and Statistical Manual of Mental Disorders*, 5th ed.; American Psychiatric Association Publishing: Washington, DC, USA, 2023.
29. Stamu-O'Brien, C.; Shivakumar, S.; Messas, T.; Kroumpouzos, G. Through the looking glass: Skin signs that help diagnose eating disorders. *Clin. Dermatol.* **2023**, *41*, 67–74. [CrossRef]

30. Johansson, A.-K.; Norring, C.; Unell, L.; Johansson, A. Eating disorders and oral health: A matched case–control study. *Eur. J. Oral Sci.* **2012**, *120*, 61–68. [CrossRef] [PubMed]
31. Paszynska, E.; Dutkiewicz, A.; Osinska, A.; Mozol-Jursza, M.; Smalc, N.; Tyszkiewicz-Nwafor, M.; Dmitrzak-Weglarz, M.; Slopien, A.; Jenerowicz, D. Anorexia Nervosa with Vomiting Episodes: Dermatological and Oral Complications. *Eur. J. Dent.* **2020**, *14*, 180–185. [CrossRef] [PubMed]
32. Uhlen, M.-M.; Tveit, A.B.; Refsholt Stenhagen, K.; Mulic, A. Self-induced vomiting and dental erosion—A clinical study. *BMC Oral Health* **2014**, *14*, 92. [CrossRef]
33. Mehler, P.S.; Brown, C. Anorexia nervosa—Medical complications. *J. Eat. Disord.* **2015**, *3*, 11. [CrossRef]
34. Mehler, P.S.; Rylander, M. Bulimia Nervosa—Medical complications. *J. Eat. Disord.* **2015**, *3*, 12. [CrossRef]
35. Miller, K.K.; Grinspoon, S.K.; Ciampa, J.; Hier, J.; Herzog, D.; Klibanski, A. Medical Findings in Outpatients With Anorexia Nervosa. *Arch. Intern. Med.* **2005**, *165*, 561–566. [CrossRef]
36. Casiero, D.; Frishman, W.H. Cardiovascular complications of eating disorders. *Cardiol. Rev.* **2006**, *14*, 227–231. [CrossRef] [PubMed]
37. Cass, K.; McGuire, C.; Bjork, I.; Sobotka, N.; Walsh, K.; Mehler, P.S. Medical Complications of Anorexia Nervosa. *Psychosomatics* **2020**, *61*, 625–631. [CrossRef] [PubMed]
38. Bahia, A.; Mascolo, M.; Gaudiani, J.L.; Mehler, P.S. PseudoBartter syndrome in eating disorders. *Int. J. Eat. Disord.* **2012**, *45*, 150–153. [CrossRef]
39. Cremonini, F.; Camilleri, M.; Clark, M.M.; Beebe, T.J.; Locke, G.R.; Zinsmeister, A.R.; Herrick, L.M.; Talley, N.J. Associations among binge eating behavior patterns and gastrointestinal symptoms: A population-based study. *Int. J. Obes.* **2009**, *33*, 342–353. [CrossRef]
40. Duriez, P.; Mastellari, T.; Viltart, O.; Gorwood, P. Clinical meaning of body temperatures in anorexia nervosa and bulimia nervosa. *Eur. Eat. Disord. Rev.* **2022**, *30*, 124–134. [CrossRef] [PubMed]
41. El Ghoch, M.; Calugi, S.; Lamburghini, S.; Dalle Grave, R. Anorexia Nervosa and Body Fat Distribution: A Systematic Review. *Nutrients* **2014**, *6*, 3895–3912. [CrossRef]
42. Fischer, M.; JeVenn, A.; Hipskind, P. Evaluation of Muscle and Fat Loss as Diagnostic Criteria for Malnutrition. *Nutr. Clin. Pract.* **2015**, *30*, 239–248. [CrossRef]
43. Holmes, S.R.M.; Sabel, A.L.; Gaudiani, J.L.; Gudridge, T.; Brinton, J.T.; Mehler, P.S. Prevalence and management of oropharyngeal dysphagia in patients with severe anorexia nervosa: A large retrospective review. *Int. J. Eat. Disord.* **2016**, *49*, 159–166. [CrossRef]
44. Quesnel, D.A.; Cooper, M.; Fernandez-del-Valle, M.; Reilly, A.; Calogero, R.M. Medical and physiological complications of exercise for individuals with an eating disorder: A narrative review. *J. Eat. Disord.* **2023**, *11*, 3. [CrossRef]
45. Roerig, J.L.; Steffen, K.J.; Mitchell, J.E.; Zunker, C. Laxative Abuse: Epidemiology, Diagnosis and Management. *Drugs* **2010**, *70*, 1487–1503. [CrossRef]
46. Santangelo, G.; Bursi, F.; Toriello, F.; Tamagni, M.E.; Fior, G.; Massironi, L.; Bertelli, S.; Fanin, A.; Gambini, O.; Carugo, S.; et al. Echocardiographic changes in anorexia nervosa: A pathophysiological adaptation or a disease? *Intern. Emerg. Med.* **2022**, *17*, 777–787. [CrossRef] [PubMed]
47. Sansone, R.A.; Sansone, L.A. Hoarseness: A sign of self-induced vomiting? *Innov. Clin. Neurosci.* **2012**, *9*, 37–41.
48. Silva, V.; Schukow, C.P.; Restini, C.B.A. Striae distensae as a diagnostic indicator for eating disorder pathologies. *Int. J. Dermatol.* **2023**, *62*, 715–722. [CrossRef]
49. Solmi, M.; Veronese, N.; Correll, C.U.; Favaro, A.; Santonastaso, P.; Caregaro, L.; Vancampfort, D.; Luchini, C.; De Hert, M.; Stubbs, B. Bone mineral density, osteoporosis, and fractures among people with eating disorders: A systematic review and meta-analysis. *Acta Psychiatr. Scand.* **2016**, *133*, 341–351. [CrossRef]
50. Tamiya, H.; Ouchi, A.; Chen, R.; Miyazawa, S.; Akimoto, Y.; Kaneda, Y.; Sora, I. Neurocognitive impairments are more severe in the binge-eating/purging anorexia nervosa subtype than in the restricting subtype. *Front. Psychiatry* **2018**, *9*, 138. [CrossRef]
51. Yahalom, M.; Spitz, M.; Sandler, L.; Heno, N.; Roguin, N.; Turgeman, Y. The significance of bradycardia in anorexia nervosa. *Int. J. Angiol.* **2013**, *22*, 83–94. [CrossRef] [PubMed]
52. Mascolo, M.; Chu, E.S.; Mehler, P.S. Abuse and clinical value of diuretics in eating disorders therapeutic applications. *Int. J. Eat. Disord.* **2011**, *44*, 200–202. [CrossRef]
53. Mascolo, M.; Dee, E.; Townsend, R.; Brinton, J.T.; Mehler, P.S. Severe gastric dilatation due to superior mesenteric artery syndrome in anorexia nervosa. *Int. J. Eat. Disord.* **2015**, *48*, 532–534. [CrossRef] [PubMed]
54. Klein, D.A.; Sylvester, J.E.; Schvey, N.A. Eating Disorders in Primary Care: Diagnosis and Management. *Am. Fam. Physician* **2021**, *103*, 22–32.
55. Vall, E.; Wade, T.D. Predictors of treatment outcome in individuals with eating disorders: A systematic review and meta-analysis. *Int. J. Eat. Disord.* **2015**, *48*, 946–971. [CrossRef]

56. Academy for Eating Disorders' (AED) Medical Care Standards Committee. *AED Eating Disorders: A Guide to Medical Care*; Academy for Eating Disorders: Wakefield, MA, USA, 2021.
57. Fichter, M.M.; Quadflieg, N. Mortality in eating disorders—Results of a large prospective clinical longitudinal study. *Int. J. Eat. Disord.* **2016**, *49*, 391–401. [CrossRef] [PubMed]
58. Crone, C.; Fochtmann, L.J.; Attia, E.; Boland, R.; Escobar, J.; Fornari, V.; Golden, N.; Guarda, A.; Jackson-Triche, M.; Manzo, L.; et al. The American Psychiatric Association Practice Guideline for the Treatment of Patients With Eating Disorders. *Am. J. Psychiatry* **2023**, *180*, 167–171. [CrossRef]
59. Andersen, D.; Baird, S.; Bates, T.; Chapel, D.L.; Cline, A.D.; Ganesh, S.N.; Garner, M.; Grant, B.L.; Hamilton, K.K.; Jablonski, K.; et al. Academy of Nutrition and Dietetics: Revised 2017 Standards of Practice in Nutrition Care and Standards of Professional Performance for Registered Dietitian Nutritionists. *J. Acad. Nutr. Diet.* **2018**, *118*, 132–140.e15. [CrossRef]
60. Fuller, S.J.; Philpot, U.; Group, W. The development of consensus-based guidelines for dietetic practice in nasogastric tube feeding under restraint for patients with anorexia nervosa using a modified Delphi process. *J. Hum. Nutr. Diet.* **2020**, *33*, 287–294. [CrossRef] [PubMed]
61. Garber, A.K.; Sawyer, S.M.; Golden, N.H.; Guarda, A.S.; Katzman, D.K.; Kohn, M.R.; Le Grange, D.; Madden, S.; Whitelaw, M.; Redgrave, G.W. A systematic review of approaches to refeeding in patients with anorexia nervosa. *Int. J. Eat. Disord.* **2016**, *49*, 293–310. [CrossRef] [PubMed]
62. Harrington, B.C.; Jimerson, M.; Haxton, C.; Jimerson, D.C. Initial evaluation, diagnosis, and treatment of anorexia nervosa and bulimia nervosa. *Am. Fam. Physician* **2015**, *91*, 46–52.
63. Hornberger, L.L.; Lane, M.A.; Lane, M.; Breuner, C.C.; Alderman, E.M.; Grubb, L.K.; Powers, M.; Upadhya, K.K.; Wallace, S.B.; Loveless, M. Identification and management of eating disorders in children and adolescents. *Pediatrics* **2021**, *147*, e2020040279. [CrossRef]
64. Linsenmeyer, W.; McGuire, J.; Patel, M.; Theobald, N.; Helein, A. Conducting a Nutrition-Focused Physical Examination on a Transgender Patient: Considerations for Registered Dietitian Nutritionists. *J. Acad. Nutr. Diet.* **2025**, S2212–S2672. [CrossRef]
65. Mittnacht, A.M.; Bulik, C.M. Best nutrition counseling practices for the treatment of anorexia nervosa: A Delphi study. *Int. J. Eat. Disord.* **2015**, *48*, 111–122. [CrossRef]
66. Peebles, R.; Hardy, K.K.; Wilson, J.L.; Lock, J.D. Are Diagnostic Criteria for Eating Disorders Markers of Medical Severity? *Pediatrics* **2010**, *125*, e1193–e1201. [CrossRef]
67. Setnick, J. Micronutrient Deficiencies and Supplementation in Anorexia and Bulimia Nervosa. *Nutr. Clin. Pract.* **2010**, *25*, 137–142. [CrossRef]
68. Watson, H.J.; Bulik, C.M. Update on the treatment of anorexia nervosa: Review of clinical trials, practice guidelines and emerging interventions. *Psychol. Med.* **2013**, *43*, 2477–2500. [CrossRef]
69. White, J.V.; Guenter, P.; Jensen, G.; Malone, A.; Schofield, M.; Academy Malnutrition Work Group; A.S.P.E.N. Malnutrition Task Force; A.S.P.E.N. Board of Directors. Consensus Statement: Academy of Nutrition and Dietetics and American Society for Parenteral and Enteral Nutrition. *J. Parenter. Enter. Nutr.* **2012**, *36*, 275–283. [CrossRef]
70. Heruc, G.; Hart, S.; Stiles, G.; Fleming, K.; Casey, A.; Sutherland, F.; Jeffrey, S.; Roberton, M.; Hurst, K. ANZAED practice and training standards for dietitians providing eating disorder treatment. *J. Eat. Disord.* **2020**, *8*, 77. [CrossRef]
71. McCuen-Wurst, C.; Ruggieri, M.; Allison, K.C. Disordered eating and obesity: Associations between binge-eating disorder, night-eating syndrome, and weight-related comorbidities. *Ann. N. Y. Acad. Sci.* **2018**, *1411*, 96–105. [CrossRef]
72. National Eating Disorders Association. Statistics and Research on Eating Disorders. Available online: https://www.nationaleatingdisorders.org/statistics-research-eating-disorders (accessed on 17 July 2024).
73. Doyle, E.A.; Quinn, S.M.; Ambrosino, J.M.; Weyman, K.; Tamborlane, W.V.; Jastreboff, A.M. Disordered Eating Behaviors in Emerging Adults With Type 1 Diabetes: A Common Problem for Both Men and Women. *J. Pediatr. Health Care* **2017**, *31*, 327–333. [CrossRef]
74. Send, S. Cultural Competence and Nutrition-focused Physical Exam. *Support Line* **2023**, *45*, 14–19.
75. Smolak, L.; Murnen, S.K.; Ruble, A.E. Female athletes and eating problems: A meta-analysis. *Int. J. Eat. Disord.* **2000**, *27*, 371–380. [CrossRef]
76. Knapp, J.; Aerni, G.; Anderson, J. Eating Disorders in Female Athletes: Use of Screening Tools. *Curr. Sports Med. Rep.* **2014**, *13*, 214–218. [CrossRef] [PubMed]
77. Mulchandani, M.; Shetty, N.; Conrad, A.; Muir, P.; Mah, B. Treatment of eating disorders in older people: A systematic review. *Syst. Rev.* **2021**, *10*, 275. [CrossRef] [PubMed]
78. Chouraqui, J.-P.; Turck, D.; Briend, A.; Darmaun, D.; Bocquet, A.; Feillet, F.; Frelut, M.-L.; Girardet, J.-P.; Guimber, D.; Hankard, R.; et al. Religious dietary rules and their potential nutritional and health consequences. *Int. J. Epidemiol.* **2020**, *50*, 12–26. [CrossRef]

79. Sala, M.; Reyes-Rodríguez, M.L.; Bulik, C.M.; Bardone-Cone, A. Race, ethnicity, and eating disorder recognition by peers. *Eat. Disord.* **2013**, *21*, 423–436. [CrossRef]
80. Kamody, R.C.; Grilo, C.M.; Udo, T. Disparities in DSM-5 defined eating disorders by sexual orientation among U.S. adults. *Int. J. Eat. Disord.* **2020**, *53*, 278–287. [CrossRef]
81. Sato, Y.; Fukudo, S. Gastrointestinal symptoms and disorders in patients with eating disorders. *Clin. J. Gastroenterol.* **2015**, *8*, 255–263. [CrossRef]
82. Elizabeth, J.; Andrea, K.; Aurelia, N. 2016 update on eating disorders in athletes: A comprehensive narrative review with a focus on clinical assessment and management. *Br. J. Sports Med.* **2016**, *50*, 154. [CrossRef]
83. Tu, C.-Y.; Meg Tseng, M.-C.; Chang, C.-H. Night eating syndrome in patients with eating disorders: Is night eating syndrome distinct from bulimia nervosa? *J. Formos. Med. Assoc.* **2019**, *118*, 1038–1046. [CrossRef]
84. Barry, M.R.; Sonneville, K.R.; Leung, C.W. Students with Food Insecurity Are More Likely to Screen Positive for an Eating Disorder at a Large, Public University in the Midwest. *J. Acad. Nutr. Diet.* **2021**, *121*, 1115–1124. [CrossRef]
85. Colton, P.A.; Olmsted, M.P.; Daneman, D.; Farquhar, J.C.; Wong, H.; Muskat, S.; Rodin, G.M. Eating disorders in girls and women with type 1 diabetes: A longitudinal study of prevalence, onset, remission, and recurrence. *Diabetes Care* **2015**, *38*, 1212–1217. [CrossRef]
86. Bahji, A.; Mazhar, M.N.; Hudson, C.C.; Nadkarni, P.; MacNeil, B.A.; Hawken, E. Prevalence of substance use disorder comorbidity among individuals with eating disorders: A systematic review and meta-analysis. *Psychiatry Res.* **2019**, *273*, 58–66. [CrossRef] [PubMed]
87. Durso, L.E.; Latner, J.D. Understanding Self-directed Stigma: Development of the Weight Bias Internalization Scale. *Obesity* **2008**, *16*, S80–S86. [CrossRef] [PubMed]
88. Pearl, R.L.; White, M.A.; Grilo, C.M. Weight bias internalization, depression, and self-reported health among overweight binge eating disorder patients. *Obesity* **2014**, *22*, E142–E148. [CrossRef] [PubMed]
89. Castellini, G.; Trisolini, F.; Ricca, V. Psychopathology of eating disorders. *J. Psychopathol.* **2014**, *20*, 461–470.
90. Academy of Nutrition and Dietetics. NCP Step 1: Nutrition Assessment. Available online: https://www.ncpro.org/pubs/2020-encpt-en/page-026 (accessed on 26 August 2024).
91. Academy of Nutrition and Dietetics. Electronic Nutrition Care Process Terminology. Available online: https://ncpro.org/?set_ga_opt_in_cookie=1&set_ga_opt_in=Save+Settings (accessed on 13 April 2025).
92. Esper, D.H. Utilization of Nutrition-Focused Physical Assessment in Identifying Micronutrient Deficiencies. *Nutr. Clin. Pract.* **2015**, *30*, 194–202. [CrossRef]
93. Pogatshnik, C.; Hamilton, C. Nutrition-focused physical examination: Skin, nails, hair, eyes and oral cavity. *Support Line* **2011**, *33*, 7–13.
94. Hammond, M.I.; Myers, E.F.; Trostler, N. Nutrition Care Process and Model: An Academic and Practice Odyssey. *J. Acad. Nutr. Diet.* **2014**, *114*, 1879–1894. [CrossRef]
95. Lebow, J.; Sim, L.A.; Kransdorf, L.N. Prevalence of a History of Overweight and Obesity in Adolescents With Restrictive Eating Disorders. *J. Adolesc. Health* **2015**, *56*, 19–24. [CrossRef]
96. Weir, C.B.; Jan, A. *BMI Classification Percentile and Cut Off Points*; StatPearls Publishing: Treasure Island, FL, USA, 2019.
97. Eiring, K.; Wiig Hage, T.; Reas, D.L. Exploring the experience of being viewed as "not sick enough": A qualitative study of women recovered from anorexia nervosa or atypical anorexia nervosa. *J. Eat. Disord.* **2021**, *9*, 142. [CrossRef]
98. Forney, K.J.; Brown, T.A.; Holland-Carter, L.A.; Kennedy, G.A.; Keel, P.K. Defining "significant weight loss" in atypical anorexia nervosa. *Int. J. Eat. Disord.* **2017**, *50*, 952–962. [CrossRef] [PubMed]
99. Academy of Nutrition and Dietetics. Nutrition-Focused Physical Findings. Available online: https://www.nutritioncaremanual.org/topic.cfm?ncm_category_id=11&lv1=64571&lv2=273972&lv3=273975&ncm_toc_id=273975&ncm_heading=Resources (accessed on 16 July 2024).
100. Detsky, A.; JR, M.; Baker, J.; Johnston, N.; Whittaker, S.; Mendelson, R.; Jeejeebhoy, K. What is subjective global assessment of nutritional status? *J. Parenter. Enter. Nutr.* **1987**, *11*, 8–13. [CrossRef]
101. Cederholm, T.; Bosaeus, I.; Barazzoni, R.; Bauer, J.; Van Gossum, A.; Klek, S.; Muscaritoli, M.; Nyulasi, I.; Ockenga, J.; Schneider, S.M.; et al. Diagnostic criteria for malnutrition—An ESPEN Consensus Statement. *Clin. Nutr.* **2015**, *34*, 335–340. [CrossRef]
102. Kilde, K.; Nitzsche, L. Nutrition-focused Physical Examination: A Head-to-Toe Approach with a Focus on Micronutrient Deficiencies. *Support Line* **2016**, *38*, 3–9.
103. Hummell, A.C.; Cummings, M. Role of the nutrition-focused physical examination in identifying malnutrition and its effectiveness. *Nutr. Clin. Pract.* **2022**, *37*, 41–49. [CrossRef]

104. Radler, D.R.; Lister, T. Nutrient Deficiencies Associated With Nutrition-Focused Physical Findings of the Oral Cavity. *Nutr. Clin. Pract.* **2013**, *28*, 710–721. [CrossRef]
105. Pinhas, L.; Morris, A.; Crosby, R.D.; Katzman, D.K. Incidence and Age-Specific Presentation of Restrictive Eating Disorders in Children: A Canadian Paediatric Surveillance Program Study. *Arch. Pediatr. Adolesc. Med.* **2011**, *165*, 895–899. [CrossRef]
106. Chen, S.T.; Ong, S.H.; Lim, P.Y.; Chong, K.S. Nutrition-Focused Physical Exam in assessing nutritional status of children with neurological impairment. *Hum. Nutr. Metab.* **2024**, *37*, 200279. [CrossRef]
107. Saha, S.; Beach, M.C.; Cooper, L.A. Patient centeredness, cultural competence and healthcare quality. *J. Natl. Med. Assoc.* **2008**, *100*, 1275–1285. [CrossRef]
108. Schwartz, D.B. Integrating Patient-Centered Care and Clinical Ethics Into Nutrition Practice. *Nutr. Clin. Pract.* **2013**, *28*, 543–555. [CrossRef]
109. Substance Abuse and Mental Health Services Administration. *SAMHSA's Concept of Trauma and Guidance for a Trauma-Informed Approach*; HHS Publication No. (SMA) 14-4884; Substance Abuse and Mental Health Services Administration: Rockville, MD, USA, 2014.
110. Substance Abuse and Mental Health Services Administration; Center for Substance Abuse Treatment. *A Treatment Improvement Protocol Trauma-Informed Care in Behavioral Health Services*; Substance Abuse and Mental Health Services Administration: Rockville, MD, USA, 2014.
111. Diemer, E.W.; White Hughto, J.M.; Gordon, A.R.; Guss, C.; Austin, S.B.; Reisner, S.L. Beyond the Binary: Differences in Eating Disorder Prevalence by Gender Identity in a Transgender Sample. *Transgend. Health* **2018**, *3*, 17–23. [CrossRef]
112. Mujallad, A. CNE SERIES. Modesty Among Muslim Women: Implications for Nursing Care. *MEDSURG Nurs.* **2016**, *25*, 169–172. [PubMed]
113. Sabin, J.A.; Riskind, R.G.; Nosek, B.A. Health Care Providers' Implicit and Explicit Attitudes Toward Lesbian Women and Gay Men. *Am. J. Public Health* **2015**, *105*, 1831–1841. [CrossRef] [PubMed]
114. Riddle, M.C.; Safer, J.D. Medical considerations in the care of transgender and gender diverse patients with eating disorders. *J. Eat. Disord.* **2022**, *10*, 178. [CrossRef] [PubMed]
115. Elliott, D.E.; Bjelajac, P.; Fallot, R.D.; Markoff, L.S.; Reed, B.G. Trauma-informed or trauma-denied: Principles and implementation of trauma-informed services for women. *J. Community Psychol.* **2005**, *33*, 461–477. [CrossRef]
116. Fallot, R.; Harris, M. *Creating Cultures of Trauma-Informed Care: A Self-Assessment and Planning Protocol*; Community Connections: Washington, DC, USA, 2009.
117. Academy for Eating Disorders. About AED. Available online: https://www.aedweb.org/about-aed/who-we-are (accessed on 8 July 2024).
118. Academy of Nutrition and Dietetics. Behavioral Health Nutrition: A Dietetic Practice Group of the Academy of Nutrition and Dietetics. Available online: https://www.bhndpg.org/bhn-knowledge-basecamp (accessed on 10 July 2024).
119. Eating Disorder Registered Dietitians and Professionals. EDRDPRO Eating Disorder Registered Dietitians and Profesionals. Available online: https://edrdpro.com/ (accessed on 26 August 2024).
120. International Federation of Eating Disorder Dietitians. Welcome to the International Federation of Eating Disorder Dietitians. Available online: https://ifedd.org/ (accessed on 10 June 2024).
121. Nasa, P.; Jain, R.; Juneja, D. Delphi methodology in healthcare research: How to decide its appropriateness. *World J. Methodol.* **2021**, *11*, 116–129. [CrossRef] [PubMed]
122. White, R.E.; Cooper, K. Grounded Theory. In *Qualitative Research in the Post-Modern Era: Critical Approaches and Selected Methodologies*; White, R.E., Cooper, K., Eds.; Springer International Publishing: Cham, Switzerland, 2022; pp. 339–385.
123. Humphrey-Murto, S.; Varpio, L.; Gonsalves, C.; Wood, T.J. Using consensus group methods such as Delphi and Nominal Group in medical education research. *Med. Teach.* **2017**, *39*, 14–19. [CrossRef]
124. Jorm, A.F. Using the Delphi expert consensus method in mental health research. *Aust. N. Z. J. Psychiatry* **2015**, *49*, 887–897. [CrossRef]
125. McMaster, C.M.; Wade, T.; Franklin, J.; Hart, S. Development of consensus-based guidelines for outpatient dietetic treatment of eating disorders: A Delphi study. *Int. J. Eat. Disord.* **2020**, *53*, 1480–1495. [CrossRef]
126. Kankam, P.K. The use of paradigms in information research. *Libr. Inf. Sci. Res.* **2019**, *41*, 85–92. [CrossRef]
127. Lee, K.M.; Hunger, J.M.; Tomiyama, A.J. Weight stigma and health behaviors: Evidence from the Eating in America Study. *Int. J. Obes.* **2021**, *45*, 1499–1509. [CrossRef]
128. Giovinazzo, S.; Sukkar, S.G.; Rosa, G.M.; Zappi, A.; Bezante, G.P.; Balbi, M.; Brunelli, C. Anorexia nervosa and heart disease: A systematic review. *Eat. Weight. Disord. Stud. Anorex. Bulim. Obes.* **2019**, *24*, 199–207. [CrossRef]
129. Klinkby Støving, R.; Bennedbæk, F.N.; Hegedüs, L.; Hagen, C. Evidence of diffuse atrophy of the thyroid gland in patients with anorexia nervosa. *Int. J. Eat. Disord.* **2001**, *29*, 230–235. [CrossRef]

130. Fleming, S.; Gill, P.; Jones, C.; Taylor, J.A.; Van den Bruel, A.; Heneghan, C.; Roberts, N.; Thompson, M. The Diagnostic Value of Capillary Refill Time for Detecting Serious Illness in Children: A Systematic Review and Meta-Analysis. *PLoS ONE* **2015**, *10*, e0138155. [CrossRef] [PubMed]
131. Bulik, C.M.; Carroll, I.M.; Mehler, P. Reframing anorexia nervosa as a metabo-psychiatric disorder. *Trends Endocrinol. Metab.* **2021**, *32*, 752–761. [CrossRef]
132. Gaudiani, J.L.; Sabel, A.L.; Mascolo, M.; Mehler, P.S. Severe anorexia nervosa: Outcomes from a medical stabilization unit. *Int. J. Eat. Disord.* **2012**, *45*, 85–92. [CrossRef] [PubMed]
133. Frølich, J.; Winkler, L.A.; Abrahamsen, B.; Bilenberg, N.; Hermann, A.P.; Støving, R.K. Fractures in women with eating disorders—Incidence, predictive factors, and the impact of disease remission: Cohort study with background population controls. *Int. J. Eat. Disord.* **2020**, *53*, 1080–1087. [CrossRef] [PubMed]
134. Legroux, I.; Cortet, B. Factors influencing bone loss in anorexia nervosa: Assessment and therapeutic options. *RMD Open* **2019**, *5*, e001009. [CrossRef]
135. Shi, E.; Oloff, L.M.; Todd, N.W. Stress injuries in the athlete. *Clin. Podiatr. Med. Surg.* **2023**, *40*, 181–191. [CrossRef]
136. Ji, M.-X.; Yu, Q. Primary osteoporosis in postmenopausal women. *Chronic Dis. Transl. Med.* **2015**, *1*, 9–13. [CrossRef]
137. Hembree, W.C.; Cohen-Kettenis, P.T.; Gooren, L.; Hannema, S.E.; Meyer, W.J.; Murad, M.H.; Rosenthal, S.M.; Safer, J.D.; Tangpricha, V.; T'Sjoen, G.G. Endocrine Treatment of Gender-Dysphoric/Gender-Incongruent Persons: An Endocrine Society Clinical Practice Guideline. *J. Clin. Endocrinol. Metab.* **2017**, *102*, 3869–3903. [CrossRef]
138. El Ghoch, M.; Alberti, M.; Capelli, C.; Calugi, S.; Dalle Grave, R. Resting Energy Expenditure in Anorexia Nervosa: Measured versus Estimated. *J. Nutr. Metab.* **2012**, *2012*, 652932. [CrossRef]
139. Thibault, R.; Genton, L.; Pichard, C. Body composition: Why, when and for who? *Clin. Nutr.* **2012**, *31*, 435–447. [CrossRef]
140. Cass, K.; Blalock, D.; Manwaring, J.; Wesselink, D.; Lundberg, C.; Mehler, P. Changes in Grip Strength, Depression, and Cognitive Functioning During Medical Stabilization for Anorexia Nervosa: Exploring the Utility of Grip Strength as a Marker of Severity. *J. Rehabil. Res. Pract.* **2024**, *5*, 14–22. [CrossRef]
141. Mehler, P.S.; Winkelman, A.B.; Andersen, D.M.; Gaudiani, J.L. Nutritional Rehabilitation: Practical Guidelines for Refeeding the Anorectic Patient. *J. Nutr. Metab.* **2010**, *2010*, 625782. [CrossRef] [PubMed]
142. Glorio, R.; Allevato, M.; De Pablo, A.; Abbruzzese, M.; Carmona, L.; Savarin, M.; Ibarra, M.; Busso, C.; Mordoh, A.; Llopis, C.; et al. Prevalence of cutaneous manifestations in 200 patients with eating disorders. *Int. J. Dermatol.* **2000**, *39*, 348. [CrossRef] [PubMed]
143. Gaudiani, J.L.; Mehler, P.S. Rare medical manifestations of severe restricting and purging: "Zebras," missed diagnoses, and best practices. *Int. J. Eat. Disord.* **2016**, *49*, 331–344. [CrossRef]
144. Gibson, D.; Watters, A.; Mehler, P.S. The intersect of gastrointestinal symptoms and malnutrition associated with anorexia nervosa and avoidant/restrictive food intake disorder: Functional or pathophysiologic?—A systematic review. *Int. J. Eat. Disord.* **2021**, *54*, 1019–1054. [CrossRef]
145. Atkinson, J.; Kavanagh, G.; McCracken, G. Pica: Recognizing the presentation in primary dental care. *Dent. Update* **2023**, *50*, 211–214. [CrossRef]
146. Jousse-Joulin, S. Chapter 17—Salivary Glands. In *Essential Applications of Musculoskeletal Ultrasound in Rheumatology*; Wakefield, R.J., D'Agostino, M.A., Eds.; W.B. Saunders: Philadelphia, PA, USA, 2010; pp. 199–206.
147. Anderson, S.; Gopi-Firth, S. Eating disorders and the role of the dental team. *Br. Dent. J.* **2023**, *234*, 445–449. [CrossRef]
148. Hetterich, L.; Mack, I.; Giel, K.E.; Zipfel, S.; Stengel, A. An update on gastrointestinal disturbances in eating disorders. *Mol. Cell. Endocrinol.* **2019**, *497*, 110318. [CrossRef]
149. Mathenge, N.; Osiro, S.; Rodriguez, I.I.; Salib, C.; Tubbs, R.S.; Loukas, M. Superior mesenteric artery syndrome and its associated gastrointestinal implications. *Clin. Anat.* **2014**, *27*, 1244–1252. [CrossRef]
150. Qualtrics$^{XM}$. Survey Basic Overview. Available online: https://www.qualtrics.com/support/survey-platform/survey-module/survey-module-overview/ (accessed on 10 June 2024).
151. Qualtrics$^{XM}$. What Is a Likert Scale. Available online: https://www.qualtrics.com/experience-management/research/likert-scale/ (accessed on 8 July 2024).
152. International Association of Eating Disorders Professionals Foundation. About the International Association of Eating Disorders Professionals Foundation. Available online: https://iaedpfoundation.com/about-us/ (accessed on 10 June 2024).
153. Trevelyan, E.G.; Robinson, P.N. Delphi methodology in health research: How to do it? *Eur. J. Integr. Med.* **2015**, *7*, 423–428. [CrossRef]
154. Jünger, S.; Payne, S.A.; Brine, J.; Radbruch, L.; Brearley, S.G. Guidance on Conducting and REporting DElphi Studies (CREDES) in palliative care: Recommendations based on a methodological systematic review. *Palliat. Med.* **2017**, *31*, 684–706. [CrossRef] [PubMed]

155. Spranger, J.; Homberg, A.; Sonnberger, M.; Niederberger, M. Reporting guidelines for Delphi techniques in health sciences: A methodological review. *Z. Evidenz Fortbild. Qual. Gesundheitswesen* **2022**, *172*, 1–11. [CrossRef] [PubMed]
156. Patricia, S.; Nathan, E.; Juan Sebastian, R.; Paula, M.; Randall, R.L.; Robert, T.; Roberto, V.; Mohit, B.; Michelle, G. Moving forward through consensus: Protocol for a modified Delphi approach to determine the top research priorities in the field of orthopaedic oncology. *BMJ Open* **2016**, *6*, e011780. [CrossRef]
157. IBM Corp. *IBM SPSS Statistics for Windows*; Version 27.0; IBM Corp.: Armonk, NY, USA, 2023.
158. Diamond, I.R.; Grant, R.C.; Feldman, B.M.; Pencharz, P.B.; Ling, S.C.; Moore, A.M.; Wales, P.W. Defining consensus: A systematic review recommends methodologic criteria for reporting of Delphi studies. *J. Clin. Epidemiol.* **2014**, *67*, 401–409. [CrossRef] [PubMed]
159. Pike, K.; Rothpletz-Puglia, P.; Byham-Gray, L. Development of a Sports Nutrition Focused Physical Examination Tool: A Delphi Study. Ph.D. Thesis, Rutgers School of Health Professions, Newark, NJ, USA, 2022.
160. Bathgate, J.R.; Pike, K.; Newkirk, M.; Rigassio Radler, D.; Byham-Gray, L.; Ziegler, J. Development of a bronchopulmonary dysplasia nutrition focused physical examination tool: A modified Delphi study. *Pediatr. Pulmonol.* **2024**, *59*, 1695–1707. [CrossRef]
161. Tylka, T.L.; Annunziato, R.A.; Burgard, D.; Daníelsdóttir, S.; Shuman, E.; Davis, C.; Calogero, R.M. The weight-inclusive versus weight-normative approach to health: Evaluating the evidence for prioritizing well-being over weight loss. *J. Obes.* **2014**, *2014*, 983495. [CrossRef]
162. Stankorb, S.M.; Rigassio-Radler, D.; Khan, H.; Touger-Decker, R. Nutrition Focused Physical Examination Practices of Registered Dietitians. *Top. Clin. Nutr.* **2010**, *25*, 335–344. [CrossRef]
163. MacKle, T.J.; Touger-Decker, R.; Maillet, J.O.S.; Holland, B.K. Registered dietitians' use of physical assessment parameters in professional practice. *J. Am. Diet. Assoc.* **2003**, *103*, 1632–1638. [CrossRef]
164. Andersen, D.; Baird, S.; Bates, T.; Chapel, D.L.; Cline, A.D.; Ganesh, S.N.; Garner, M.; Grant, B.L.; Hamilton, K.K.; Jablonski, K.; et al. Academy of Nutrition and Dietetics: Revised 2017 Scope of Practice for the Registered Dietitian Nutritionist. *J. Acad. Nutr. Diet.* **2018**, *118*, 141–165. [CrossRef]

*nutrients*

MDPI

*Article*

# Association Between Body Image and Physical Activity, Sociodemographic, and Morphological Variables in Adult Women

**Andrzej Knapik [1], Ryszard Plinta [1] and Rafał Gnat [2,*]**

[1] Department of Adapted Physical Activity and Sport, Faculty of Health Sciences, Medical University of Silesia, 40-751 Katowice, Poland; aknapik@sum.edu.pl (A.K.); rplinta@sum.edu.pl (R.P.)
[2] Institute of Physiotherapy and Health Sciences, Academy of Physical Education, 40-065 Katowice, Poland
* Correspondence: rafal.gnat@interia.pl; Tel.: +48-515957646

**Abstract: Background/Objectives**: Studying non-pathological determinants of body image (BI) among adult women is essential to provide a holistic understanding of the factors that shape BI and thus to promote positive mental health, support tailored interventions, address societal impacts, and ultimately facilitate women's healthier relationships with their bodies. The data available on this particular topic are sparse. The importance of BI for well-being prompted the authors to study its relationships with sociodemographic (age, education level, professional, marital, material status), morphological (body mass index—BMI) and behavioural (habitual physical activity) variables. **Methods**: A cross-sectional study of a cohort of 740 volunteer women was conducted with the one-time measurement of the dependent variables—individual components of BI—using the standard Body Esteem Scale (BES) questionnaire. **Results**: The BMI was a factor differentiating all domains of the BES. Among the analysed sociodemographic variables, intergroup comparisons showed significant effects of education and material status in the following BES domains: sexual attractiveness and weight concern. The physical condition domain showed differences between the categories of professional and material status. The level of physical activity differentiated respondents in the domains of weight concern and physical condition. **Conclusions**: Adult women are generally critical about their bodies. A factor that adversely affects the BES is an excessive BMI. Sociodemographic factors influence BI to a lesser degree. Physical activity shows a correlation with the BES domain of physical condition, particularly among young women.

**Keywords:** body image; physical activity; self-esteem

## 1. Introduction

Body image (BI) is a multidimensional construct encompassing an individual's perceptions of and attitudes towards their own body, including its physical appearance and functional aspects [1,2]. In simpler terms, it refers to "the picture we have in our minds of the size, shape, and form of our bodies" [3]. Body image can change over time, influenced by various factors, such as individual, biological, social, cultural, and historical elements, which exert differing levels of impact at different periods. The diversity of influences on BI has led to the development of various conceptualisations of the term. However, the synthesis of these approaches suggests that BI's multidimensional nature encompasses perceptual, affective, and behavioural dimensions [4,5].

Body image also encompasses attitudes and interactions with others. A common stereotype links physical attractiveness with positive personal traits, a perception not limited to Western culture but prevalent worldwide [6]. Researchers adopting a sociocultural approach suggest that cultural values shape the perception of others according to prevailing standards of attractiveness and associated expectations. This perception of others, in turn, influences how individuals view themselves [7].

Body image can take on either a positive or negative nature. Positive BI involves an individual valuing their uniqueness, showing affection for their body, and respecting it [8,9]. Such individuals also accept aspects of their appearance that do not conform to socially imposed standards [10–12]. Body acceptance is strongly linked to overall self-esteem [13–15]. From an interpersonal perspective, empirical studies indicate that individuals with positive BI tend to associate with people who accept their bodies and exhibit a positive attitude towards them [16]. Conversely, negative BI is associated with psychological distress [17]. Its documented consequences extend beyond a diminished perception of physical attractiveness [18] to include low overall self-esteem [19], symptoms of depression [20], sexual dysfunction [21], and eating disorders [19,22,23]. It is understood that the development of negative BI results from the interplay between individual personality traits and sociocultural influences [17,24].

In modern times, the media exerts significant pressure in terms of shaping BI, often promoting slender figures. This can create challenges for individuals who perceive themselves as differing from these ideals [25–30]. As BI is a key element of self-identity, feelings about one's body can profoundly influence how individuals perceive themselves and their capabilities [31]. It has a notable impact on self-esteem, quality of life, and overall well-being [32–34].

The literature review highlights a predominant focus among researchers on the negative aspects of BI, often linked to various diseases, disorders, or adverse social conditions. In contrast, the positive dimensions of BI and its associations with non-pathological factors are explored less frequently, despite their significant relevance to women's mental, emotional, and physical well-being. Women with positive BI are more likely to experience higher self-esteem, reduced anxiety, and greater overall life satisfaction. Additionally, positive BI fosters resilience against societal pressures, unrealistic beauty standards, and age-related physical changes, enabling women to embrace their individuality and feel confident across different areas of life [8,10–14]. Given the importance of BI and its complex relationships with well-being, general functioning, and quality of life, it is vital to examine BI in relation to sociodemographic (e.g., age, education level, professional, marital, and material status), morphological (e.g., body mass index—BMI), and behavioural (e.g., physical activity) variables.

## 2. Materials and Methods

### 2.1. Design

This was a cross-sectional study, categorising participants according to their habitual physical activity, body mass index (BMI), age, educational level, and professional, marital, and material status. The cohort consisted of 740 volunteer women. Dependent variables—individual components of BI, namely sexual attractiveness, weight concern, and physical condition—were assessed through one-time measurement using the standardised Body Esteem Scale (BES) questionnaire (Figure 1).

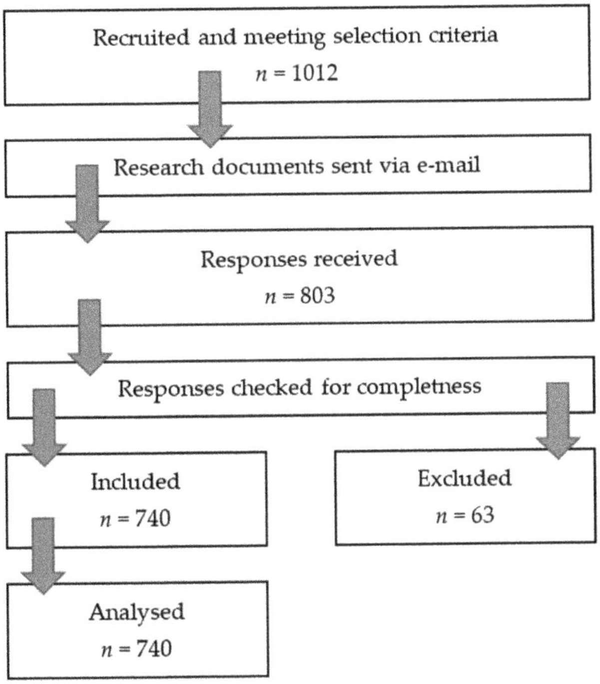


**Figure 1.** Flowchart depicting the consecutive stages of the procedure.

### 2.2. *Participants*

A total of 1012 women were recruited for the study, and 740 completed questionnaires (73%) meeting the completeness criteria were collected. Participants were recruited from the Silesian Voivodship in Southern Poland. The study employed a convenience sampling method: trained recruiters provided information to individuals within their networks—family, friends, student groups, colleagues, and members of fitness or sports clubs—and requested their email addresses. Links to the research questionnaires were subsequently sent to these addresses. This approach ensured voluntary and anonymous participation. Due to the planned sample size, the only established selection criteria were gender and age. The authors were solely interested in women's BI. The lower age limit of 18 years was set in compliance with legal requirements for adulthood, while the upper limit of 59 years was chosen as it precedes the WHO-defined threshold for old age (60 years). Other variables, such as eating disorders, menstrual cycle phases, comorbid conditions, etc., were not considered, as they were not related to the stated aim of the study. It was assumed that, given the large sample size, their impact would be significantly limited. Basic demographic data are summarised in Table 1. All procedures were conducted in accordance with the Declaration of Helsinki as revised in 2013.

**Table 1.** Demographic characteristics in the total group of participants.

| Characteristic | Mean ± Std. Dev. (Min–Max) |
|---|---|
| Age (years) | 27.73 ± 8.66 (19–59) |
| Body height (m) | 1.67 ± 6.45 (150–197) |
| Body mass (kg) | 62.69 ± 10.58 (42–128) |
| Body mass index (kg/m$^2$) | 22.31 ± 3.38 (16.18–41.86) |

### 2.3. Method

This study utilised a questionnaire comprising a demographic section and standardised scales. In the demographic section, participants provided their age, height, and weight, which were used to calculate their BMI. Additional sociodemographic data were collected, including the education level (basic professional, high school, university), professional status (student, employed, unemployed), marital status (single, in a relationship), and self-assessed material status (low, average, high). These factors were treated as independent variables.

The questionnaire section included the Body Esteem Scale (BES) [35] to assess BI-related self-esteem (the dependent variable) and the Subjective Experience of Work Load (SEWL) [36,37] to evaluate physical activity (PA) levels. The BES consists of 35 items representing various body parts and functions. Participants rate their feelings towards each item on a scale of 1 to 5, where 1 indicates "definitely negative", 2 "moderately negative", 3 "neutral", 4 "moderately positive", and 5 "definitely positive". The items are grouped into three domains: sexual attractiveness, weight concern, and physical condition. The score for each domain is calculated as the average of the ratings for its respective items, with higher scores indicating better self-esteem.

The SEWL comprises 16 items divided into three parts, including one focused on work-related items, one addressing sports, and the last addressing leisure-time PA. Points are allocated based on the physical effort associated with these activities. The remaining closed-ended statements pertain to the frequency of PA and the perception of physical exertion. The responses are scored, and calculation algorithms are applied to determine the "amount" of PA across three domains: work-related, sports, and leisure time. The combined score of these three domains represents the overall habitual PA. The SEWL outcomes were also treated as independent variables.

### 2.4. Statistical Analysis

In certain analyses, the respondents were divided into the following age groups: 18–29, 30–39, 40–49, and 50–59 years. Physical activity was also categorised: women who engaged in regular PA at least once a week, for a minimum of 1.5 h, over at least one year were classified as active; those who did not meet these criteria were classified as inactive. Body mass index categories included underweight, normal weight, overweight, and obese.

Descriptive statistics (numbers, percentages, and medians/mean values) were calculated for all variables. The internal consistency of the questionnaires was assessed using Cronbach's alpha coefficients.

Relationships between the BES domains and other quantitative variables were examined using Pearson's correlation coefficients. For univariate analyses, non-parametric tests were applied, including the Mann–Whitney test and Kruskal–Wallis ANOVA with a post hoc analysis. Multivariate analyses were conducted using backward stepwise regression. The critical $p$-value threshold was set at 0.05.

## 3. Results

### 3.1. Internal Consistency

The results obtained for the individual domains of the BES demonstrated strong internal consistency, with Cronbach's alpha coefficients as follows: sexual attractiveness—0.83, weight concern—0.91, and physical condition—0.91. Similarly, the SEWL questionnaire exhibited excellent internal consistency, with a Cronbach's alpha of 0.92.

*3.2. Body Mass Index*

Negative, weak-to-moderate correlations were observed between BMI and all individual BES domains. These trends were most pronounced in the 18–29 and 30–39 age groups. The strongest correlation was identified between weight concern and BMI (Table 2, Figure 2).

The performed categorisation of BMI returned a total of 51 (6.89%; mean 17.89 ± 0.48) underweight, 572 (77.30%; mean 21.48 ± 1.65) normal, 91 (12.30%; mean 26.99 ± 1.45) overweight, and 26 (3.51%; mean 34.33 ± 5.55) obese women (Kruskal–Wallis ANOVA and all post hoc comparisons $p < 0.001$).

The analysis of the BMI differences (Kruskal–Wallis ANOVA) showed that BMI was a factor differentiating BES in all domains: sexual attractiveness ($p < 0.01$), weight concern ($p < 0.0001$), and physical condition ($p < 0.0001$). Post hoc comparisons showed a number of differences (Table 3, Figure 2).

**Table 2.** Correlations between Body Esteem Scale domains and body mass index. Age groups (years) are also considered.

| Variable | Age Group | | | | | | | | Total | |
|---|---|---|---|---|---|---|---|---|---|---|
| | 18–29 | | 30–39 | | 40–49 | | 50–59 | | | |
| | *r* | *p* | *r* | *p* | *r* | *p* | *r* | *p* | *r* | *p* |
| SA | −0.02 | ns | −0.34 | *** | −0.05 | ns | −0.12 | ns | −0.12 | ** |
| WC | −0.42 | *** | −0.54 | *** | −0.30 | * | −0.26 | ns | −0.42 | *** |
| PC | −0.14 | ** | −0.37 | *** | −0.50 | *** | −0.33 | ns | −0.24 | *** |

SA—sexual attractiveness; WC—weight concern; PC—physical condition; r—Pearson's correlation coefficient; ns—non-significant; * $p < 0.05$; ** $p < 0.01$; *** $p < 0.001$.

**Table 3.** Body Esteem Scale and body mass index categories. Age groups (years) are also considered.

| Variable | BMI | Age Group | | | | | | | | Total | |
|---|---|---|---|---|---|---|---|---|---|---|---|
| | | 18–29 | | 30–39 | | 40–49 | | 50–59 | | | |
| | | Me | *p* | Me | *p* | Me | *p* | Me | *p* | Me | *p* |
| SA | a: U | 3.85 | | 3.54 | | 4.62 | | 3.69 | | 3.85 | |
| | b: N | 3.92 | | 4.08 | | 3.62 | | 3.85 | | 3.92 | |
| | c: OV | 3.92 | | 3.85 | | 3.58 | | 3.58 | | 3.77 | a–d * |
| | d: OB | 3.69 | ns | 3.27 | b–d ** | 3.73 | ns | 2.69 | ns | 3.62 | b–d ** |
| WC | | | a–b * | | | | | | | | a–b * |
| | a: U | 4.00 | a–c ** | 3.60 | | 4.30 | | 3.40 | | 4.00 | a–c *** |
| | b: N | 3.50 | a–d *** | 3.40 | a–d * | 3.60 | | 3.50 | | 3.50 | a–d *** |
| | c: OV | 2.50 | b–c *** | 2.90 | b–c * | 2.85 | | 3.60 | | 2.80 | b–c *** |
| | d: OB | 2.20 | b–d * | 2.00 | b–d *** | 3.05 | ns | 2.30 | ns | 2.30 | b–d *** |
| PC | a: U | 3.67 | | 4.50 | | 4.33 | | 4.22 | | 3.78 | |
| | b: N | 3.89 | | 4.00 | | 4.00 | | 3.78 | | 3.89 | a–d *** |
| | c: OV | 3.50 | | 3.44 | | 3.39 | | 3.89 | | 3.56 | b–c ** |
| | d: OB | 3.06 | ns | 2.83 | b–d * | 3.50 | ns | 2.83 | ns | 3.17 | b–d *** |

BMI—body mass index; U—underweight; N—normal; OV—overweight; OB—obese; SA—sexual attractiveness; WC—weight concern; PC—physical condition; Me—median; ns—non-significant; * $p < 0.05$; ** $p < 0.01$; *** $p < 0.001$ (post hoc for Kruskal–Wallis ANOVA).

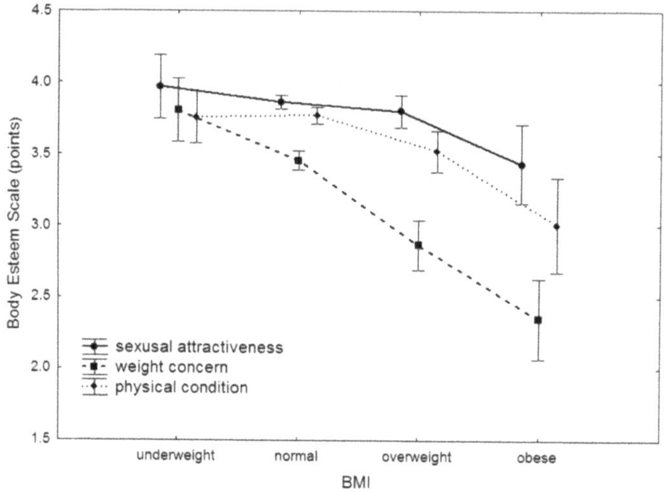


**Figure 2.** Results obtained in the Body Esteem Scale domains (mean and 95% confidence interval (whiskers)) for individual body mass index (BMI) categories in the total group of respondents.

### 3.3. Sociodemographic Variables

Intergroup comparisons of the analysed sociodemographic variables revealed that education and material status significantly influenced the sexual attractiveness domain of the BES. Differences in the weight concern domain were observed only between the 18–29 and 30–39 age groups. Meanwhile, the physical condition domain showed differences between categories of professional status as well as material status (Table 4).

**Table 4.** Body Esteem Scale domains and sociodemographic variables.

| Variable | Category | n (%) | SA | | WC | | PC | |
|---|---|---|---|---|---|---|---|---|
| | | | Me | p | Me | p | Me | p |
| Education Level (n = 740) | a: basic professional | 62 (8.38 | 3.54 | a–b *** a–c *** | 3.30 | ns | 3.89 | ns |
| | b: high school | 220 (29.73) | 3.92 | | 3.50 | | 3.67 | |
| | c: university | 458 (61.89) | 3.92 | | 3.40 | | 3.78 | |
| Professional Status (n = 740) | d: student | 345 (46.62) | 3.85 | ns | 3.40 | ns | 3.67 | d–e * |
| | e: working | 296 (40.00) | 3.85 | | 3.50 | | 3.89 | |
| | f: not working | 99 (13.38) | 3.77 | | 3.40 | | 3.88 | |
| Marital Status (n = 656) | g: single | 174 (26.52) | 3.77 | ns | 3.30 | ns | 3.78 | ns |
| | h: in relationship | 482 (73.48) | 3.92 | | 3.40 | | 3.67 | |
| Material Status (n = 581) | i: low | 160 (27.54) | 3.96 | i–j ** | 3.50 | ns | 3.89 | j–k * |
| | j: average | 223 (38.38) | 3.77 | | 3.40 | | 3.67 | |
| | k: high | 198 (34.08) | 3.92 | | 3.60 | | 3.89 | |
| Age (years) (n = 740) | l: 18–29 | 551 (74.46) | 3.92 | | 3.50 | | 3.78 | |
| | m: 30–39 | 106 (14.32) | 3.85 | | 3.10 | | 3.67 | |
| | n: 40–49 | 50 (6.76) | 3.62 | | 3.40 | | 3.67 | |
| | o: 50–59 | 33 (4.46) | 3.69 | ns | 3.50 | a–b * | 3.78 | ns |

SA—sexual attractiveness; WC—weight concern; PC—physical condition; Me—median; ns—non-significant; * $p < 0.05$; ** $p < 0.01$; *** $p < 0.001$ (marital status—Mann–Whitney test, other variables—post hoc for Kruskal–Wallis ANOVA).

### 3.4. *Physical Activity*

The correlations between the individual BES domains and the SEWL questionnaire outcomes were weak to moderate, positive, and occasionally significant (Table 5). These correlations were particularly evident in the overall participant group, the 18–29 age category, and within the physical condition domain of the BES.

**Table 5.** Body Esteem Scale and physical activity (SEWL total score). Age groups (years) are also considered.

| Variable | Age Group | | | | | | | | Total | |
|---|---|---|---|---|---|---|---|---|---|---|
| | 18–29 | | 30–39 | | 40–49 | | 50–59 | | | |
| | *r* | *p* | *r* | *p* | *r* | *p* | *r* | *p* | *r* | *p* |
| SA | 0.13 | ** | 0.02 | ns | 0.16 | ns | 0.35 | * | 0.13 | ** |
| WC | 0.12 | * | 0.09 | ns | 0.20 | ns | 0.16 | ns | 0.13 | ** |
| PC | 0.43 | *** | 0.32 | ** | 0.34 | * | 0.32 | ns | 0.40 | *** |

SA—sexual attractiveness; WC—weight concern; PC—physical condition; r—Pearson's correlation coefficient; ns—non-significant; * $p < 0.05$; ** $p < 0.01$; *** $p < 0.001$.

The SEWL-score-based categorisation of PA identified 538 active women (72.70%; mean score 9.22 ± 1.80) and 202 inactive women (27.30%; mean score 7.99 ± 1.92) (Mann–Whitney test, $p < 0.001$). Comparing the participants based on the defined activity criterion revealed differences in the weight concern and physical condition domains of the BES. These differences were evident in the total participant group and within the 30–39 age subgroup. In the 18–29 age subgroup, differences were observed solely in the physical condition domain (Table 6, Figure 3).

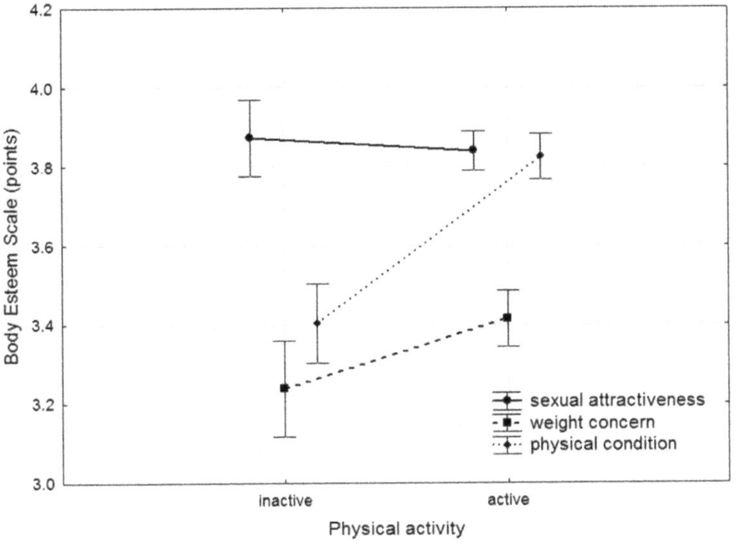


**Figure 3.** Results obtained in the Body Esteem Scale domains (mean and 95% confidence interval (whiskers)) for individual physical activity categories in the total group of respondents.

**Table 6.** Body Esteem Scale and Subjective Experience of Work Load scale categories. Age groups (years) are also considered.

| Variable | Activity | Age Group | | | | | | | | Total | |
|---|---|---|---|---|---|---|---|---|---|---|---|
| | | 18–29 | | 30–39 | | 40–49 | | 50–59 | | | |
| | | Me | *p* | Me | *p* | Me | *p* | Me | *p* | Me | *p* |
| SA | A | 3.85 | ns | 3.77 | ns | 3.62 | ns | 3.65 | ns | 3.85 | ns |
| | IA | 3.92 | | 4.00 | | 3.62 | | 3.69 | | 3.92 | |
| WC | A | 3.50 | ns | 3.10 | * | 3.40 | ns | 3.50 | ns | 3.50 | * |
| | IA | 3.40 | | 3.00 | | 3.90 | | 3.40 | | 3.30 | |
| PC | A | 3.89 | *** | 3.89 | ** | 3.67 | ns | 3.89 | ns | 3.89 | *** |
| | IA | 3.33 | | 3.33 | | 3.67 | | 3.33 | | 3.33 | |

A—active women; IA—inactive women; SA—sexual attractiveness; WC—weight concern; PC—physical condition; Me—median; ns—non-significant; * $p < 0.05$; ** $p < 0.01$; *** $p < 0.001$ (Mann–Whitney test).

### 3.5. Regression Analysis

The backward stepwise regression analysis identified several significant predictors of the BES score. Body mass index emerged as a significant predictor across all BES domains: sexual attractiveness ($\beta = -0.14$; $p < 0.001$), weight concern ($\beta = -0.44$; $p < 0.0001$), and physical condition ($\beta = -0.30$; $p < 0.0001$). As reflected by the regression β-coefficient, BMI was consistently negatively associated with the scores in all three domains. Conversely, all other identified predictors exhibited positive correlations with the BES scores. Professional status was a significant predictor in the weight concern ($\beta = 0.11$; $p < 0.05$) and physical condition ($\beta = 0.13$; $p < 0.001$) domains. Education level was found to be significant in the sexual attractiveness domain ($\beta = 0.12$; $p < 0.01$). Additionally, within the physical condition domain, material status ($\beta = 0.09$; $p < 0.05$) and PA ($\beta = 0.26$; $p < 0.0001$) were identified as significant predictors (Figure 4).

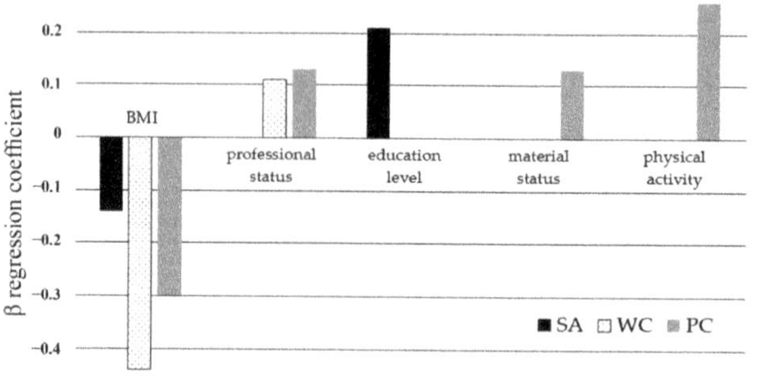


**Figure 4.** Summary of the identified significant predictors of the outcomes in the individual Body Esteem Scale domains: sexual attractiveness (SA), weight concern (WC), and physical condition (PC). Presented are β-coefficients of backward stepwise regression. Non-significant factors were omitted. BMI—body mass index.

## 4. Discussion

The analysis of the results highlights a rather critical self-assessment of BI within the participant group. The medians (Table 3) and means (Figure 2) ranged between intermediate and moderately positive. On one hand, this suggests a realistic approach to the self-evaluation of BI; on the other, it may reflect a perceived gap between one's current state and personal expectations. This phenomenon is characteristic of highly developed socioeconomic environments, where physical appearance is highly valued, and the female ideal demands continuous effort towards maintaining and improving one's body [38–40].

No significant differences in sexual attractiveness were observed between active and inactive women (Table 6). This phenomenon may be explained by historical and social factors, particularly concerning traditional gender roles [41]. For men, the functional aspects of the body, such as physical fitness and activity, have historically played a far greater role in determining attractiveness [42]. Researchers interpret this as reflecting men's long-standing role as protectors, where physical fitness and activity were essential in ensuring safety. Consequently, an active and fit man was perceived as a more desirable partner, and, even today, these traits remain key motivators for male PA [43,44]. Historically, women's roles in most cultures were markedly different. Indicators of sexual attractiveness were, and still often are, linked to body regions perceived as crucial for reproduction, such as the breasts, abdomen, pelvic area, and face [45,46]. Features like the waist-to-hip ratio and breast size are particularly significant in determining reproductive attractiveness in women [47]. Unfortunately, anthropometric data related to these factors were not collected in this study. However, it can be speculated that women with an underweight or normal BMI may have had more favourable proportions in these areas, as a higher BMI is often associated with abdominal obesity [48]. This interpretation, of course, applies only to one specific aspect of sexual attractiveness. The significant differences in sexual attractiveness observed between the BMI categories lend further support to this argument (Table 3, Figure 2).

An intriguing finding is the absence of a statistically significant correlation between age and sexual attractiveness ($p > 0.05$). Previous studies suggest that attributes linked to fertility, such as body shape, facial features, and voice pitch, significantly influence a woman's perceived attractiveness [49]. While the medians for sexual attractiveness were marginally higher in age groups I and II compared to groups III and IV, the hypothesis that age negatively impacts sexual attractiveness was not supported by this study. This aligns with the notion that physical attractiveness and sexual attractiveness, the latter being more biologically rooted, are distinct constructs [50]. This perspective is somewhat consistent with the observation that critical BI evaluations peak during the perimenopausal period but remain relatively stable during the pre- and postmenopausal stages [51]. Such patterns may reflect both the increased lifespan of modern humans—extending all individual life stages—and the growing influence of non-biological factors, such as evolving social and moral norms. Sexual activity, no longer solely tied to procreation, has become a recognised element of contemporary quality of life [52]. This is further supported by the observed effects of education and material status on sexual attractiveness (Table 3 and regression analysis).

In contemporary society, weight concern has become a significant component of BI. Self-perception is shaped by numerous factors, including emotional, social, cultural, and interpersonal influences. Increasingly, the media play a pivotal role in this process by promoting the ideal of the "perfect body" [53]. A slim figure is often associated with success, health, and sexual attractiveness [54]. From a public health perspective, this trend has both positive and negative implications. On the positive side, weight concern may contribute to mitigating the global obesity epidemic to some extent [55]. However, among young women and girls with a healthy BMI, it can also lead to risky behaviours, such as excessively restrictive diets, overly intense exercise routines, or even the use of weight loss drugs [56]. The prevailing "cult of thinness" is evident across all three domains of the BES (Figure 2). While the decline in BES mean values varies with increasing BMI, this trend is particularly pronounced in the weight concern domain.

When considering PA, differences in weight concern between active and inactive women were observed only in the second age group, i.e., women in their thirties. It appears that, during this stage of life, women regard PA as the most effective method of weight control. Comparisons of the weight concern scores across the categories of education level, professional status, marital status, and material status revealed no significant differences. Notably, among the three BES domains analysed, weight concern was the most critically assessed, reflecting, in many cases, an excessive focus on body weight. This may be a consequence of the previously mentioned "cult of thinness". Research suggests that women are generally more self-critical of their body weight than men [57]. However, this observation does not hold true across all ethnic groups and cultural settings [58,59]. The regression analysis identified professional status as a partial predictor of weight concern. This finding may be linked to the professional roles and societal expectations of women in Western cultures, although further research is needed to explore this relationship in greater depth.

The results indicate that the physical condition of underweight and normal-weight women is nearly equivalent. However, as the BMI increases, the physical condition declines sharply, reaffirming the well-established fact that excess body weight negatively impacts functional capabilities. When considering PA, differences between active and inactive women were evident across the entire group of respondents. However, when age was factored in, the impact of activity on physical condition was more pronounced in younger women. This suggests that the importance placed on physical condition diminishes with age. Nevertheless, given the ongoing dynamics of social and cultural changes, this issue may require further attention. Previous studies highlight that PA is a positive factor in shaping and enhancing BI [60].

The physical condition domain of the BES revealed significant differences between students and working women. This may be explained by the temporary pressures that young women face during their studies, which can restrict their ability to maintain an active lifestyle, thereby negatively impacting their physical condition. A notable difference was also observed between women reporting an average and high material status. However, the relationship between material status and BI remains ambiguous in the context of other studies. Some reports suggest that individuals with lower financial status tend to be dissatisfied with their body shape [61], while others argue that dissatisfaction with BI is more prevalent among those with higher socioeconomic status [62]. This discrepancy is likely influenced by cultural variations.

The data presented offer a deeper understanding of the role of various non-pathological factors in shaping women's self-perception of their bodies. From a broader perspective, this knowledge is essential in developing targeted health promotion strategies that encourage positive BI and, in turn, enhance public mental health outcomes. The findings also highlight the need to expand future research to explore the intricate interplay between pathological and non-pathological influences on BI. Investigating more specific populations could provide valuable insights. For example, women with conditions that affect body image, or adolescents, who are particularly susceptible to social pressures during their formative years, often face lasting mental health implications. By focusing on these groups, researchers can gain a more nuanced understanding of the complex interrelations among societal, psychological, and biological factors, thereby paving the way for tailored interventions that address diverse and specific needs.

The limitations of the study include reduced generalisability, as local social, economic, cultural, and religious factors, typical of a Central European population with a culture similar to that of Western Europe—predominantly Christian and politically leaning towards conservatism and centre-right ideologies—may have influenced BI and should be taken into account. Additionally, the number of respondents across the different age, BMI, and PA categories was largely uneven, which may have affected the statistical analysis. However, our primary objective was to gather a large cohort of participants, and all categorisations were performed retrospectively, making them of secondary importance compared to the main goal. Due to this same reason, such variables as eating disorders, menstrual cycle phases, comorbid conditions, etc., were not considered as selection criteria when categorising the variables. According to the authors, these important issues should be the subject of separate research using appropriate tools for that purpose. However, it was assumed that BI remains relatively stable in cases of chronic psychological disorders and that the Body Esteem Scale demonstrates a certain resilience to temporary changes in body image associated, for example, with menstrual cycle phases.

Readers should also note the use of subjective measurement tools. Nevertheless, the typical limitations associated with subjectivity were mitigated through the implementation of an anonymous approach and the use of standardised questionnaires specifically designed to address the challenges of subjective self-assessment. It is also important to note the limitations of the cross-sectional design, which does not allow for the formulation of cause-and-effect relationships.

## 5. Conclusions

Adult women tend to be highly critical of their bodies. An excessive BMI appears to be a key factor negatively associated with all three domains of the BES. Sexual attractiveness is linked to the level of education but shows no significant relationship with age or PA levels. Professional status is moderately related to BI and self-esteem in the domain of physical condition, while PA shows a correlation with physical condition, particularly among young women.

**Author Contributions:** Conceptualisation, A.K., R.P. and R.G.; methodology, A.K. and R.G.; software, R.G.; validation, A.K. and R.G.; formal analysis, R.G.; investigation, A.K. and R.P.; resources, R.P.; data curation, R.G.; writing—original draft preparation, A.K., R.P. and R.G.; writing—review and editing, A.K., R.P. and R.G.; visualisation, R.G.; supervision, A.K.; project administration, A.K. All authors have read and agreed to the published version of the manuscript.

**Funding:** All expenses were covered by the Medical University of Silesia, Katowice, Poland, based on agreement BNW-1-162/N/3/Z.

**Institutional Review Board Statement:** The study was conducted in accordance with the Declaration of Helsinki and approved by the Institutional Ethics Committee of the Academy of Physical Education in Katowice (8/2020 25 September 2020).

**Informed Consent Statement:** Informed consent was obtained from all subjects involved in the study.

**Data Availability Statement:** The data presented in this study are available on request from the corresponding author due to legal reasons.

**Conflicts of Interest:** The authors declare no conflicts of interest.

## Abbreviations

The following abbreviations are used in this manuscript:

| | |
|---|---|
| BI | Body Image |
| BMI | Body Mass Index |
| BES | Body Esteem Scale |
| SEWL | Subjective Experience of Work Load |
| PA | Physical Activity |

## References


1. Abbott, B.D.; Barber, B.L. Differences in functional and aesthetic body image between sedentary girls and girls involved in sports and physical activity: Does sport type make a difference? *Psychol. Sport Exerc.* **2011**, *12*, 333–342. [CrossRef]
2. Alleva, J.M. Shapeshifting: Focusing on body functionality to improve body image. *J. Aesthetic Nurs.* **2017**, *6*, 300–301. [CrossRef]
3. Jackson, K.L.; Janssen, I.; Appelhans, B.M.; Kazlauskaite, R.; Karavolos, K.; Dugan, S.A.; Avery, E.A.; Shipp-Johnson, K.J.; Powell, L.H.; Kravitz, H.M. Body image satisfaction and depression in midlife women: The Study of Women's Health Across the Nation (SWAN). *Arch. Womens Ment. Health* **2014**, *17*, 177–187. [CrossRef]
4. Cash, T.F.; Theriault, J.E.; Annis, N.M. Body image in an interpersonal context: Adult attachment, fear of intimacy, and social anxiety. *J. Soc. Clin. Psychol.* **2004**, *23*, 89–102. [CrossRef]
5. Krawczyk, R.; Menzel, J.; Thompson, J.K. Methodological issues in the study of body image and appearance. In *The Oxford Handbook of the Psychology of Appearance*; Rumsey, N., Harcourt, D., Eds.; Oxford University Press: Oxford, UK, 2012; pp. 605–619. [CrossRef]
6. Pop, C.L. Physical Activity, Body Image, and Subjective Well-Being. In *Well-Being and Quality of Life—Medical Perspective*; Mollaoglu, M., Ed.; Intech Open: Rijeka, Croatia, 2017. [CrossRef]
7. Jackson, L.A. Physical attractiveness: A sociocultural perspective. In *Body Images: A Handbook of Theory, Research, and Clinical Practice*; Cash, T.F., Pruzinsky, T., Eds.; Guilford Press: New York, NY, USA, 2004; pp. 13–21.
8. Avalos, L.; Tylka, T.L.; Wood-Barcalow, N. The Body Appreciation Scale: Development and psychometric evaluation. *Body Image* **2005**, *2*, 285–297. [CrossRef]
9. Wood-Barcalow, N.L.; Tylka, T.L.; Augustus-Horvath, C.L. "But I like my body": Positive body image characteristics and a holistic model for young-adult women. *Body Image* **2010**, *7*, 106–116. [CrossRef] [PubMed]
10. Tylka, T.L. Positive psychology perspectives on body image. In *Encyclopedia of Body Image and Human Appearance*; Cash, T.F., Ed.; San Diego Academic Press: San Diego, CA, USA, 2012; Volume 2, pp. 657–663.
11. Tylka, T.L.; Wood-Barcalow, N.L. What is and what is not positive body image? Conceptual foundations and construct definition. *Body Image* **2015**, *14*, 118–129. [CrossRef] [PubMed]
12. Swami, V.; Weis, L.; Barron, D.; Furnham, A. Positive body image is positively associated with hedonic (emotional) and eudaimonic (psychological and social) well-being in british adults. *J. Soc. Psychol.* **2018**, *158*, 541–552. [CrossRef] [PubMed]
13. Swami, V.; Airs, N.; Chouhan, B.; Leon, M.A.P.; Towell, T. Are there ethnic differences in positive body image among female British undergraduates? *Eur. Psychol.* **2009**, *14*, 288–296. [CrossRef]
14. Swami, V.; Campana, A.N.; Coles, R. Acceptance of cosmetic surgery among British female university students: Are there ethnic differences? *Eur. Psychol.* **2012**, *17*, 55–62. [CrossRef]
15. Tylka, T.L.; Kroon Van Diest, A.M. The Intuitive Eating Scale—2: Item refinement and psychometric evaluation with college women and men. *J. Couns. Psychol.* **2013**, *60*, 137–153. [CrossRef] [PubMed]
16. Alleva, J.M.; Medoch, M.M.; Priestley, K.; Philippi, J.L.; Hamaekers, J.; Salvinho, E.N.; Sanne, H.; Custers, M. "I appreciate your body, because..." Does promoting positive body image to a friend affect one's own positive body image? *Body Image* **2021**, *3*, 134–138. [CrossRef]
17. Allen, M.S.; Walter, E.E. Personality and body image: A systematic review. *Body Image* **2016**, *19*, 79–88. [CrossRef] [PubMed]
18. Bessenoff, G.R. Can the media affect us? Social comparison, self-discrepancy, and the thin ideal. *Psychol. Women Q.* **2006**, *30*, 239–251. [CrossRef]
19. Johnson, F.; Wardle, J. Dietary restraint, body image dissatisfaction, and psychological distress: A prospective analysis. *J. Abnorm. Psychol.* **2005**, *114*, 119. [CrossRef] [PubMed]
20. Przezdziecki, A.; Sherman, K.A.; Baillie, A.; Taylor, A.; Foley, E.; Stalgis-Bilinski, K. My changed body: Breast cancer, body image, distress and self-compassion. *Psycho-Oncology* **2013**, *22*, 1872–1879. [CrossRef]
21. Woertman, L.; van den Brink, F. Body image and female sexual functioningand behavior: A review. *J. Sex Res.* **2012**, *49*, 184–211. [CrossRef]

22. Levine, M.P.; Piran, N. The role of body image in the prevention of eating disorders. *Body Image* **2004**, *1*, 57–70. [CrossRef]
23. Menzel, J.E.; Schaefer, L.M.; Burke, N.L.; Mayhew, L.L.; Brannick, M.T.; Thompson, J.K. Appearance-related teasing, body dissatisfaction, and disorder edeating: A meta-analysis. *Body Image* **2010**, *7*, 261–270. [CrossRef] [PubMed]
24. Dionne, M.M.; Davis, C. Body image and personality. In *Encyclopedia of Body Image and Human Appearance*; Cash, T.F., Ed.; Elsevier: Amsterdam, The Netherlands, 2012; pp. 136–140.
25. Grabe, S.; Ward, L.M.; Hyde, J.S. The role of the media in body image concerns among women: A meta-analysis of experimental and correlational studies. *Psychol. Bull.* **2008**, *134*, 460–476. [CrossRef]
26. Bell, B.T.; Dittmar, H. Does media type matter? The role of identification in adolescent girls' media consumption and the impact of different thin-ideal media on body image. *Sex Roles* **2011**, *65*, 478. [CrossRef]
27. Ferguson, C.J.; Munoz, M.E.; Contreras, S.; Velasquez, K. Mirror, mirror on the wall: Peer competition, television influences, and body image dissatisfaction. *J. Soc. Clin. Psychol.* **2011**, *30*, 458–483. [CrossRef]
28. Levine, M.P.; Chapman, K. Media influences on body image. In *Body Image: A Handbook of Science, Practice, and Prevention*, 2nd ed.; Cash, T.F., Smola, L.K., Eds.; Guilford Press: New York, NY, USA, 2011; pp. 101–109.
29. McCabe, M.P.; McGreevy, S.J. Role of media and peers on body change strategies among adult men: Is body size important? *Eur. Eat. Disord. Rev.* **2011**, *19*, 438–446. [CrossRef]
30. Abraczinskas, M.; Fisak, B.; Barnes, R.D. The relation between parental influence, body image, and eating behaviors in a nonclinical female sample. *Body Image* **2012**, *9*, 93–100. [CrossRef]
31. Chrisler, J.C.; Ghiz, L. Body image issues of older women. *Women Ther.* **1993**, *14*, 67–75. [CrossRef]
32. Mercurio, A.E.; Landry, L.J. Self-objectification and well-being: The impact of self-objectification on women's overall sense of self-worth and life satisfaction. *Sex Roles* **2008**, *58*, 458–466. [CrossRef]
33. Choma, B.L.; Shove, C.; Busseri, M.A.; Sadava, S.W.; Hosker, A. Assessing the role of body image coping strategies as mediators or moderators of the links between self-objectification, body shame, and well-being. *Sex Roles* **2009**, *61*, 699–713. [CrossRef]
34. Albawardi, N.M.; AlTamimi, A.A.; AlMarzooqi, M.A.; Alrasheed, L.; Al-Hazzaa, H.M. Associations of body dissatisfaction with lifestyle behaviors and socio-demographic factors among saudi females attending fitness centers. *Front. Psychol.* **2021**, *12*, 611472. [CrossRef] [PubMed]
35. Mendelson, B.K.; Mendelson, M.J.; White, D.R. Body Esteem Scale for adolescents and adults. *J. Personal. Assess.* **2001**, *76*, 90–106. [CrossRef]
36. Baecke, J.A.; Burema, J.; Frijters, J.E. A short questionnaire for the measurement of habitual physical activity in epidemiological studies. *Am. J. Clin. Nutr.* **1982**, *36*, 936–942. [CrossRef]
37. Hertogh, E.M.; Monninkhof, E.M.; Schouten, E.G.; Peeters, P.H.; Schuit, A.J. Validity of the Modified Baecke Questionnaire: Comparison with energy expenditure according to the doubly labeled water method. *Int. J. Behav. Nutr. Phys. Act.* **2008**, *5*, 30. [CrossRef]
38. Ahern, A.L.; Bennett, K.M.; Kelly, M.; Hetherington, M.M. A qualitative exploration of young women's attitudes towards the thin ideal. *J. Health Psychol.* **2011**, *16*, 70–79. [CrossRef]
39. Swami, V.; Smith, J.-M. How not to feel good naked? The effects of television programs that use 'real women' on female viewers' body image and mood. *J. Soc. Clin. Psychol.* **2012**, *31*, 151–168. [CrossRef]
40. Swami, V.; Tran, U.S.; Stieger, S.; Voracek, M. Associations between women's body image and happiness: Results of the YouBeauty.com Body Image Survey (YBIS). *J. Happiness Stud.* **2015**, *16*, 705–718. [CrossRef]
41. Diekman, A.B.; Goodfriend, W. Rolling with the changes: A role congruity perspective on gender norms. *Psychol. Women Q.* **2006**, *30*, 369–383. [CrossRef]
42. Hőnekopp, J.; Rudolph, U.; Beier, L.; Liebert, A.; Mϋller, C. Physical attractiveness of face and body as indicators of physical fitness in men. *Evol. Hum. Behav.* **2007**, *28*, 106–111. [CrossRef]
43. Henss, R. Waist-to-hip ratio and female attractiveness. Evidence from photographic stimuli and methodological considerations. *Personal. Individ. Difer.* **2000**, *28*, 501–513. [CrossRef]
44. Agliata, D.; Tantleff-Dunn, S. The impact of media exposure on males' body image. *J. Soc. Clin. Psychol.* **2004**, *23*, 7–22. [CrossRef]
45. Platek, S.M.; Singh, D. Optimal Waist-to-Hip Ratios in Women Activate Neural Reward Centers in Men. *PLoS ONE* **2010**, *5*, e9042. [CrossRef] [PubMed]
46. Garza, R.; Heredia, R.R.; Cieslicka, A.B. Male and Female Perception of Physical Attractiveness: An Eye Movement Study. *Evol. Psychol.* **2016**, *14*, 1474704916631614. [CrossRef]
47. Ladabaum, U.; Mannalithara, A.; Myer, P.A.; Singh, G. Obesity, abdominal obesity, physical activity, and caloric intake in US adults: 1988 to 2010. *Am. J. Med.* **2014**, *127*, 717–727. [CrossRef]
48. Feinberg, D.R.; Jones, B.C.; DeBruine, L.M.; Moore, F.R.; Law Smith, M.J.; Cornwell, E.; Tiddeman, B.P.; Boothroyd, L.G.; Perrett, D.I. The voice and face of woman: One ornament that signals quality? *Evol. Hum. Behav.* **2005**, *26*, 398–408. [CrossRef]

49. Bovet, J.; Barkat-Defradas, M.; Durand, V.; Faurie, C.; Raymond, M. Women's attractiveness is linked to expected age at menopause. *J. Evol. Biol.* **2018**, *31*, 229–238. [CrossRef] [PubMed]
50. Cutler, W.B.; Genovese, E. Pheromones, sexual attractiveness and quality of life in menopausal women. *Climacteric* **2009**, *5*, 112–121. [CrossRef]
51. Séjourné, N.; Got, F.; Solans, C.; Raynal, P. Body image, satisfaction with sexual life, self-esteem, and anxiodepressive symptoms: A comparative study between premenopausal, perimenopausal, and postmenopausal women. *J. Women Aging* **2019**, *31*, 18–29. [CrossRef] [PubMed]
52. Jiannine, L.M. An investigation of the relationship between physical fitness, self-concept, and sexual functioning. *J. Educ. Health Promot.* **2018**, *7*, 57. [CrossRef] [PubMed]
53. Graff Low, K.; Charanasomboon, S.; Brown, C.; Hiltunen, G.; Long, K.; Reinhalter, K.; Jones, H. Internalization of the thin ideal, weight, and body image concerns. *Soc. Behav. Personal. Int. J.* **2003**, *31*, 81–90. [CrossRef]
54. Wronka, I.; Suliga, E.; Pawlińska-Chmara, R. Perceived and desired body weight among female university students in relation to BMI-based weight status and socio-economic factors. *Ann. Agric. Environ. Med.* **2013**, *20*, 533–538. [PubMed]
55. World Health Organization. Obesity and Overweight. 2021. Available online: https://www.who.int/news-room/fact-sheets/detail/obesity-and-overweight (accessed on 15 September 2024).
56. Zaborskis, A.; Petronyte, G.; Sumskas, L.; Kuzman, M.; Iannotti, R.J. Body image and weight control among adolescents in Lithuania, Croatia, and the United States in the context of global obesity. *Croat. Med. J.* **2008**, *49*, 233–242. [CrossRef]
57. Park, E. Overestimation and underestimation: Adolescents' weight perception in comparison to BMI-based weight status and how it varies across socio-demographic factors. *J. Sch. Health* **2011**, *81*, 57–64. [CrossRef] [PubMed]
58. Metcalf, P.A.; Scragg, R.K.R.; Willoughby, P.; Finau, S.; Tipene-Leach, D. Ethnic differences in perceptions of body size in middle-aged European, Maori and Pacific People living in New Zealand. *Int. J. Obes.* **2000**, *24*, 593–599. [CrossRef]
59. Rahman, M.; Berenson, A.B. Self-perception of weight and its association with weight-related behaviors in young, reproductive-aged women. *Obstet. Gynecol.* **2010**, *116*, 1274–1280. [CrossRef] [PubMed]
60. Hausenblas, A.A.; Fallon, E.A. Exercise and body image: A meta-analysis. *Psychol. Health* **2006**, *21*, 33–47. [CrossRef]
61. Silva, D.A.S.; Nahas, M.V.; Sousa, T.F.; Del Duca, G.F.; Peres, K.G. Prevalence and associated factors with body image dissatisfaction among adults in southern Brazil: A population-based study. *Body Image* **2011**, *8*, 427–431. [CrossRef] [PubMed]
62. Holmqvist, K.; Frisen, A. Body dissatisfaction across cultures: Findings and research problems. *Eur. Eat. Disord. Rev.* **2010**, *18*, 133–146. [CrossRef]

*nutrients*

MDPI

*Article*

# Predisposing Potential Risk Factors for Severe Anorexia Nervosa in Adolescents

**Elena Bozzola [1,*,†], Flavia Cirillo [1,†], Cristina Mascolo [1], Livia Antilici [1], Umberto Raucci [2], Benedetta Guarnieri [1], Annamaria Ventricelli [1], Elettra De Santis [1], Giulia Spina [1], Massimiliano Raponi [3], Alberto Villani [1,†] and Maria Rosaria Marchili [1,†]**

1 Pediatric Unit, Bambino Gesù Children's Hospital, IRCCS, 00165 Rome, Italy; flavia.cirillo@opbg.net (F.C.); cristina.mascolo@opbg.net (C.M.); livia.antilici@opbg.net (L.A.); benedetta.guarnieri@opbg.net (B.G.); annamaria.ventricelli@opbg.net (A.V.); elettra.desantis@opbg.net (E.D.S.); giulia.spina@opbg.net (G.S.); alberto.villani@opbg.net (A.V.); mrosaria.marchili@opbg.net (M.R.M.)

2 Pediatric Emergency Unit, Bambino Gesù Children's Hospital, IRCCS, 00165 Rome, Italy; umberto.raucci@opbg.net

3 Sanitary Direction, Bambino Gesù Children's Hospital, IRCCS, 00165 Rome, Italy; massimiliano.raponi@opbg.net

* Correspondence: elena.bozzola@opbg.net

† These authors contributed equally to this work.

**Abstract:** Background: Anorexia nervosa is a serious eating disorder that mainly affects children and adolescents. Most patients present with extreme body dissatisfaction and an obsessive focus on body weight and food. Anorexia nervosa is a complex and multifactorial condition characterised by biological, psychological, and social factors. However, studies that have explored the cumulative risk that predisposes to anorexia nervosa are limited. This study aims to explore the potential risk factors for a severe form of the disease in patients affected by anorexia nervosa and to identify whether they may interact and reinforce each other, contributing to the severity of the disorder. Methods: For this study, we enrolled children and adolescents under 18 years of age hospitalised at IRCCS Bambino Gesù Pediatric Hospital, Rome, Italy, for anorexia nervosa from 1 December 2022 to 31 August 2024, identifying and analysing potential risk factors. Elevated shape and weight concerns were found in all patients. Psychiatric and neurodevelopment comorbidities were identified in 76 patients (51.35%), life stress events in 69 (46.62%), and a family history of eating and weight control behaviours in 39 (26.35%). Out of the sample size, 20.27% of patients did not live in a traditionally structured family. This study used the Kiddie-SADS-Present and Lifetime Version interview, the Coddington Life Events Scales, and the Trauma Symptom Checklist for Children questionnaires. Results: Patients with an extreme or severe index of anorexia nervosa are more likely to have multiple predisposing factors. In detail, four predisposing factors were found in 18.6% of patients with an extreme severity index, in 15.5% of those with a severe score, and in 10.3 and 10.6% of those with a moderate and mild score, respectively. Conclusions: Cumulative potential risk factors are more likely to be found in cases of severe course disease and patients hospitalised for anorexia. Prompt identification of predisposing factors and an effective plan of action are required to avoid a severe course disorder.

**Keywords:** anorexia nervosa; adolescents; body image; cumulative potential risk factors

## 1. Introduction

Eating disorders in children and adolescents encompass a range of behavioural conditions marked by profound and persistent disruptions in eating habits, along with distressing thoughts and emotions. After the COVID-19 pandemic, the number of minors affected by anorexia nervosa, who limit food intake and/or engage in excessive physical activity even when the individual is already underweight, dramatically increased [1–3]. This is a potentially life-threatening condition that affects the quality of life of children and adolescents, as well as their families, and has one of the highest mortality rates among psychiatric disorders [4]. Anorexia nervosa is more common in females than in males. The prevalence in the general population is approximately 12 times greater in females than males (1.42% in females and 0.12% in males) [4]. Most eating disorders involve extreme body dissatisfaction and an obsessive focus on body weight and food, resulting in dangerous dietary routines that negatively affect nutritional intake, causing adverse effects on the growth and development of children and adolescents [4].

The principal cause of anorexia nervosa has not been found yet. The aetiology of anorexia nervosa includes both genetic and environmental contributions as potential predisposing factors. So, eating disorders, particularly anorexia nervosa, may be defined as complex and multifactorial conditions characterised by a combination of biological, psychological, and social factors [4–12]. Support for genetic involvement is based on aggregates in families and twin studies [6,8,13]. An influence vulnerability not only to anorexia nervosa but also to psychiatric disorders commonly comorbid in patients or relatives has been observed. These include anxiety disorders, obsessive–compulsive disorder, major depression, and substance use disorders, as well as bulimia nervosa [14]. As for environmental contributions, well-established potential risk factors for anorexia nervosa include sociocultural pressures for thinness, elevated shape and weight concerns, dietary restraint, exercise, and family history of eating and weight control behaviour [15–17].

Childhood abuse is associated with psychiatric problems and may contribute to onset of anorexia nervosa [12].

Patients with anorexia nervosa who are not medically stable should first be hospitalised as they are at risk of mortality stemming from both the physical and psychiatric complications. To help inform the decision regarding inpatient hospitalisation, clinicians can use the severity index for anorexia nervosa in the American Psychiatric Association's Diagnostic and Statistical Manual of Mental Disorders, Fifth Edition, Text Revision (DSM-5-TR) [18]. According to this document, the severity index is based upon the patient's body mass index (BMI) and is divided as follows:

- Mild-BMI 17 to 18.49 $kg/m^2$;
- Moderate-BMI 16 to 16.99 $kg/m^2$;
- Severe-BMI 15 to 15.99 $kg/m^2$;
- Extreme-BMI $< 15$ $kg/m^2$.

As previously demonstrated, high acute hospitalisation cost is associated with anorexia nervosa in paediatric patients. Three key factors, namely, comorbidities, enteral feeding, and guarding, correlate to a prolonged hospital stay and cost [19].

A prompt identification of predisposing potential risk factors and prompt diagnosis is fundamental to start an early intervention, to prevent medical complications due to prolonged anorexia nervosa, and to reintroduce the child/adolescent to social life.

The aim of this study is to explore potential risk factors that may predispose patients to hospitalisation from both psychological and medical perspectives. The main objective of our analysis is to identify whether there is a correlation between the presence of cumulative potential risk factors and the severity of anorexia nervosa.

## 2. Material and Methods

For the purpose of this study, we designed a cross-sectional study, including children and adolescents aged younger than 18 years, admitted with a diagnosis of anorexia nervosa to IRCCS Bambino Gesù Children's Hospital, Rome, Italy, which is a reference third level centre for pediatric eating disorders in Italy. The study period ranged from 1 December 2022 to 31 August 2024. According to the DSM-5-TR, the diagnosis of anorexia nervosa requires each of the following:

- Restriction of energy intake that leads to a low body weight, given the patient's age, sex, developmental trajectory, and physical health;
- Intense fear of gaining weight or becoming fat, or persistent behaviour that prevents weight gain, despite being underweight;
- Distorted perception of body weight and shape, undue influence of weight and shape on self-worth, or denial of the medical seriousness of one's low body weight [18].

Patients were excluded if they did not meet the inclusion criteria. As for the enrolled patients, the severity index has been calculated and the predisposing potential risk factors have been noted, including familiarity, life stress events, psychiatric comorbidities. According to the BMI, the severity index was calculated for every patient and the sample size divided into 4 groups.

Potential risk factors were identified by the psychologist during interviews with the parents, following the Coddington Life Events Scales (CLES) questionnaire—both the child (CLES–C) and the adolescent (CLES–A) versions—and previous studies [4,20].

Information was entered into a database specifically created for data collection. The various potential risk factors we identified, which will be further discussed in the final analysis, pertain to psychosocial potential risk factors, family structure information, the presence of psychiatric disorders in the family, and, with regard to anorexic patients, all psychological potential risk factors related to weight and body image, along with psychiatric comorbidities. Psychosocial potential risk factors, for example, include abuse and mistreatment, bullying, witnessing violence, moving house, substance abuse within the family, parental separation, domestic accidents, and school-related conflicts. Experiencing the death of a relative or a friend also represents a potential risk factor. Regarding family structure, we consider whether the patient lives with both parents, with one parent, or if the parents are divorced. With respect to the presence of past or current psychiatric disorders in the family, we observe severe intellectual disability, language disorders, schizophrenia, mood disorders, anxiety disorders, obsessive–compulsive disorder, and conduct and impulse control disorders. The presence of eating disorders is also noted. Finally, concerning the psychiatric comorbidities associated with anorexia, we highlight, in particular, intellectual disabilities, pragmatic communication disorder, mood disorders, anxiety disorders, obsessive–compulsive disorder, trauma-related disorders, and impulse control disorders.

The psychological tests used were the Kiddie-SADS-Present and Lifetime Version (K-SADS-PL) and the Trauma Symptom Checklist for Children (TSCC) [21,22]. The K-SADS-PL is one of the most valuable tools currently available for diagnosing psychiatric disorders through a semi-structured interview with a child or adolescent in collaboration with their parents. Interview Supplement 5 provides structured questions to assess the presence of eating disorders by exploring specific behaviours, thoughts regarding eating and weight, body image. The supplement provides information from both the child and parents on the duration and intensity of symptoms and helps to identify the potential risk of medical complications. The Trauma Symptom Checklist for Children (TSCC) is a psychological assessment tool designed to evaluate the presence and severity of psychological symptoms related to traumatic experiences in children and adolescents. TSCC is widely used in clinical and research settings to identify emotional and behavioural disturbances resulting from trauma, such as abuse, violence, or natural disasters. It is a self-report questionnaire typically administered to children and adolescents aged 8 to 16. It provides a quantitative assessment of post-traumatic symptoms and is composed of 54 items that cover six distinct symptom domains, including anxiety, depression, anger, post-traumatic stress, dissociation, and sexual concerns [22].

The diagnostic process for anorexia nervosa, according to the criteria of the DSM-5-TR, focused on the three main diagnostic criteria, with a specific emphasis on the issues related to body image distortion through the K-SADS-PL supplement [18,23].

*Statistical Data Analysis*

For the statistical analysis, we used SPSS Statistics (version 26.0). All tests were conducted at a significance level of $\alpha = 0.05$. To explore the relationships between BMI and the number of potential risk factors, we performed a Kruskal–Wallis test and Spearman's Rank Correlation, as appropriate.

## 3. Results

According to the inclusion criteria, 165 patients were admitted to Bambino Gesù Children's Hospital for eating disorders. Out of them, 17 were excluded because they did not meet the diagnostic criteria for anorexia nervosa. As for the remaining 149, 148 were affected by anorexia nervosa and 1 by atypical anorexia nervosa.

General data are presented in Table 1.

**Table 1.** Demographic data of enrolled patients. Continuous data are expressed in number (%) and median (IQR).

| Demographic Data | Findings |
| --- | --- |
| Female | 142 (95.9) |
| Age, years | 15.0 (13.6–16.0) |
| Length of Hospitalisation, days | 18 (11–28.7) |

Potential predisposing factors were evaluated for any patient, finding elevated shape and weight concerns in all of them and a family history of eating and weight control behaviours in 39 (26.35%).

Psychiatric and neurodevelopment comorbidities were identified in 76 (51.35%) and life stress events in 69 (46.62%) patients. Out of the sample size, 20.27% of patients did not live in a traditional structured family. Table 2 summarises the results.

In our sample size, one potential predisposing factor was found in 29.72% (44), two in 32.43% (48), three in 22.29 (33), and four in 15.54% (23).

According to the severity index, we divided patients into four groups, namely: Group A (Mild-BMI 17 to 18.49 kg/m$^2$), made up of 29 patients (19.59%); Group B (Moderate-BMI 16 to 16.99 kg/m$^2$), made up of 20 (13.51%); Group C (Severe-BMI 15 to 15.99 kg/m$^2$), including 26 (17.56%); and Group D (Extreme-BMI < 15 kg/m$^2$), composed of 73 (49.32%).

Patients affected by an extreme or severe index were more likely to have more than one potential predisposing factor. Four potential predisposing factors were found in 18.6% of patients with an extreme severity index, 15.5% of those with a severe score, and in 10.3

and 10.6% of those with a moderate and mild score, respectively. Figure 1 summarises the results.

**Table 2.** Potential predisposing factors to anorexia.

| Potential Predisposing Factors | Number of Patients |
|---|---|
| Shape and weight concerns | 148 (100%) |
| A family history of eating and weight control behaviours | 39 (26.35%) classified as follows:<br>1 severe intellectual disability<br>1 language disorder<br>2 schizophrenia spectrum disorders<br>2 bipolar disorders<br>14 depressive disorders<br>3 anxiety disorders<br>1 obsessive–compulsive disorder<br>6 anorexia nervosa<br>1 unspecified disruptive, impulse-control, and conduct disorder<br>2 substance-related and addictive disorders<br>1 bulimia<br>1 neurocognitive disorder |
| Psychiatric and neurodevelopment comorbidities | 76 (51.35%) classified as follows:<br>2 severe intellectual disability disorders<br>1 asocial (pragmatic) communication disorder<br>4 bipolar disorders<br>28 depressive disorders<br>27 anxiety disorders<br>3 obsessive–compulsive disorders<br>7 trauma- and stressor-related disorders<br>4 unspecified disruptive, impulse-control, and conduct disorders |
| Life stress events | 69 (46.62%) classified as follows:<br>8 childhood maltreatment or abuse<br>19 bullying<br>6 exposure to violence<br>11 experiencing relative/friend death<br>3 experiencing relative illness<br>1 substance abuse<br>1 moving house<br>2 separation from parents<br>2 teacher–student conflicts<br>1 domestic accident |
| Family structure | 118 living with both parents<br>28 living with divorced parents<br>2 single-parent family |

Psychological tests were consistent with self-esteem, body dissatisfaction, and emotional dysregulation.

We conducted a Kruskal–Wallis test to compare the groups based on disease severity (mild, moderate, severe, and extreme) with respect to the number of potential risk factors, but the results did not reach statistical significance ($p = 0.810$).

We conducted a Spearman's rank correlation analysis to assess the relationship between BMI and the number of potential risk factors. The analysis revealed a weak negative correlation (rho = −0.034) which was not statistically significant ($p = 0.340$).

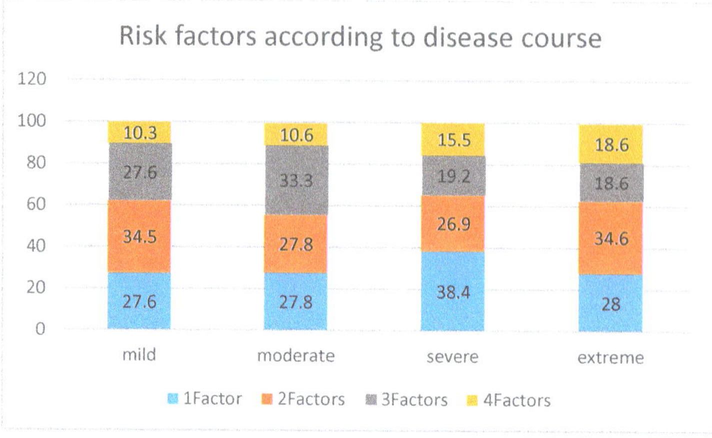


**Figure 1.** Number of potential risk factors according to disease course in patients.

## 4. Discussion

Potential risk factors, either psychological, social, and biological, were searched and identified in paediatric patients hospitalised for anorexia nervosa. When these factors co-occur or accumulate, the impact seems to become of concern, potentially intensifying the effect. In our study, the coexistence of four or more potential risk factors seems to correlate to a more serious form of eating disorder. In fact, the percentage of patients with four potential risk factors was almost double in the extreme subgroup than in mild and moderate ones. Nevertheless, comparing groups based on disease severity (mild, moderate, severe, and extreme) with respect to the number of potential risk factors, no statistical significance was reached. This may suggest that while an accumulation of potential risk factors is observed in more severe cases descriptively, this pattern does not achieve statistical support in the current sample size, likely due to the designed model of the study. In fact, given the retrospective nature of the study, we did not calculate the simple size, which actually corresponds to all eligible patients hospitalised during the study period. Patients with a lower BMI corresponding to a severe or extreme case were more likely to be hospitalised. Mild or moderate cases are generally treated as outpatients, receiving reduced assistance in terms of exams, interviews, and psychological support.

Furthermore, we decided not to involve outpatient subjects because we preferred to have a CLES questionnaire performed in the same context (hospital) and by the same operators. In fact, although the interviews allow for greater flexibility of assessment, they are also influenced by the clinical judgment and the professional experience of the psychologist.

All the patients participating in this study presented with shape and weight concerns. According to self-esteem theory, one of the primary psychological vulnerabilities for the development of anorexia nervosa is low self-esteem [24]. Research suggests that low self-esteem is linked to body dissatisfaction rather than to an actual overweight condition, thus becoming a strong predictor of dysfunctional eating behaviours. This potential risk factor is exacerbated by significant concerns about physical appearance and body weight, which act as key mediators for the onset of an eating disorder. Therefore, low self-esteem is one of the primary potential psychological risk factors to consider when addressing eating disorders. Eating disorders, especially anorexia nervosa, are frequently associated with difficulties in emotional regulation. This is supported by emotional regulations and interpersonal theories; individuals with eating disorders often use food control as a dysfunctional coping strategy for managing stress and interpersonal conflicts. Consequently, restrictive eating may be employed as a means to avoid confronting distressing emotional experiences, both intrapersonal and interpersonal, in an effort to regain a sense of control [24].

In the sample size, more than half of the patients were diagnosed with psychiatric and neurodevelopment comorbidities. Individuals with eating disorders frequently exhibit anxiety, depression, or personality traits related to perfectionism and impulsivity [25]. In anorexia nervosa, anxiety and perfectionism can intensify concerns about weight and body image, predisposing individuals to the disorder and serving as maintenance factors. The literature suggests that individuals diagnosed with anorexia nervosa may exhibit deficits in cognitive processes such as attention, working memory, and problem-solving. These impairments can negatively impact self-regulation abilities, increasing the likelihood of dysfunctional eating behaviours. According to the executive function theory, individuals with anorexia nervosa may struggle to assess the long-term consequences of their actions, thereby reinforcing restrictive eating patterns and supporting dysfunctional eating behaviours as control and coping strategies [24].

A life stress event was found in almost half of the patients, confirming that it is a potential risk factor for anorexia nervosa. Several studies emphasise the significant role of childhood trauma as a potential risk factor for developing eating disorders, particularly anorexia nervosa [26–30]. Various forms of abuse, neglect, and severe family dysfunction significantly increase the likelihood of developing eating disorders. The accumulation of traumatic experiences in a person's life exponentially raises the risk of such disorders. Furthermore, childhood trauma not only serves as a potential risk factor but also as a maintenance factor. Adverse experiences in childhood can alter the neurobiological stress response, compromising the hypothalamic–pituitary–adrenal axis and leading to emotional dysregulation [28]. This, in turn, can reinforce the use of food control as a dysfunctional coping strategy for stress, even into adulthood [27,29–31].

Among potential social risk factors, family structure and organisation should be investigated. Dysfunctional family dynamics can be a significant potential risk factor for the development of anorexia nervosa. For example, high parental expectations, poor communication, dysfunctional relationships such as intrafamily conflict, and mental health disorders within the family may all contribute to the onset of an eating disorder [32]. Specifically, insecure attachment or dysfunctional mother–child interactions, particularly during feeding, as well as emotional neglect, are all identified as significant potential risk factors for the development of anorexia nervosa in offspring [33].

Patients affected by an extreme or severe index were more likely to have more potential predisposing factors.

Our study supports the hypothesis of cumulative potential risk factors in the development and maintenance of anorexia nervosa acting in an additive manner [34]. Each potential risk factor, such as low self-esteem, emotional dysregulation, and childhood trauma, may individually contribute to the vulnerability to the disorder. However, when these factors co-occur or accumulate, the impact seems to become more significant, potentially intensifying the effect. In detail, the coexistence of four or more potential risk factors seems to correlate with a more serious form of eating disorder. The percentage of patients with four potential risk factors was almost double in the extreme subgroup compared to mild and moderate ones. Multiple interacting potential risk factors may overwhelm an individual's psychological resilience and coping mechanisms, thereby increasing the likelihood of engaging in dysfunctional eating behaviours. For example, an individual with a history of trauma, combined with low self-esteem and difficulties in emotional regulation, may be at higher risk compared to someone exposed to a single factor. Moreover, these cumulative potential risks may not only precipitate the onset of anorexia nervosa but also play a critical role in the chronicity of the disorder as they reinforce maladaptive coping strategies, such as restrictive eating, to manage underlying emotional or psychological distress.

This perspective is supported by evidence from the literature indicating that multiple adverse experiences, particularly those occurring during formative years, can significantly elevate the risk of developing eating disorders. The interplay between these factors suggests a more complex aetiological pathway whereby cumulative stressors and vulnerabilities disrupt key psychological processes, contributing to the difficulty in achieving long-term recovery.

Understanding the potential risk factors and their potential contribution to the onset of anorexia nervosa may be useful for identifying at-risk groups and providing screening and prevention programs, as well as effective interventions. Despite the available body of literature testing environmental, genetic, and family potential risk factors, due to the contrasting results, no convincing evidence supports a specific potential risk factor [25,35,36]. In the literature, eating disorders seem to be associated with multiple potential risk factors [25]. The value of the present study is that we speculate that the co-occurrence of more potential risk factors rather than a specific potential risk factor is the key to interpreting the severity of presentation. Moreover, it provides a methodological direction for future studies such as large-scale collaborative studies and data sharing to validate preliminary evidence. Better knowledge of the processes and factors that lead to anorexia is useful to define prevention strategies and implement targeted early interventions.

Moreover, our work is part of a broader landscape of research on potential risk factors, from a cumulative perspective, for anorexia nervosa. Several studies broaden this perspective by highlighting the weight of emotional dysregulation and family dynamics, the impact of social and cultural norms, and core potential risk factors of anorexia such as low BMI and body dissatisfaction [37,38]. The literature emphasises the need for integrated approaches to understand the interactions between biological, psychological, and social factors.

A potential limitation of this study is our not having examined a causal link between the variables analysed. Moreover, we did not investigate the role of the media and social media in the onset and course of the disease. In fact, a problematic use of social media and the internet can serve as potential risk factors by promoting unrealistic aesthetic ideals that foster body dissatisfaction and a desire for thinness. Social media platforms may amplify social comparison, exposing individuals to idealised body images that reinforce concerns about weight and body shape [25,39]. Furthermore, the use of weight control and dieting applications can contribute to the maintenance of restrictive eating behaviours. Experiences of cyberbullying should also be considered as potential risk factors in the online environment [39,40].

## 5. Conclusions

The co-presence of multiple potential risk factors may correlate to the severity of anorexia nervosa, both from a medical and a psychological point of view.

In detail, the presence of four or more potential risk factors is associated with a more severe form of anorexia nervosa, even if no statistical significance could be identified, likely due to the study design. The preliminary results may be of relevance for clinical practice in order to guide prevention strategies, intercepting patients at risk of hospitalisation and severe course disease early. Further studies, including large-scale collaborative studies and data sharing, should be planned to validate our preliminary evidence.

**Author Contributions:** Supervision, M.R.M. and A.V. (Alberto Villani); conceptualisation, F.C. and E.B.; data curation, B.G. and E.D.S.; methodology, U.R. and A.V. (Annamaria Ventricelli); analysis,

C.M. and M.R.; statistical analysis, L.A.; critically review, G.S. All authors have read and agreed to the published version of the manuscript.

**Funding:** This work was supported also by the Italian Ministry of Health with "current Research funds".

**Institutional Review Board Statement:** Ethical review and approval were waived for this study because the consent for the use of medical data for research purposes was obtained at hospital admission as per the hospital regulation. No further consent was asked as the collected data were part of the clinical assessment as per the Bambino Gesù Children's Hospital protocol.

**Informed Consent Statement:** Informed consent was obtained from the parents of all subjects involved in the study.

**Data Availability Statement:** Data can be collected at Cirillo's room and are available on reasonable request by contacting Cirillo (flavia.cirillo@opbg.net).

**Conflicts of Interest:** The authors declare no conflicts of interest.

## List of Abbreviations

| | |
|---|---|
| AN | Anorexia nervosa |
| ED | Eating disorders |
| DSM-5-TR | Diagnostic and Statistical Manual of Mental Disorders, Fifth Edition, Text Revision |
| BMI | Body mass index |
| CLES | Coddington Life Events Scales |
| K-SADS-PL | Kiddie-SADS-Present and Lifetime Version |
| TSCC | Trauma Symptom Checklist for Children |

## References

1. Spettigue, W.; Obeid, N.; Erbach, M.; Feder, S.; Finner, N.; Harrison, M.E.; Isserlin, L.; Robinson, A.; Norris, M.L. The impact of COVID-19 on adolescents with eating disorders: A cohort study. *J. Eat. Disord.* **2021**, *9*, 65. [CrossRef]
2. Vyver, E.; Katzman, D.K. Anorexia nervosa: A paediatric health crisis during the COVID-19 pandemic. *Paediatr. Child. Health* **2021**, *26*, 317–318. [CrossRef] [PubMed]
3. Schlissel, A.C.; Richmond, T.K.; Eliasziw, M.; Leonberg, K.; Skeer, M.R. Anorexia nervosa and the COVID-19 pandemic among young people: A scoping review. *J. Eat. Disord.* **2023**, *11*, 122. [CrossRef]
4. Guarda, A. Eating Disorders: Overview of Epidemiology, Clinical Features, and Diagnosis. Uptodate. 2024. Available online: https://www.uptodate.com/contents/eating-disorders-overview-of-epidemiology-clinical-features-and-diagnosis (accessed on 10 October 2024).
5. Steinhausen, H.; Jakobsen, H.; Helenius, D.; Munk-Jørgensen, P.; Strober, M. A nation-wide study of the family aggregation and risk factors in anorexia nervosa over three generations. *Int. J. Eat. Disord.* **2015**, *48*, 1–8. [CrossRef] [PubMed]
6. Watson, H.J.; Palmos, A.B.; Hunjan, A.; Baker, J.H.; Yilmaz, Z.; Davies, H.L. Genetics of eating disorders in the genome-wide era. *Psychol. Med.* **2021**, *51*, 2287. [CrossRef]
7. Strober, M.; Freeman, R.; Lampert, C.; Diamond, J.; Kaye, W. Controlled family study of anorexia nervosa and bulimia nervosa: Evidence of shared liability and transmission of partial syndromes. *Am. J. Psychiatry* **2000**, *157*, 393. [CrossRef]
8. Bulik, C.M.; Sullivan, P.F.; Tozzi, F.; Furberg, H.; Lichtenstein, P.; Pedersen, N.L. Prevalence, heritability, and prospective risk factors for anorexia nervosa. *Arch. Gen. Psychiatry* **2006**, *63*, 305. [CrossRef] [PubMed]
9. Duncan, L.; Yilmaz, Z.; Gaspar, H.; Walters, R.; Goldstein, J.; Anttila, V.; Bulik-Sullivan, B.; Ripke, S.; Eating Disorders Working Group of the Psychiatric Genomics Consortium; Thornton, L.; et al. Significant Locus and Metabolic Genetic Correlations Revealed in Genome-Wide Association Study of Anorexia Nervosa. *Am. J. Psychiatry* **2017**, *174*, 850. [CrossRef] [PubMed]
10. Watson, H.J.; Yilmaz, Z.; Thornton, L.M.; Hübel, C.; Coleman, J.R.; Gaspar, H.A.; Bryois, J.; Hinney, A.; Leppä, V.M.; Mattheisen, M.; et al. Genome-wide association study identifies eight risk loci and implicates metabo-psychiatric origins for anorexia nervosa. *Nat. Genet.* **2019**, *51*, 1207. [CrossRef] [PubMed]
11. Hilbert, A.; Pike, K.M.; Goldschmidt, A.B.; Wilfley, D.E.; Fairburn, C.G.; Dohm, F.-A.; Walsh, B.T.; Weissman, R.S. Risk factors across the eating disorders. *Psychiatry Res.* **2014**, *220*, 500. [CrossRef]

12. Molendijk, M.L.; Hoek, H.W.; Brewerton, T.D.; Elzinga, B.M. Childhood maltreatment and eating disorder pathology: A systematic review and dose-response meta-analysis. *Psychol. Med.* **2017**, *47*, 1402. [CrossRef] [PubMed]
13. Steiger, H.; Booij, L. Eating Disorders, Heredity and Environmental Activation: Getting Epigenetic Concepts into Practice. *J. Clin. Med.* **2020**, *9*, 1332. [CrossRef] [PubMed]
14. Sirufo, M.M.; Magnanimi, L.M.; Ginaldi, L.; De Martinis, M. Anorexia nervosa and autoimmune comorbidities: A bidirectional route? *CNS Neurosci. Ther.* **2022**, *28*, 1921–1929. [CrossRef]
15. Bryant, E.; Aouad, P.; Hambleton, A.; Touyz, S.; Maguire, S. 'In an otherwise limitless world, I was sure of my limit.'† Experiencing Anorexia Nervosa: A phenomenological metasynthesis. *Front. Psychiatry* **2022**, *13*, 894178. [CrossRef]
16. Mento, C.; Silvestri, M.C.; Muscatello, M.R.A.; Rizzo, A.; Celebre, L.; Praticò, M.; Zoccali, R.A.; Bruno, A. Psychological Impact of Pro-Anorexia and Pro-Eating Disorder Websites on Adolescent Females: A Systematic Review. *Int. J. Environ. Res. Public Health* **2021**, *18*, 2186. [CrossRef]
17. McAdams, C.J.; Efseroff, B.; McCoy, J.; Ford, L.; Timko, C.A. Social Processing in Eating Disorders: Neuroimaging Paradigms and Research Domain Organizational Constructs. *Curr. Psychiatry Rep.* **2022**, *24*, 777–788. [CrossRef] [PubMed]
18. American Psychiatric Association. *Diagnostic and Statistical Manual of Mental Disorders*, 5th ed.; Text Revision (DSM-5-TR); American Psychiatric Association: Washington, DC, USA, 2022; pp. 391–398.
19. Marchili, M.; Bozzola, E.; Guolo, S.; Marchesani, S.; Spina, G.; Mascolo, C.; Vicari, S.; De Jacobis, I.T.; Raponi, M.; Villani, A. Pediatric acute hospitalization for anorexia nervosa: An economic evaluation. *Ital. J. Pediatr.* **2024**, *27*, 33. [CrossRef]
20. Coddington, R.D. *Coddington Life Events Scales (CLES) Technical Manual*; Multi-Health Systems: Toronto, ON, Canada, 1999.
21. Kaufman, J.; Birmaher, B.; Brent, D.; Rao, U.; Flynn, C.; Moreci, P.; Williamson, D.; Ryan, N. Schedule for affective disorders and schizophrenia for school-age children-present and lifetime version (K-SADS-PL): Initial reliability and validity data. *J. Am. Acad. Child. Adolesc. Psychiatry* **1997**, *36*, 77–80. [CrossRef]
22. Briere, J. *Trauma Symptom Checklist for Children (TSCC): Professional Manual*; Psychological Assessment Resources: Odessa, FL, USA, 2004.
23. Mohammadi, M.R.; Mostafavi, S.A.; Hooshyari, Z.; Khaleghi, A.; Ahmadi, N.; Molavi, P.; Armani Kian, A.; Safavi, P.; Delpisheh, A.; Talepasand, S.; et al. Prevalence, correlates and comorbidities of feeding and eating disorders in a nationally representative sample of Iranian children and adolescents. *Int. J. Eat. Disord.* **2020**, *53*, 349–361. [CrossRef] [PubMed]
24. Zanella, E.; Lee, E. Integrative review on psychological and social risk and prevention factors of eating disorders including anorexia nervosa and bulimia nervosa: Seven major theories. *Heliyon* **2022**, *8*, e11422. [CrossRef]
25. Barakat, S.; McLean, S.A.; Bryant, E.; Le, A.; Marks, P.; Touyz, S.; Maguire, S. Risk factors for eating disorders: Findings from a rapid review. *J. Eat. Disord.* **2023**, *11*, 8. [CrossRef]
26. Yoon, C.Y.; Mason, S.M.; Loth, K.; Jacobs, D.R. Adverse childhood experiences and disordered eating among middle-aged adults: Findings from the coronary artery risk development in young adults study. *Prev. Med.* **2022**, *162*, 107124. [CrossRef] [PubMed]
27. Ziobrowski, H.N.; Buka, S.L.; Austin, S.B.; Duncan, A.E.; Simone, M.; Sullivan, A.J.; Horton, N.J.; Field, A.E. Child and adolescent maltreatment patterns and risk of eating disorder behaviors developing in young adulthood. *Child. Abus. Negl.* **2021**, *120*, 105225. [CrossRef] [PubMed]
28. Marciello, F.; Monteleone, A.M.; Cascino, G.; Patriciello, G.; Pellegrino, F.; Fiorenza, G.; Monteleone, P. Early traumatic experiences and eating disorders: A focus on the endogenous stress response system. *J. Psychopathol.* **2020**, *26*, 77–84.
29. Kovács-Tóth, B.; Oláh, B.; Kuritárné Szabó, I.; Túry, F. Adverse childhood experiences increase the risk for eating disorders among adolescents. *Front. Psychol.* **2022**, *13*, 1063693. [CrossRef] [PubMed]
30. Chu, J.; Raney, J.H.; Ganson, K.T.; Wu, K.; Rupanagunta, A.; Testa, A.; Jackson, D.B.; Murray, S.B.; Nagata, J.M. Adverse childhood experiences and binge-eating disorder in early adolescents. *J. Eat. Disord.* **2022**, *10*, 168. [CrossRef] [PubMed]
31. Lie, S.Ø.; Bulik, C.M.; Andreassen, O.A.; Rø, Ø.; Bang, L. Stressful life events among individuals with a history of eating disorders: A case-control comparison. *BMC Psychiatry* **2021**, *21*, 501. [CrossRef]
32. Doom, J.R.; Deer, L.K.; Mickel, T.; Infante, A.; Rivera, K.M. Eating behaviors as pathways from early childhood adversity to adolescent cardiometabolic risk. *Health Psychol.* **2024**, *43*, 448–461. [CrossRef]
33. Del Casale, A.; Adriani, B.; Modesti, M.N.; Virzì, S.; Parmigiani, G.; Vento, A.E.; Speranza, A.M. Anorexia nervosa and familial risk factors: A systematic review of the literature. *Curr. Psychol.* **2023**, *42*, 25476–25484. [CrossRef]
34. Stice, E.; Desjardins, C.D. Interactions between risk factors in the prediction of onset of eating disorders: Exploratory hypothesis generating analyses. *Behav. Res. Ther.* **2018**, *105*, 52–62. [CrossRef]
35. Solmi, M.; Radua, J.; Stubbs, B.; Ricca, V.; Moretti, D.; Busatta, D.; Carvalho, A.F.; Dragioti, E.; Favaro, A.; Monteleone, A.M.; et al. Risk factors for eating disorders: An umbrella review of published meta-analyses. *Braz. J. Psychiatry* **2021**, *43*, 314–323. [CrossRef] [PubMed]

36. McKay, M.T.; Cannon, M.; Chambers, D.; Conroy, R.M.; Coughlan, H.; Dodd, P.; Healy, C.; O'donnell, L.; Clarke, M.C. Childhood trauma and adult mental disorder: A systematic review and meta-analysis of longitudinal cohort studies. *Acta Psychiatr. Scand.* **2021**, *143*, 189–205. [CrossRef] [PubMed]
37. Bourdier, L.; Orri, M.; Carre, A.; Gearhardt, A.N.; Romo, L.; Dantzer, C.; Berthoz, S. Are emotionally driven and addictive-like eating behaviors the missing links between psychological distress and greater body weight? *Appetite* **2018**, *120*, 536–546. [CrossRef]
38. Jahrami, H.; Saif, Z.; Faris, M.A.; Levine, M.P. The relationship between risk of eating disorders, age, gender and body mass index in medical students: A meta-regression. *Eat. Weight. Disord.* **2019**, *24*, 169–177. [CrossRef]
39. Bozzola, E.; Spina, G.; Agostiniani, R.; Barni, S.; Russo, R.; Scarpato, E.; Di Mauro, A.; Di Stefano, A.V.; Caruso, C.; Corsello, G.; et al. The Use of Social Media in Children and Adolescents: Scoping Review on the Potential Risks. *Int. J. Environ. Res. Public Health* **2022**, *19*, 9960. [CrossRef]
40. Ioannidis, K.; Taylor, C.; Holt, L.; Brown, K.; Lochner, C.; Fineberg, N.A.; Corazza, O.; Chamberlain, S.R.; Roman-Urrestarazu, A.; Czabanowska, K. Problematic usage of the internet and eating disorder and related psychopathology: A multifaceted, systematic review and meta-analysis. *Neurosci. Biobehav. Rev.* **2021**, *125*, 569–581. [CrossRef] [PubMed]

*nutrients*

MDPI

*Article*

# Anorexia Nervosa in Polish Children and Adolescents in the Context of the COVID-19 Pandemic—An Observational Single Centre Study

**Katarzyna Anna Dyląg [1,2,*], Kamil Paweł Skowron [1], Magdalena Kurnik-Łucka [1], Łukasz Drzewiecki [2], Katarzyna Przybyszewska [2], Magdalena Król-Dykas [1], Paulina Dumnicka [3], Zuzanna Gawlik [4], Jakub Gawlik [5], Sebastian Sikora [6] and Krzysztof Gil [1]**

[1] Department of Pathophysiology, Jagiellonian University Medical College, 31-121 Krakow, Poland; kamil.skowron@uj.edu.pl (K.P.S.); magdalena.kurnik@uj.edu.pl (M.K.-Ł.); magdalena.krol@doctoral.uj.edu.pl (M.K.-D.); krzysztof.m.gil@uj.edu.pl (K.G.)

[2] St. Louis Children Hospital, 31-503 Krakow, Poland; lukasz.drzewiecki@dzieciecyszpital.pl (Ł.D.); katarzyna.przybyszewska@dzieciecyszpital.pl (K.P.)

[3] Department of Medical Biochemistry, Jagiellonian University Medical College, 31-034 Krakow, Poland; paulina.dumnicka@uj.edu.pl

[4] Hospital of the Brothers Hospitallers of Saint John of God, 31-061 Krakow, Poland; zuzanna.nowak@alumni.uj.edu.pl

[5] University Hospital, 30-688 Krakow, Poland; jkgawlik@gmail.com

[6] SWPS University, 31-864 Krakow, Poland; ssikora@swps.edu.pl

* Correspondence: katarzyna.anna.dylag@uj.edu.pl

**Abstract:** Introduction: Anorexia nervosa (AN) is a psychiatric disorder with a high mortality rate and significant prevalence in the paediatric population. Preliminary reports during the COVID-19 pandemic suggested an increased incidence of AN among children and adolescents. The objective of this study was to analyse new cases of AN before, during, and after the pandemic, with a particular focus on the physical manifestations of the disease. Methods: This single-centre, retrospective study included new AN cases from the hospital database of a regional paediatric hospital (a reference centre for AN) between 2013 and 2023. Data analysed included the duration of the disease, body mass index (BMI), weight loss, length of hospitalisation, laboratory markers (leukopenia, anaemia, hypoproteinaemia, hyperferritinaemia, hypophosphataemia, dyslipidaemia, and hypothyroxinaemia) and clinical features of malnutrition (such as amenorrhea bradycardia, pericardial effusion, and cortical/subcortical atrophy). Results: This study was conducted in a Polish regional hospital. We analysed 228 hospitalized female patients aged 10 to 18 years diagnosed with AN, focusing on clinical characteristics, biochemical markers, and the impact of the COVID-19 pandemic. The COVID-19 pandemic was shown to have a significant impact, with longer hospitalisations observed during and after the pandemic and a lower BMI on admission post-pandemic compared to pre-pandemic. In addition, nutritional treatment became more prevalent over time while biochemical markers such as anaemia, hypothyroidism, hypophosphataemia, and dyslipidaemia were statistically more common post-pandemic. Conclusions: This study demonstrates a significant impact of the COVID-19 pandemic on the clinical course and hospitalisation patterns of paediatric patients with AN. These findings suggest that the pandemic may have exacerbated disease severity and altered treatment approaches, emphasizing the need for enhanced clinical management and follow-up strategies for AN in the paediatric population during such health crises.

**Keywords:** anorexia nervosa; eating disorder; adolescents; COVID-19; SARS-CoV-2; coronavirus

## 1. Introduction

Anorexia nervosa (AN) is an eating disorder characterized by progressive weight loss, intense fear of gaining weight, and distorted perception of body weight and shape [1].

The prevalence of AN differs significantly between countries, Europe and North America reporting a significantly higher prevalence than Asia and South America [2,3]. In Europe, the prevalence of AN is estimated to be 1–4% of women [4,5]. Mortality in patients with AN is significantly higher than in the general population [6]. According to a meta-analysis of global eating disorder mortality rates conducted by Arcelus et al., the standardized mortality ratio (SMR) was 5.9 (95% CI 4.2–8.3), indicating an approximately six-fold greater risk [7]. A recent study revealed that, after 5 years of follow-up, the SMR of patients with anorexia nervosa and severe nutritional problems was 15.9 [8]. The course of relapsing-remitting disease is observed among most patients [9,10]. AN occurs among children, adolescents, and adults, and a gradual decrease in the age of the first onset of the disease has been observed [11,12]. AN symptoms include limiting of energy consumption in relation to needs, resulting in a markedly reduced body weight considering age and sex, developmental trajectory and physical well-being [13]. In DSM-5, the severity of anorexia nervosa is classified into four stages according to the individual's BMI: extreme (BMI $< 15$ kg/m$^2$), severe (BMI 15–15.99 kg/m$^2$), moderate (BMI 16–16.99 kg/m$^2$) and mild (BMI $\geq 17$ kg/m$^2$) [6]. Furthermore, the DSM-5 classification distinguishes a subtype of anorexia nervosa, atypical anorexia nervosa, which is diagnosed when all criteria for anorexia nervosa are met, except that, despite significant weight loss, the individual's weight is within or above the normal range. Bozzola et al. in 2024 summerised the most important risk factors and comorbidities such as persistent functional abdominal discomfort or autoimmune/autoinflammatory disorders. Furthermore, evidence indicates the influence of genetic predisposition, along with brain structure and pathways, on the development of anorexia nervosa. Finally, the gastrointestinal microbiota has also been identified as a potential risk factor for the development of anorexia nervosa due to intricate direct and indirect interactions between the gut and the brain [14].

The first reports suggesting that the COVID-19 pandemic contributed to a significant increase in cases of anorexia nervosa in the pediatric age group were published in July 2020 [15]. The authors suggested that the circumstances created by the pandemic caused an increase in both morbidity and severity of AN [16]. Since then, it has been well established that the COVID-19 pandemic affected both patients with pre-existing eating disorders, causing an increase in severity and incidence of the symptoms [17,18] differences in the disease characteristics [19,20] as well as a worsening of general well-being [21]. Furthermore, an increased comorbidity of psychiatric disorders was observed among individuals with AN. Most of the authors concentrated on the mental health of AN patients in the context of a pandemic. However, some studies analyzed the pediatric population in terms of increased occurrence of new cases of AN and physical determinants of health in patients with AN who required hospital treatment.

Unlike most European countries, Poland was not badly affected by the first wave of pandemic regarding COVID-19 morbidity and mortality due to the early introduction of lockdown [22,23]. However, the emotional impact of severe restrictions was high [24–26]. The second and third pandemic waves brought a rapid and tragic increase in COVID-19 mortality [22]. Furthermore, the Polish mental health system, including child and adolescent psychiatry services, is believed to have been inefficient for years [27–30]. In the time of the COVID-19 pandemic, access to psychiatric and psychological services in Poland, as in other European countries, was limited [27,30,31], yet an increased incidence of mental health services was observed [27,30,32]. Although it is known that the COVID-19 pandemic influenced body image, self-esteem, and predisposition to eating disorders among Polish women [33] the influence of the pandemic on the incidence of AN in children and adolescents in Poland has not yet been studied.

The objective of this study was to investigate the trend of morbidity in AN before, during and after the COVID-19 pandemic and to assess the severity of malnutrition and the presence of other biomarkers of malnutrition in AN in these groups.

## 2. Materials and Methods

### 2.1. Study Design

The study was a retrospective analysis conducted at St. Louis Children's Hospital, ul. Strzelecka 2, 31-503 Krakow, Poland, using data extracted from medical records of 228 patients hospitalised over a 10-year period between 2013–2023 that were included in the study. The inclusion criteria were a newly made diagnosis of anorexia nervosa (coded F50.00 in the ICD-10 classification) in a patient treated in an inpatient service and of female sex. The exclusion criteria were other causes of malnutrition (i.e., comorbid gastrointestinal disorder) and male sex. During the study group selection process, 17 male patients with an AN diagnosis (corresponding to 6.9% of all AN cases) and were excluded from further analysis. Anorexia nervosa affects only a very small percentage of adolescent males, and the clinical presentation of eating disorders in males may differ from that in females, often involving less restrictive methods of weight control and a focus on muscle gain rather than weight loss [34]. In our study, we included only female cases to increase the homogeneity of the study population, as the number of male cases was too small for comparative analysis and would only lead to biased results. The mean age of the participants was 13.9 $\pm$ 1.6. The duration of the disease, the body mass index (BMI) on admission, the extent of weight loss, the length of hospitalisation, and laboratory biomarkers of malnutrition (leucopenia, anaemia, hypoproteinaemia, hyperferritinaemia, hypophosphataemia, dyslipidaemia, and hypothyroxinaemia) as well as clinical ones (bradycardia, pericardial effusion, and cortical/subcortical atrophy) were analysed. The study population was divided into three study groups: before, during, and after the SARS-CoV-2 pandemic (pre-COVID, peri-COVID, and post-COVID, respectively). The epidemic state in Poland, introduced by the Ministry of Health and lasting from 20 March 2020 to 15 May 2022, was taken as the duration of the pandemic.

### 2.2. Statistical Analysis

Statistical analyses were performed using IBM SPSS. Quantitative data were shown as mean and standard deviation (SD) or median (interquartile range), depending on the distribution (normal or different from normal in the Shapiro–Wilk test). Qualitative data were presented as the number of patients in a category and the percentage of the respective group, with the exclusion of patients with a lack of data. A one-way ANOVA was selected to compare the means of normally distributed continuous variables across the three groups of patients admitted before, during, and after the COVID-19 pandemic. The Kruskal–Wallis Test was employed for non-normally distributed continuous data, providing a non-parametric alternative to ANOVA. Pearson's Chi-Square Test was used for categorical variables, such as patient categories or demographic factors, to test for associations or differences in proportions across the groups. A paired $t$-test was used for comparing the means of dependent populations, such as on admission and at discharge, within the same group of patients, provided the data were normally distributed. Wilcoxon Signed Rank Test served as a non-parametric alternative to the paired $t$-test. A $p$-value less than 0.05 was considered statistically significant. Data integrity was ensured through a rigorous cleaning and validation process. All datasets were reviewed for accuracy, with duplicate records and incomplete entries excluded to prevent bias. All steps of data handling and analysis were documented for reproducibility, and statistical assumptions were checked to ensure valid results.

### 2.3. Ethical Considerations

The study was approved by the Ethics Committee of the Regional Board of Physicians with the approval number of OIL/KBL/15/2024 (6 May 2024). Patient confidentiality was maintained by data pseudonymization. Each patient was assigned a unique, randomly generated identification number to replace any personally identifiable information. Direct identifiers, such as names, personal ID numbers, and contact information, were removed from the dataset. Once the data needed for the analysis were collected, the information linking patient identifiers to their original identifiers was stored in an encrypted database, accessible only to the principal investigator for data verification. This process ensured compliance with relevant ethical and regulatory standards.

## 3. Results

### 3.1. General Characteristics

The study included 228 hospitalised female patients diagnosed with anorexia nervosa, with an age range of 10 to 18 years. In Figure 1 we present a distribution of cases over time. Most of those included in the study exhibited a severe form of the disease, as defined by the BMI criteria outlined in the fifth edition of the Diagnostic and Statistical Manual of Mental Health (DSM-5) (Figure 2). BMI values increased slightly during hospitalisation, with a mean difference between BMI at discharge and admission of 0.457 ± 0.813 ($p < 0.001$). There was a negative correlation between BMI and ferritin levels and patients with a lower BMI at admission and a higher ferritin exhibited a longer hospitalisation period (r = −0.437 and r = 0.333, respectively, $p < 0.001$). The duration of the disease showed a positive correlation with the duration of amenorrhoea (r = 0.318, $p < 0.001$). Among biochemical indicators of the severity of malnutrition, hyperferritinaemia and hypothyroidism were the most commonly observed (Table 1). Younger age was associated with a higher prevalence of hypophosphataemia (r = −0.366, $p < 0.001$).

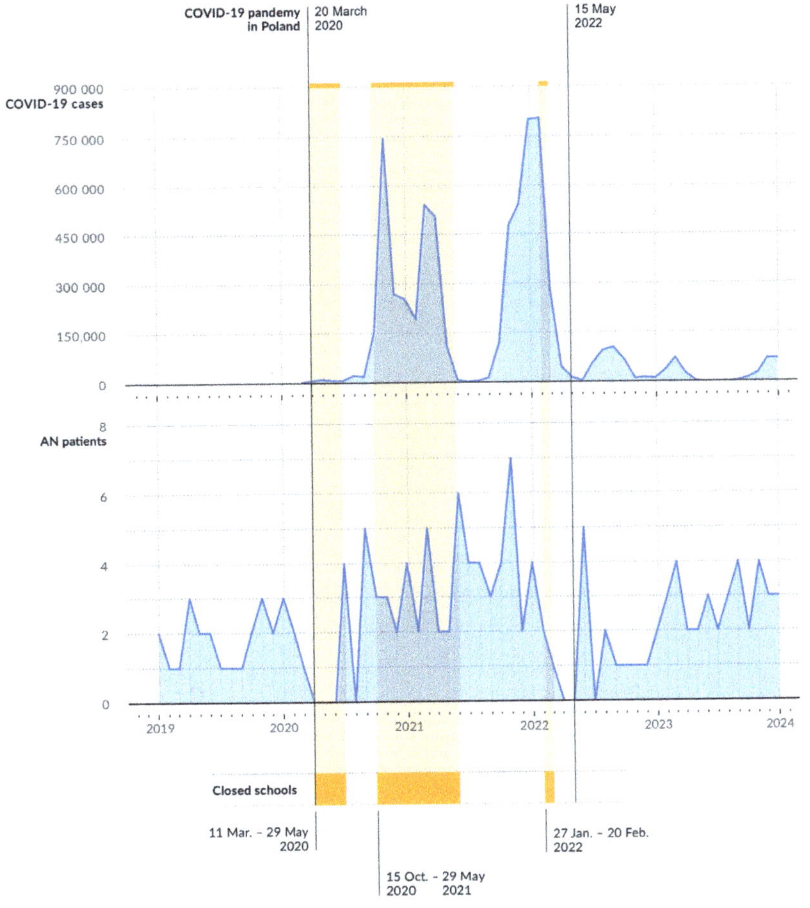


**Figure 1.** Number of cases of anorexia nervosa on the timeline juxtaposed with COVID-19 morbidity.

**Table 1.** General characteristics of the study groups. WBC, white blood cells; PLT, platelets. Bold—statistically significant difference between groups. Data are presented as mean ± SD or median (interquartile range).

| Variable | All Patients ($n$ = 228) | Pre-COVID ($n$ = 113) | Peri-COVID ($n$ = 71) | Post-COVID ($n$ = 44) | $p$ |
|---|---|---|---|---|---|
| Age (years) | 13.9 ± 1.6 | 14 (2) | 14 (2) | 14 (2) | 0.531 |
| Hospitalisation (days) | 5 (8) | 4 (5) | 6 (6.5) | 9 (13.25) | **<0.001** |
| Nutritional treatment | 30.7% | 17.7% | 35.2% | 56.8% | **<0.001** |
| Height (cm) | 160 (10) | 159 (12) | 161 (10) | 161 (9.8) | 0.121 |
| Weight at admission (kg) | 35.9 ± 6.4 | 36.2 ± 7.3 | 35.9 ± 5.4 | 35.4 ± 5.5 | 0.766 |
| BMI at admission (kg/m$^2$) | 14.0 ± 1.7 | 14.4 ± 1.9 | 14.1 ± 1.4 | 13.7 ± 1.6 | **0.047** |
| Weight at discharge (kg) | 36.6 ± 5.7 | 36.1 ± 6.6 | 36.7 ± 4.9 | 37.7 ± 4.9 | 0.324 |
| BMI at discharge (kg/m$^2$) | 14.5 ± 1.4 | 14.5 ± 1.6 | 14.4 ± 1.2 | 14.5 ± 1.1 | 0.822 |
| Amenorrhoea (months) | 4 (5) | 6 (9) | 3 (5) | 4 (2.25) | **0.018** |
| Duration of the disease (months) | 7 (8) | 6 (8) | 7 (8) | 6.5 (7.75) | 0.816 |
| Weight loss during the disease (kg) | 11 (8) | 11 (8) | 11 (8.5) | 10 (7.1) | 0.701 |
| WBC ($\times 10^3/\mu$L) | 5.22 ± 1.72 | 5.45 ± 1.65 | 5.13 ± 2.1 | 4.75 ± 1.1 | **0.013** |
| PLT ($\times 10^3/\mu$L) | 225.5 ± 54.6 | 225.1 ± 54.5 | 223.8 ± 57.3 | 229.3 ± 51.5 | 0.918 |
| Ferritin (µg/L) | 122.3 (108.7) | 110.5 (85.3) | 133.4 (115.2) | 122 (103.9) | 0.295 |
| Leukopenia, $n$ (%) | 28.8% | 21.6% | 38% | 31.8% | 0.051 |
| Anaemia, $n$ (%) | 14.2% | 9% | 14.1% | 27.3% | **0.013** |
| Hypoproteinaemia, $n$ (%) | 5.5% | 5.7% | 8.6% | 0% | 0.145 |
| Hyperferritinaemia, $n$ (%) | 71.1% | 71.4% | 68.8% | 74.4% | 0.817 |
| Hypothyroidism, $n$ (%) | 80.6% | 71.3% | 89.4% | 86% | **0.011** |
| Pericardial fluid, $n$ (%) | 27% | 32.8% | 26% | 18.9% | 0.319 |
| Bradycardia, $n$ (%) | 38.5% | 42.3% | 38.2% | 29.3% | 0.244 |
| Cortical-subcortical atrophy, $n$ (%) | 26.1% | 35.4% | 19.3% | 19.4% | 0.082 |
| Hypophosphataemia, $n$ (%) | 15% | 4.3% | 26.1% | 20.9% | **<0.001** |
| Dyslipidaemia, $n$ (%) | 43.3% | 31.6% | 42.4% | 61% | **0.015** |

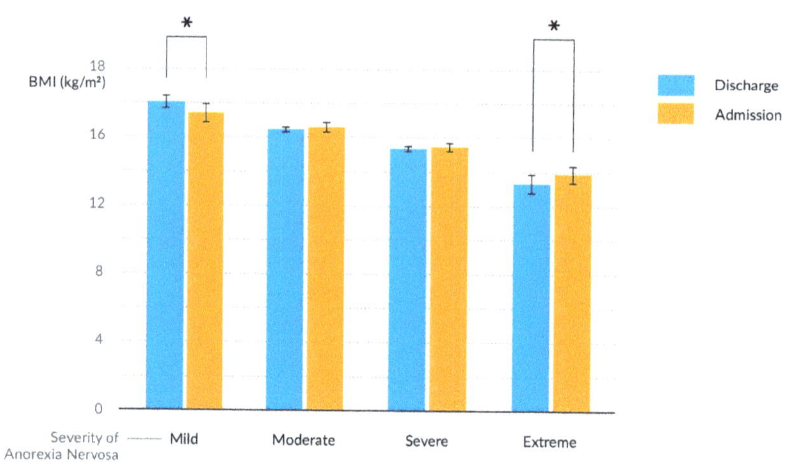


**Figure 2.** Quantitative division of the study population into disease severity categories based on DMS-V criteria based on BMI values and change in BMI during hospitalisation in each category. * $p < 0.05$.

### 3.2. Trends Across the COVID Timeline

The analysis included a population divided into three study groups: pre-COVID, peri-COVID, and post-COVID. The groups were homogeneous with respect to age, duration of the disease, reported weight loss during the course of the disease, and BMI achieved at discharge. During and after the pandemic, it was found that the duration of hospitalisation was longer than that observed prior to the pandemic ($p < 0.001$, Figure 3). In the cohort studied before the advent of the pandemic, the duration of amenorrhoea was significantly longer than in the peri-COVID group ($p = 0.015$). The BMI values on admission were comparable during and after the pandemic, although they were lower after the pandemic compared to before ($p = 0.042$).

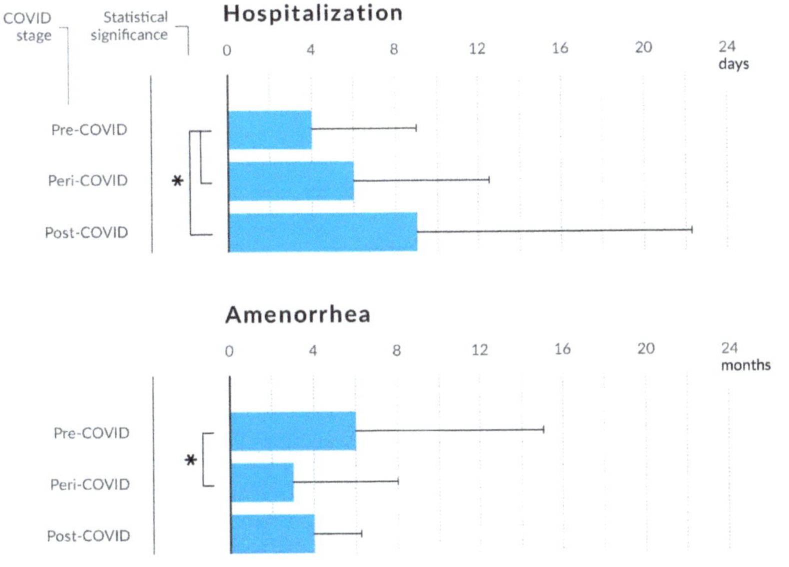


**Figure 3.** Duration of hospitalisation in days and duration of amenorrhoea in months reported on the day of admission. * Statistically significant difference between variables ($p < 0.05$).

However, the prevalence of anaemia, decreased thyroid hormone levels, hypophosphataemia, and dyslipidaemia was higher among patients after COVID than in the pre-COVID group. Patients during the pandemic exhibited a transition period of change, as indicated by the aforementioned characteristics. Only with regard to nutritional therapy and the incidence of hypophosphataemia were there discernible statistical differences when comparing these patients with pre-pandemic patients. The percentage of patients who received nutritional treatment gradually increased from the pre-pandemic period to the post-pandemic period (Figure 4). Leukocyte, platelet, and ferritin levels were found to be comparable between the study groups. However, the prevalence of anaemia, decreased thyroid hormone levels, hypophosphataemia, and dyslipidaemia was higher among patients after COVID than in the pre-COVID group. Patients during the pandemic exhibited a transition period of change, as indicated by the aforementioned characteristics. Only with regards to nutritional therapy and the incidence of hypophosphataemia were there discernible statistical differences when comparing these patients with pre-pandemic patients.

Data on hospitalisations on an annual basis reveal a notable increase in 2021, followed by a precipitous decline in the following year. In 2023, the upward trajectory observed during the previous decade was resumed. Annual trends in AN hospitalisations are presented in Figure 5. The age distribution of hospitalised patients has remained relatively stable over time (Figure 6).

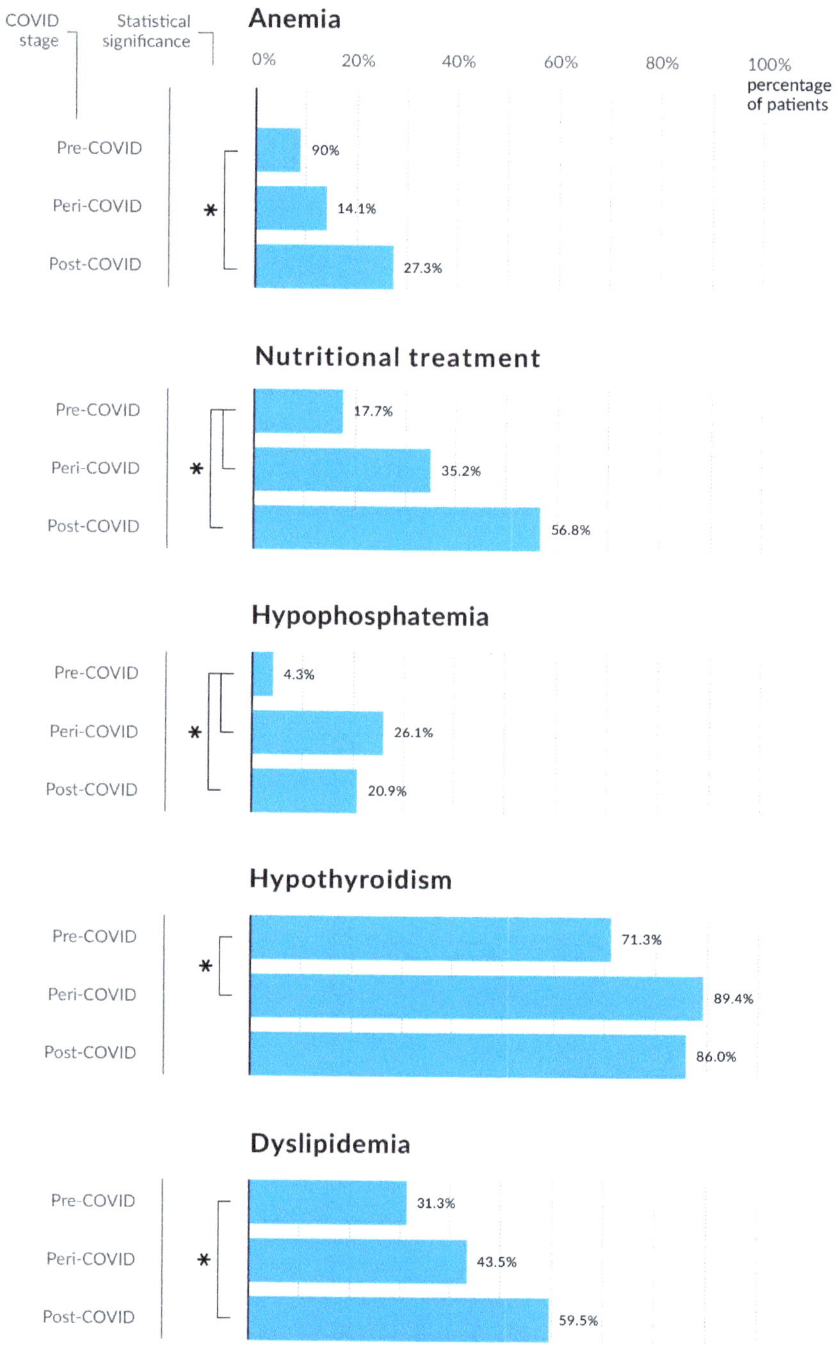


**Figure 4.** Selected characteristics that demonstrate the major differences between the study groups. * Statistically significant difference between variables ($p < 0.05$).

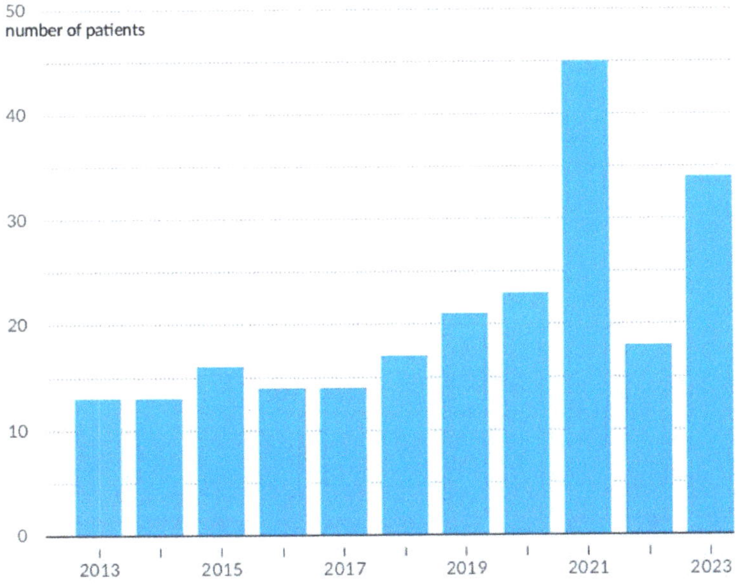


**Figure 5.** Annual number of AN hospitalisations from 2013 to 2023.

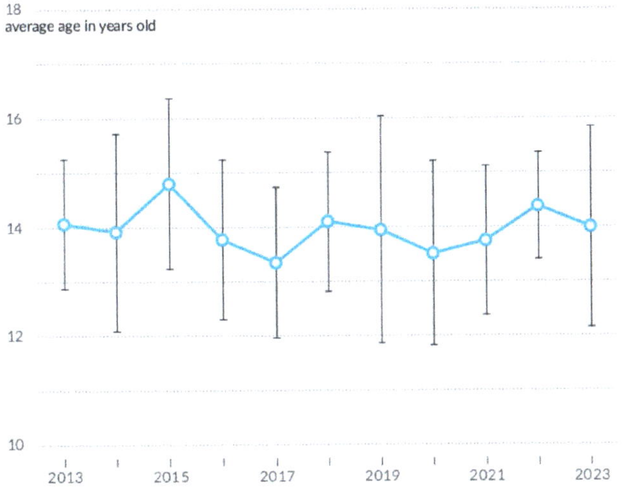


**Figure 6.** Annual average age of AN patients hospitalised between 2013–2023.

## 4. Discussion

In our study, we observed a rapid increase in hospitalizations due to anorexia nervosa during the COVID-19 pandemic with an increase in hospitalization duration, nutritional treatment requirement, and duration of amenorrhea. This result is consistent with reports from pediatric centers in other countries. Haripersad et al. [15] and Hansen were the first to alarm the medical community about the increase in new hospitalisations due to AN during the first lockdown; however, due to the time when they were published, their reports were lacking more in-depth analysis. Lin et al. [35] established a significant increase in hospital admissions due to AN accompanied by an extension of hospitalisation. Otto et al. [36] reported the same trend, which observed a double incidence in AN during

COVID-19 compared to before COVID-19 years. Matthews et al. [37] suggested that the odds of hospitalization for patients with AN were eight times higher during the pandemic than in previous years. Agostino et al. [38] established that not only the incidence of AN increased with pandemic, but also patients diagnosed with AN before pandemic were characterized with fewer markers of disease severity than patients diagnosed during pandemic. Similarly, Vyver et al. [39] demonstrated that AN morbidity doubled during the COVID-19 period. Girardi et al. [40], Hurtado et al. [41], Rafferty at al. [42], Silber et al. [43], Hyam et al. [44], Wong et al. [45] and Giraldo et al. [46] documented an increased AN-related admission rate in their respective countries. Schlapfer et al. showed that an increase in admissions due to AN was associated with a prolonged admission time and nutritional interventions. Goldberg et al. [47] concluded that the increase in admissions was not accompanied by an increase in the severity of the disease. In 2023 Schlissel et al. presented a review in which they observed during COVID-19 pandemia increased incidence of AN and AAN among adolescents. They concluded that overall the pandemic has exacerbated symptoms specific to eating disorders, hence straining an already overextended healthcare system [18]. Increases were also observed in other non-Western nations, including Pakistan and Singapore, but in these cases, investigators did not distinguish between forms of eating disorders [48,49]. To date, there have been no Polish data on trends in new admissions and the severity of AN in light of the pandemic. However, our findings confirm general trends throughout the world [15,18,20,36–44,47,50–52]. To our knowledge, this is also the first study to analyze post-pandemic trends in the incidence of AN. In our sample, we observed that the trend of increased morbidity from AN continues in the years after COVID-19. Hurtado et al. analyzed peripandemic and postpandemic years together and their observations are consistent with ours [41], however, this trend should be carefully researched in the near future. Interestingly, Herpetz-Dahlman et al. indicated that after pandemic admission rates due to AN remained high among children with a decrease among adolescents [53]. The trend should be carefully studied in the future.

Anorexia nervosa is a psychiatric condition with a significant impact on the biological mechanisms of the human body [54]. In our peri-pandemic and post-pandemic cohorts, we report a higher frequency of anemia, hypothyroidism, hypophosphatemia, and dyslipidemia compared to the pre-pandemic cohort. These parameters offer a more precise reflection of patients' somatic state, thereby providing a more nuanced understanding of their physical condition that extends beyond the limitations of a BMI-based approach to disease severity assessment [55]. One way of gaining a deeper insight is to monitor the negative effects of malnutrition on the functioning of the endocrine system, causing disorders of the hypothalamic-pituitary-adrenal, gonadal or thyroid axis that can mostly be interpreted as response reaction to systemic illness which may also simultaneously serve as adaptive mechanisms. For instance, it appears that the reduced levels of thyroid hormones in individuals with anorexia nervosa have an adaptive role in response to malnutrition to slow down metabolism and thus reduce the body's energy expenditure (resting energy expenditure) [56]. It is often accompanied by increases in serum ACTH and cortisol levels (maintaining euglycemia) and decreases in serum gonadotropin and sex hormone concentrations (reducing energy expenditure), as observed in malnourished patients. Hypothyroidism, together with hypoinsulinemia, increased endogenous cholesterol synthesis and elevated cortisol levels, may contribute to dyslipidaemia. Additionally, severe malnutrition and metabolic disorders can result in anaemia, which is primarily caused by bone marrow hypoplasia. It can thus be inferred that the elevated prevalence of the aforementioned disorders observed in the peri- and/or post-COVID-19 patient population indicates a more severe somatic presentation of the AN than was observed in the pre-pandemic period, despite only slight differences in BMI between those groups. Few researchers have investigated the influence of pandemic on malnutrition biomarkers in patients with AN. Stra [57] face et al. observed that hospitalized patients had leukocytopenia, neutropenia, hypovitaminosis, and hormonal disturbances; however, the authors only illustrated the COVID-19 cohort without comparing it with the prepandemic cohort.

Hurtado et al. included sex hormones and vitamin B12 levels in the analysis, but no differences were observed between the two analyzed groups analyzed [41]. Goldberg et al. reported a higher incidence of leucopenia among patients hospitalised during the pandemic, but did not confirm the differences in terms of hypophosphatemia and thyroid hormones levels [47]. Both pre-COVID-19 and post-COVID-19 AN patients were similar in laboratory malnutrition parameters according to Girardi et al. [40]. However, none of the authors reported differences in BMI between the groups. These findings raise a question about methods to estimate the severity of the disease in AN.

The disease severity criteria commonly used in DSM-5, which are based on patient BMI values, are increasingly being criticized for their lack of real-world applicability in terms of translating into the somatic or psychological state of patients [58,59]. In relation to this, efforts are being made to identify a new, more clinically useful indicator of disease severity. From a psychopathological point of view, methods that assess the severity of the cardinal characteristics of AN, such as the drive for thinness and overestimation of weight and shape, are being used, which are gaining increasing support in the scientific literature [60,61]. However, similar to the assessment of BMI, these methods do not provide a reliable reflection of the biological manifestations of the disease such as endocrine or electrolyte disturbances, which also directly influence therapeutic decisions. This is particularly relevant given the considerable prevalence of systemic complications in this population, including anemia, hyponatraemia and hypokalaemia, or elevated liver enzymes, which cannot be directly attributed only to the degree of undernutrition [62]. However, new reports are periodically published on potential new indicators of disease severity. One such example is a report which suggests that the evaluation of a patient's body weight history in relation to their body weight upon admission is associated with nutritional biomarkers and can be a useful method for classifying the severity of disease [63].

There are various interpretations of the reasons behind the increase in admissions due to new AN during the pandemic. Sociocultural factors are of great importance in body image and are known to be essential determinants of the development of anorexia nervosa [64–66] In fact, stress of any kind should be considered as a precipitating factor for the onset of AN [65]. Our study specifically identifies social isolation related to the COVID-19 pandemic as another important factor in the development and aggravation of AN. During the COVID-19 pandemic, educational institutions were closed and most extracurricular activities for children and adolescents that usually take place outside the family and in group settings were canceled. Teens experienced an extended period of physical isolation from their peers, teachers, extended family, and community connections. Social distancing and school closures exacerbate mental health issues of children and adolescents, who are already more susceptible to such problems than adults [67]. In addition, quarantine experiences were often associated with decreased psychological well-being and the emergence of psychological symptoms and emotional disorders, including sadness, anxiety, sleeplessness, and post-traumatic symptoms [68]. It should be mentioned that Graell et al. reported that people with a history of eating disorders have been among the most severely affected [69]. Moreover, 41% of adolescents receiving therapeutic care exhibited a resurgence of eating disorder symptoms following the lockdown, especially among those with diminished self-directedness and fewer effective coping mechanisms [70]. Furthermore, lock-down and isolation due to COVID-19 pandemic increased the use of internet and social-media, which are already known to affect loneliness [71,72]. Women, in particular, rely on social networks, which can negatively influence their perceptions of body image [57,58]. However, the long-lasting consequences of social restraint for newly diagnosed and relapsing patients with AN are unknown. Moreover, the concept of control has been considered a crucial factor in the development of AN. It has been established that people with AN tend to control their caloric intake in response to the inability to control the uncertain aspects of life [73,74]. It is difficult to imagine something more uncontrolled than the rapidly developing worldwide pandemic and the sense of instability can be considered a precipitating factor in many of the new cases of AN. Jarvers et al. documented that loss of personal control and alexithymia contributed to the symptomatology of AN during the pandemic [75].

Furthermore, the fact that medical institutions provided care predominantly by telehealth methods seems to be an important factor. Telehealth, being an effective tool that helps provide better care to patients, has several limitations, especially in terms of surveillance of patients with mental health conditions [76]. Therefore, AN diagnoses may have been delayed, resulting in a more severe clinical picture. Patients with already diagnosed AN reported dissatisfaction with telehealth services [77] and a decrease in treatment adherence was observed [78], which might suggest that in this particular group direct communication with physicians and therapists could be crucial [79]. Additionally, children and adolescents were home schooled during lockdowns. Teachers and school psychologists could not serve as whistleblowers in the context of individual weight loss. Furthermore, the crucial role of family structure in the pathophysiology of AN is undisputable [80,81]. During lockdowns, all family members were closed in a limited space, which could have increased the impact of the family and increased the outbreak of the disease. There are various interpretations of the reasons for the increase in AN cases observed during the pandemic. Most interpretations take into account psychosocial factors. It should be mentioned that direct biological mechanisms that lead to an increase in susceptibility to AN cannot be excluded on the basis of current scientific data. However, if such trends of AN related to post-COVID morbidity are maintained, especially in light of the continuously unstable epidemiological situation, such mechanisms must be searched for.

According to the latest report of the World Health Organization, Polish children have one of the lowest rates of mental well-being and one of the highest rates of attempted suicide in Europe [82]. Our results further demonstrate such a defect in the Polish healthcare system in the context of pediatric mental health and AN, which has only been aggravated by the COVID-19 pandemic. On the other hand, such pandemic-related unpreparedness was not uncommon [15,36,39–42,50–52,79,83,84], which could argue for the intersection between biological and psychological mechanisms related to SARS-CoV-2 infection [85]. Thus, our results also imply the global need for dedicated health policies, which could be easily implemented into virtual (e.g., telehealth) clinical practice in the time of possible future pandemics, military conflicts or climate-related migration and disasters. These strategies should be established for a better and earlier diagnosis of AN, among other psychiatric disorders, to prevent hospitalization. The multilevel prevention system to support youth mental well-being should be introduced across different health promotion settings, including the implementation of feasible lifestyle recommendations combined with practical interventions. Special attention should be paid to up-to-date sleep and nutritional recommendations. The latest research suggests causative links between sleep duration, screen time and mental health [86]. And the bidirectional relationship between diet quality and mental health in children and adolescents has also been highlighted [87,88].

The main strength of our study is its comprehensive approach and especially the analysis of the biological aspects of AN in the context of a pandemic. Furthermore, we included the post-pandemic perspective in the analysis. However, several limitations of our study must be acknowledged. This is a single-center retrospective study that may contribute to some selection bias. However, in the region, St. Louis Children's Hospital is the only service that provides pediatric and psychiatric services, and the vast majority of cases of AN are hospitalized at this facility. Only patients who required hospitalization were included in the study; What can limit conclusions about the trend of morbidity to more serious cases, the overall increase in cases of AN remains unknown. We also limited our analysis to only female patients. This decision was motivated by the fact that the characteristics of AN are more homogeneous among women. Despite limitations, the study offers a novel approach in analyzing trends among AN patients in the context of a pandemic. Further research is needed to fill the gaps in knowledge on interpretations of trends in the post-pandemic context. Our study also advocates for well-planned and

AN-dedicated countermeasures in the event of future pandemics [89], which are believed to be inevitable [90], and the social restrictions associated with them.

## 5. Conclusions

The study provides an insight into the trajectory of AN in the context of the COVID-19 pandemic, and, as previously documented by other researchers, it highlights a notable increase in the number of AN cases during the SARS-CoV-2 pandemic. However, an extremely important finding seems to be that the pandemic not only increased the incidence (or manifestation) of AN, but also worsened its clinical presentation, as evidenced by longer hospital stays, an increased need for nutritional interventions, and higher rates of anaemia, hypothyroidism, hypophosphataemia, and dyslipidaemia after the pandemic. The observed changes in these parameters, which may be regarded as malnutrition-related biomarkers, suggest that the disease severity in the peri-pandemic period is greater despite the presence of subtle differences in BMI values. Given the reported trends, it is crucial to consider the potential long-term effects of the pandemic on mental health, particularly in vulnerable populations such as adolescents. The study underscores the need for enhanced awareness, early intervention, and targeted support strategies to mitigate the impact of future pandemics on eating disorders like AN.

**Author Contributions:** Conceptualization, K.A.D. and K.P.S.; methodology, K.A.D. and P.D.; validation Ł.D.; formal analysis, K.P.S. and P.D.; investigation, K.A.D.; data curation, Z.G., M.K.-D. and J.G.; writing—original draft preparation, K.A.D. and K.P.S.; writing—review and editing. M.K.-Ł. and K.G.; visualization, S.S.; supervision, K.G. and K.P. All authors have read and agreed to the published version of the manuscript.

**Funding:** This research received no external funding.

**Institutional Review Board Statement:** The study was approved by the Ethics Committee of the Regional Board of Physicians. The number of approvals—OIL/KBL/15/2024 (6 May 2024).

**Informed Consent Statement:** The study was performed retrospectively and hospital database was the source of the data. The data were analysed in an anonymized form.

**Data Availability Statement:** The data presented in this study are available on request from the corresponding author due to privacy reasons.

**Acknowledgments:** The authors would like to acknowledge the St. Louis Children Hospital who enabled this study by granting access to the hospital database.

**Conflicts of Interest:** The authors declare no conflicts of interest.

## References


1. Bell, C.C. Task Force on DSM-IV. Diagnostic and Statistical Manual of Mental Disorders: DSM-IV. *JAMA* **1994**, *272*, 828–829. [CrossRef]
2. Galmiche, M.; Déchelotte, P.; Lambert, G.; Tavolacci, M.P. Prevalence of Eating Disorders over the 2000–2018 Period: A Systematic Literature Review. *Am. J. Clin. Nutr.* **2019**, *109*, 1402–1413. [CrossRef] [PubMed]
3. Erskine, H.E.; Whiteford, H.A.; Pike, K.M. The Global Burden of Eating Disorders. *Curr. Opin. Psychiatry* **2016**, *29*, 346–353. [CrossRef] [PubMed]
4. Keski-Rahkonen, A.; Mustelin, L. Epidemiology of Eating Disorders in Europe: Prevalence, Incidence, Comorbidity, Course, Consequences, and Risk Factors. *Curr. Opin. Psychiatry* **2016**, *29*, 340–345. [CrossRef] [PubMed]
5. Hoek, H.W. Review of the Worldwide Epidemiology of Eating Disorders. *Curr. Opin. Psychiatry* **2016**, *29*, 336–339. [CrossRef]
6. Fichter, M.M.; Quadflieg, N. Mortality in Eating Disorders—Results of a Large Prospective Clinical Longitudinal Study. *Int. J. Eat. Disord.* **2016**, *49*, 391–401. [CrossRef]
7. Arcelus, J.; Mitchell, A.J.; Wales, J.; Nielsen, S. Mortality Rates in Patients with Anorexia Nervosa and Other Eating Disorders. A Meta-Analysis of 36 Studies. *Arch. Gen. Psychiatry* **2011**, *68*, 724–731. [CrossRef]
8. Guinhut, M.; Godart, N.; Benadjaoud, M.A.; Melchior, J.C.; Hanachi, M. Five-Year Mortality of Severely Malnourished Patients with Chronic Anorexia Nervosa Admitted to a Medical Unit. *Acta Psychiatr. Scand.* **2021**, *143*, 130–140. [CrossRef]
9. Keel, P.K.; Dorer, D.J.; Franko, D.L.; Jackson, S.C.; Herzog, D.B. Postremission Predictors of Relapse in Women with Eating Disorders. *Am. J. Psychiatry* **2005**, *162*, 2263–2268. [CrossRef]

10. Carter, J.C.; Mercer-Lynn, K.B.; Norwood, S.J.; Bewell-Weiss, C.V.; Crosby, R.D.; Woodside, D.B.; Olmsted, M.P. A Prospective Study of Predictors of Relapse in Anorexia Nervosa: Implications for Relapse Prevention. *Psychiatry Res.* **2012**, *200*, 518–523. [CrossRef]
11. Favaro, A.; Caregaro, L.; Tenconi, E.; Bosello, R.; Santonastaso, P. Time Trends in Age at Onset of Anorexia Nervosa and Bulimia Nervosa. *J. Clin. Psychiatry* **2009**, *70*, 1715–1721. [CrossRef] [PubMed]
12. Herpertz-Dahlmann, B.; Dahmen, B. Children in Need—Diagnostics, Epidemiology, Treatment and Outcome of Early Onset Anorexia Nervosa. *Nutrients* **2019**, *11*, 1932. [CrossRef] [PubMed]
13. Zipfel, S.; Giel, K.E.; Bulik, C.M.; Hay, P.; Schmidt, U. Anorexia Nervosa: Aetiology, Assessment, and Treatment. *Lancet Psychiatry* **2015**, *2*, 1099–1111. [CrossRef] [PubMed]
14. Bozzola, E.; Barni, S.; Marchili, M.R.; Hellmann, R.; Giudice, E.D.; De Luca, G.; Cupertino, V. Anorexia Nervosa in Children and Adolescents: An Early Detection of Risk Factors. *Ital. J. Pediatr.* **2024**, *50*, 221. [CrossRef]
15. Haripersad, Y.V.; Kannegiesser-Bailey, M.; Morton, K.; Skeldon, S.; Shipton, N.; Edwards, K.; Newton, R.; Newell, A.; Stevenson, P.G.; Martin, A.C. Outbreak of Anorexia Nervosa Admissions during the COVID-19 Pandemic. *Arch. Dis. Child.* **2021**, *106*, e15. [CrossRef]
16. Linardon, J.; Messer, M.; Rodgers, R.F.; Fuller-Tyszkiewicz, M. A Systematic Scoping Review of Research on COVID-19 Impacts on Eating Disorders: A Critical Appraisal of the Evidence and Recommendations for the Field. *Int. J. Eat. Disord.* **2022**, *55*, 3. [CrossRef]
17. Devoe, D.J.; Han, A.; Anderson, A.; Katzman, D.K.; Patten, S.B.; Soumbasis, A.; Flanagan, J.; Paslakis, G.; Vyver, E.; Marcoux, G.; et al. The Impact of the COVID-19 Pandemic on Eating Disorders: A Systematic Review. *Int. J. Eat. Disord.* **2023**, *56*, 5–25. [CrossRef]
18. Schlissel, A.C.; Richmond, T.K.; Eliasziw, M.; Leonberg, K.; Skeer, M.R. Anorexia Nervosa and the COVID-19 Pandemic among Young People: A Scoping Review. *J. Eat. Disord.* **2023**, *11*, 122. [CrossRef]
19. Özbaran, N.B.; Yılancıoğlu, H.Y.; Tokmak, S.H.; Yuluğ Taş, B.; Çek, D.; Bildik, T. Changes in the Psychosocial and Clinical Profiles of Anorexia Nervosa Patients during the Pandemic. *Front. Psychiatry* **2023**, *14*, 1207526. [CrossRef]
20. Datta, N.; Van Wye, E.; Citron, K.; Matheson, B.; Lock, J.D. The COVID-19 Pandemic and Youth with Anorexia Nervosa: A Retrospective Comparative Cohort Design. *Int. J. Eat. Disord.* **2023**, *56*, 263–268. [CrossRef]
21. Rossi, L.; Silva, C.; Charvin, I.; Da Fonseca, D.; Bat-Pitault, F. Sleep and Emotionality in Adolescents with Anorexia Nervosa during the COVID-19 Pandemic. *Eur. Eat. Disord. Rev.* **2023**, *31*, 462–473. [CrossRef] [PubMed]
22. Orlewska, K.; Kozieł, D.; Klusek, J.; Orlewska, E. Burden of COVID-19 Mortality and Morbidity in Poland in 2020. *Int. J. Environ. Res. Public Health* **2022**, *19*, 5432. [CrossRef] [PubMed]
23. Kundzewicz, Z.W.; Ebi, K.L.; Duszyński, J. Lessons from the COVID-19 Pandemic: Mortality Impacts in Poland versus European Union. *Risk Anal.* 2023; *in press*. [CrossRef] [PubMed]
24. Kowal, M.; Coll-Martín, T.; Ikizer, G.; Rasmussen, J.; Eichel, K.; Studzińska, A.; Koszałkowska, K.; Karwowski, M.; Najmussaqib, A.; Pankowski, D.; et al. Who Is the Most Stressed During the COVID-19 Pandemic? Data from 26 Countries and Areas. *Appl. Psychol. Health Well Being* **2020**, *12*, 946–966. [CrossRef]
25. Sekścińska, K.; Trzcińska, A.; Pankowski, D.; Pisula, E.; Wytrychiewicz-Pankowska, K. Financial Factors and Psychological Distress during the COVID-19 Pandemic in Poland. *Int. J. Environ. Res. Public Health* **2022**, *19*, 1798. [CrossRef]
26. Bujek-Kubas, I.C.; Mojs, E. Environmental Stress and the Quality of Life Connected with COVID-19 among People in Poland and the Netherlands. *Int. J. Occup. Med. Environ. Health* **2021**, *34*, 177–188. [CrossRef]
27. Wójtowicz-Szefler, M.; Grzankowska, I.; Deja, M. The Mental Condition of Polish Adolescents during the COVID-19 Pandemic and War in Ukraine. *Front. Public Health* **2023**, *11*, 1257384. [CrossRef]
28. Grzejszczak, J.; Strzelecki, D.; Gabryelska, A.; Kotlicka-Antczak, M. Evaluation of COVID-19 Effect on Mental Health, Self-Harm, and Suicidal Behaviors in Children and Adolescents Population. *J. Clin. Med.* **2024**, *13*, 744. [CrossRef]
29. Kluczyńska, S.; Sokołowska, E. Mental Health of Children and Adolescents in Poland—Challenges and Recommendations. *Educ. Psychol.* **2023**, *69*, 67–77. [CrossRef]
30. Killaspy, H.; Cardoso, G.; White, S.; Wright, C.; Caldas de Almeida, J.M.; Turton, P.; Taylor, T.L.; Schützwohl, M.; Schuster, M.; Cervilla, J.A.; et al. Quality of Care and Its Determinants in Longer Term Mental Health Facilities across Europe; a Cross-Sectional Analysis. *BMC Psychiatry* **2016**, *16*, 31. [CrossRef]
31. Sokół-Szawłowska, M. Change in the Form of Work of Psychologists after the Outbreak of the COVID-19 Pandemic in Poland, and the Subjective Mental State of This Professional Group. *Int. J. Occup. Saf. Ergon.* **2023**, *29*, 1552–1557. [CrossRef] [PubMed]
32. Paczkowska, A.; Nowakowska, E.; Michalska, J.; Ficner, H.; Tyrchniewicz, J.; Szymański, A.; Wańkowicz, K.; Namyślak, K.; Gala, A.; Sulimiera Michalak, S.; et al. Impact of the COVID-19 Pandemic and Lockdown on Mental Health among the General Population in Poland and on the Availability of Psychological Support. *Psychol. Health Med.* **2023**, *28*, 179–188. [CrossRef] [PubMed]
33. Czepczor-Bernat, K.; Swami, V.; Modrzejewska, A.; Modrzejewska, J. COVID-19-Related Stress and Anxiety, Body Mass Index, Eating Disorder Symptomatology, and Body Image in Women from Poland: A Cluster Analysis Approach. *Nutrients* **2021**, *13*, 1384. [CrossRef] [PubMed]
34. Nagata, J.M.; Ganson, K.T.; Murray, S.B. Eating Disorders in Adolescent Boys and Young Men: An Update. *Curr. Opin. Pediatr.* **2020**, *32*, 476–481. [CrossRef]

35. Lin, J.A.; Hartman-Munick, S.M.; Kells, M.R.; Milliren, C.E.; Slater, W.A.; Woods, E.R.; Forman, S.F.; Richmond, T.K. The Impact of the COVID-19 Pandemic on the Number of Adolescents/Young Adults Seeking Eating Disorder-Related Care. *J. Adolesc. Health* **2021**, *69*, 660. [CrossRef]
36. Otto, A.K.; Jary, J.M.; Sturza, J.; Miller, C.A.; Prohaska, N.; Bravender, T.; Van Huysse, J. Medical Admissions Among Adolescents with Eating Disorders During the COVID-19 Pandemic. *Paediatrics* **2021**, *148*, e2021052201. [CrossRef]
37. Matthews, A.; Kramer, R.A.; Peterson, C.M.; Mitan, L. Higher Admission and Rapid Readmission Rates among Medically Hospitalized Youth with Anorexia Nervosa/Atypical Anorexia Nervosa during COVID-19. *Eat. Behav.* **2021**, *43*, 101573. [CrossRef]
38. Agostino, H.; Burstein, B.; Moubayed, D.; Taddeo, D.; Grady, R.; Vyver, E.; Dimitropoulos, G.; Dominic, A.; Coelho, J.S. Trends in the Incidence of New-Onset Anorexia Nervosa and Atypical Anorexia Nervosa Among Youth During the COVID-19 Pandemic in Canada. *JAMA Netw. Open* **2021**, *4*, E2137395. [CrossRef]
39. Vyver, E.; Han, A.X.; Dimitropoulos, G.; Patten, S.B.; Devoe, D.J.; Marcoux-Louie, G.; Katzman, D.K. The COVID-19 Pandemic and Canadian Paediatric Tertiary Care Hospitalizations for Anorexia Nervosa. *J. Adolesc. Health* **2023**, *72*, 344–351. [CrossRef]
40. Girardi, M.; Assalone, C.; Maines, E.; Genovese, A.; Naselli, A.; Fovino, L.N.; Soffiati, M.; Franceschi, R. Disease Characteristics and Psychiatric Comorbidities in Adolescents with Anorexia Nervosa Hospitalized During COVID-19 Pandemic. *Front. Biosci.* **2022**, *14*, 28. [CrossRef]
41. Murray Hurtado, M.; Martín Rivada, Á.; Pestano García, S.; Ramallo Fariña, Y. Influence of the COVID-19 Pandemic on the Incidence of Eating Disorders. *Anal. Pediatr.* **2024**, *101*, 21–28. [CrossRef]
42. Rafferty, C.; O'Donnell, A.; Campbell, S.; Sun, B.; King, J.; Ali, Z.; Lynch, D.; Barrett, E.; Richardson, S.; Clifford, M.; et al. Eating Disorders and COVID-19—Different or Just More? *Ir. J. Med. Sci.* **2024**, *193*, 1939–1944. [CrossRef] [PubMed]
43. Silber, A.S.; Platte, S.; Kumar, A.; Arora, S.; Kadioglu, D.; Schmidt, M.; Storf, H.; Chiocchetti, A.G.; Freitag, C.M. Admission Rates and Clinical Profiles of Children and Youth with Eating Disorders Treated as Inpatients before and during the COVID-19 Pandemic in a German University Hospital. *Front. Public Health* **2023**, *11*, 1281363. [CrossRef] [PubMed]
44. Hyam, L.; Richards, K.L.; Allen, K.L.; Schmidt, U. The Impact of the COVID-19 Pandemic on Referral Numbers, Diagnostic Mix, and Symptom Severity in Eating Disorder Early Intervention Services in England. *Int. J. Eat. Disord.* **2023**, *56*, 269–275. [CrossRef]
45. Wong, J.W.S.; Skeldon, S.; Gill, E.; Taylor, E.; Billingham, W.; Stevenson, P.G.; Shipton, N.; Ferguson, G.; Collins, R.; Newton, R.; et al. The Increase in Medical Admissions with Anorexia Nervosa during the COVID-19 Pandemic in Western Australia. *Int. J. Eat. Disord.* **2023**, *56*, 1661–1666. [CrossRef]
46. Vázquez-Giraldo, P.; Muñoz-Sanjosé, A.; López-Cuadrado, T. Hospital Admissions for Eating Disorders in Children and Adolescents in Spain: A Population-Based Study. *Int. J. Eat. Disord.* **2024**, *57*, 2299–2305. [CrossRef]
47. Goldberg, L.; Ziv, A.; Vardi, Y.; Hadas, S.; Zuabi, T.; Yeshareem, L.; Gur, T.; Steinling, S.; Scheuerman, O.; Levinsky, Y. The Effect of COVID-19 Pandemic on Hospitalizations and Disease Characteristics of Adolescents with Anorexia Nervosa. *Eur. J. Pediatr.* **2022**, *181*, 1767. [CrossRef]
48. Malik, G.M.A.; Faheem, S.; Muhammad, F. Rise of Eating Disorders and Obesity in Young Adults during the COVID-19 Pandemic. *J. Pak. Med. Assoc.* **2022**, *72*, 399. [CrossRef]
49. Gordon, C.M.; Katzman, D.K. Lessons Learned in Caring for Adolescents with Eating Disorders: The Singapore Experience. *J. Adolesc. Health* **2020**, *67*, 5–6. [CrossRef]
50. Touyz, S.; Lacey, H.; Hay, P. Eating Disorders in the Time of COVID-19. *J. Eat. Disord.* **2020**, *8*, 19. [CrossRef]
51. Schlapfer, L.; Gettis, M.A.; Dutreuil, V.; Cherven, B. Hospitalization Patterns for Adolescents with Eating Disorders during COVID-19. *J. Eat. Disord.* **2023**, *11*, 137. [CrossRef] [PubMed]
52. Taquet, M.; Geddes, J.R.; Luciano, S.; Harrison, P.J. Incidence and Outcomes of Eating Disorders during the COVID-19 Pandemic. *Br. J. Psychiatry* **2021**, *220*, 262–264. [CrossRef] [PubMed]
53. Herpertz-Dahlmann, B.; Eckardt, S.; Dempfle, A. A Never-Ending Story: The COVID-19 Pandemic and the Increase of Hospital Admissions for Typical and Atypical Anorexia Nervosa in Children, Adolescents and Young Adults in the Post-Pandemic Era in Germany. *Eur. Psychiatry* **2024**, *67*, e77. [CrossRef] [PubMed]
54. Skowron, K.; Kurnik-łucka, M.; Dadański, E.; Bętkowska-Korpała, B.; Gil, K. Backstage of Eating Disorder—About the Biological Mechanisms behind the Symptoms of Anorexia Nervosa. *Nutrients* **2020**, *12*, 1–32. [CrossRef]
55. De Filippo, E.; Marra, M.; Alfinito, F.; Di Guglielmo, M.L.; Majorano, P.; Cerciello, G.; De Caprio, C.; Contaldo, F.; Pasanisi, F. Hematological Complications in Anorexia Nervosa. *Eur. J. Clin. Nutr.* **2016**, *70*, 1305–1308. [CrossRef]
56. Calcaterra, V.; Magenes, V.C.; Siccardo, F.; Hruby, C.; Basso, M.; Conte, V.; Maggioni, G.; Fabiano, V.; Russo, S.; Veggiotti, P.; et al. Thyroid Dysfunction in Children and Adolescents Affected by Undernourished and Overnourished Eating Disorders. *Front. Nutr.* **2023**, *10*, 1205331. [CrossRef]
57. Straface, E.; Jacobis, I.T.D.; Capriati, T.; Pretelli, I.; Grandin, A.; Mascolo, C.; Vona, R.; Gambardella, L.; Cittadini, C.; Villani, A.; et al. The Impact of the COVID-19 Pandemic on Eating Disorders Risk and Symptoms: A Retrospective Study. *Ital. J. Pediatr.* **2023**, 49. [CrossRef]
58. Machado, P.P.P.; Grilo, C.M.; Crosby, R.D. Evaluation of the DSM-5 Severity Indicator for Anorexia Nervosa. *Eur. Eat. Disord. Rev.* **2017**, *25*, 221–223. [CrossRef]
59. Reas, D.L.; Rø, Ø. Investigating the DSM-5 Severity Specifiers Based on Thinness for Adults with Anorexia Nervosa. *Int. J. Eat. Disord.* **2017**, *50*, 990–994. [CrossRef]

60. Dang, A.B.; Kiropoulos, L.; Castle, D.J.; Jenkins, Z.; Phillipou, A.; Rossell, S.L.; Krug, I. Assessing Severity in Anorexia Nervosa: Do the DSM-5 and an Alternative Severity Rating Based on Overvaluation of Weight and Shape Severity Differ in Psychological and Biological Correlates? *Eur. Eat. Disord. Rev.* **2023**, *31*, 447–461. [CrossRef]
61. Krug, I.; Binh Dang, A.; Granero, R.; Agüera, Z.; Sánchez, I.; Riesco, N.; Jimenez-Murcia, S.; Menchón, J.M.; Fernandez-Aranda, F. Drive for Thinness Provides an Alternative, More Meaningful, Severity Indicator than the DSM-5 Severity Indices for Eating Disorders. *Eur. Eat. Disord. Rev.* **2021**, *29*, 482–498. [CrossRef] [PubMed]
62. Miller, K.K.; Grinspoon, S.K.; Ciampa, J.; Hier, J.; Herzog, D.; Klibanski, A. Medical Findings in Outpatients with Anorexia Nervosa. *Arch. Intern. Med.* **2005**, *165*, 561–566. [CrossRef] [PubMed]
63. Garber, A.K.; Cheng, J.; Accurso, E.C.; Adams, S.H.; Buckelew, S.M.; Kapphahn, C.J.; Kreiter, A.; Le Grange, D.; MacHen, V.I.; Moscicki, A.B.; et al. Weight Loss and Illness Severity in Adolescents with Atypical Anorexia Nervosa. *Paediatrics* **2019**, *144*, e20192339. [CrossRef] [PubMed]
64. Pacciardi, B.; Cargioli, C.; Belli, S.; Cotugno, B.; Augusto, S.; Di Paolo, L.; Mauri, M. Is Osteoporosis Risk in Anorexia Nervosa Underestimated? A Case Report Series. *Open J. Intern. Med.* **2013**, *3*, 115–118. [CrossRef]
65. Chadda, R.; Malhotra, S.; Asad, A.G.; Bambery, P. Socio-cultural factors in anorexia nervosa. *Indian J. Psychiatry* **1987**, *29*, 107.
66. Garner, D.M.; Garfinkel, P.E. Socio-Cultural Factors in the Development of Anorexia Nervosa. *Psychol. Med.* **1980**, *10*, 647–656. [CrossRef]
67. Meherali, S.; Punjani, N.; Louie-Poon, S.; Rahim, K.A.; Das, J.K.; Salam, R.A.; Lassi, Z.S. Mental Health of Children and Adolescents amidst COVID-19 and Past Pandemics: A Rapid Systematic Review. *Int. J. Environ. Res. Public Health* **2021**, *18*, 3432. [CrossRef]
68. Longobardi, C.; Morese, R.; Fabris, M.A. COVID-19 Emergency: Social Distancing and Social Exclusion as Risks for Suicide Ideation and Attempts in Adolescents. *Front. Psychol.* **2020**, *11*, 551113. [CrossRef]
69. Graell, M.; Morón-Nozaleda, M.G.; Camarneiro, R.; Villaseñor, Á.; Yáñez, S.; Muñoz, R.; Martínez-Núñez, B.; Miguélez-Fernández, C.; Muñoz, M.; Faya, M. Children and Adolescents with Eating Disorders during COVID-19 Confinement: Difficulties and Future Challenges. *Eur. Eat. Disord. Rev.* **2020**, *28*, 864–870. [CrossRef]
70. Baenas, I.; Caravaca-Sanz, E.; Granero, R.; Sánchez, I.; Riesco, N.; Testa, G.; Vintró-Alcaraz, C.; Treasure, J.; Jiménez-Murcia, S.; Fernández-Aranda, F. COVID-19 and Eating Disorders during Confinement: Analysis of Factors Associated with Resilience and Aggravation of Symptoms. *Eur. Eat. Disord. Rev.* **2020**, *28*, 855–863. [CrossRef]
71. Young, K.S. Internet Addiction: The Emergence of a New Clinical Disorder. *Cyberpsychol. Behav.* **1998**, *1*, 237–244. [CrossRef]
72. Kraut, R.; Patterson, M.; Lundmark, V.; Kiesler, S.; Mukophadhyay, T.; Scherlis, W. Internet Paradox. A Social Technology That Reduces Social Involvement and Psychological Well-Being? *Am. Psychol.* **1998**, *53*, 1017–1031. [CrossRef] [PubMed]
73. Golan, M. Eating and Control Styles Axis in Mentalisation-Based Psychotherapy in Eating Disorders: A Randomised Clinical Trial. *Front. Psychiatry* **2022**, *13*, 774382. [CrossRef] [PubMed]
74. Froreich, F.V.; Vartanian, L.R.; Grisham, J.R.; Touyz, S.W. Dimensions of Control and Their Relation to Disordered Eating Behaviours and Obsessive-Compulsive Symptoms. *J. Eat. Disord.* **2016**, *4*, 14. [CrossRef]
75. Jarvers, I.; Ecker, A.; Schleicher, D.; Kandsperger, S.; Otto, A.; Brunner, R. Anorexia Nervosa during COVID-19: Loss of Personal Control and Alexithymia as Important Contributors to Symptomatology in Adolescent Girls. *J. Eat. Disord.* **2023**, *11*, 180. [CrossRef]
76. Greenwood, H.; Krzyzaniak, N.; Peiris, R.; Clark, J.; Scott, A.M.; Cardona, M.; Griffith, R.; Glasziou, P. Telehealth Versus Face-to-Face Psychotherapy for Less Common Mental Health Conditions: Systematic Review and Meta-Analysis of Randomized Controlled Trials. *JMIR Ment. Health* **2022**, *9*, e31780. [CrossRef]
77. Fernández-Aranda, F.; Munguía, L.; Mestre-Bach, G.; Steward, T.; Etxandi, M.; Baenas, I.; Granero, R.; Sánchez, I.; Ortega, E.; Andreu, A.; et al. COVID Isolation Eating Scale (CIES): Analysis of the Impact of Confinement in Eating Disorders and Obesity-A Collaborative International Study. *Eur. Eat. Disord. Rev.* **2020**, *28*, 871–883. [CrossRef]
78. Rodríguez Guarin, M.; Gempeler Rueda, J.; Castro, S.M.; Ospina, M.M.; Villanueva Betancourth, C.; Amórtegui, J.P.; Vázquez, L. Feasibility and Acceptability of Comprehensive Virtual Treatment in Eating Disorders: Perspectives from Patients, Parents and Therapists during the COVID-19 Pandemic. *Rev. Colomb. Psiquiatr.* **2023**, *52*, 337–344. [CrossRef]
79. Yaffa, S.; Adi, E.L.; Itai, P.; Marit, J.M.; Doron, G.; Daniel, S. Treatment of Eating Disorders in Adolescents during the COVID-19 Pandemic: A Case Series. *J. Eat. Disord.* **2021**, *9*, 17. [CrossRef]
80. Batista, M.; Antić, L.Ž.; Žaja, O.; Jakovina, T.; Begovac, I. Predictors of eating disorder risk in anorexia nervosa adolescents. *Acta Clin. Croat.* **2018**, *57*, 399–410. [CrossRef]
81. Gale, C.J.; Cluett, E.R.; Laver-Bradbury, C. A Review of the Father-Child Relationship in the Development and Maintenance of Adolescent Anorexia and Bulimia Nervosa. *Issues Compr. Pediatr. Nurs.* **2013**, *36*, 48–69. [CrossRef] [PubMed]
82. "Giving Mental Health the Attention It Deserves"—Poland Adopts WHO Tool to Boost Efforts to Address Mental Health Needs. Available online: https://www.who.int/europe/news/item/10-10-2023-giving-mental-health-the-attention-it-deserves---poland-adopts-who-tool-to-boost-efforts-to-address-mental-health-needs (accessed on 16 November 2024).
83. Spettigue, W.; Obeid, N.; Erbach, M.; Feder, S.; Finner, N.; Harrison, M.E.; Isserlin, L.; Robinson, A.; Norris, M.L. The Impact of COVID-19 on Adolescents with Eating Disorders: A Cohort Study. *J. Eat. Disord.* **2021**, *9*, 65. [CrossRef] [PubMed]
84. Hansen, S.J.; Stephan, A.; Menkes, D.B. The Impact of COVID-19 on Eating Disorder Referrals and Admissions in Waikato, New Zealand. *J. Eat. Disord.* **2021**, *9*, 105. [CrossRef] [PubMed]

85. Tizenberg, B.N.; Brenner, L.A.; Lowry, C.A.; Okusaga, O.O.; Benavides, D.R.; Hoisington, A.J.; Benros, M.E.; Stiller, J.W.; Kessler, R.C.; Postolache, T.T. Biological and Psychological Factors Determining Neuropsychiatric Outcomes in COVID-19. *Curr. Psychiatry Rep.* **2021**, *23*, 10. [CrossRef]
86. Alonzo, R.; Hussain, J.; Stranges, S.; Anderson, K.K. Interplay between Social Media Use, Sleep Quality, and Mental Health in Youth: A Systematic Review. *Sleep Med. Rev.* **2021**, *56*, 101414. [CrossRef]
87. Shawon, M.S.R.; Rouf, R.R.; Jahan, E.; Hossain, F.B.; Mahmood, S.; Gupta, R.D.; Islam, M.I.; Al Kibria, G.M.; Islam, S. The Burden of Psychological Distress and Unhealthy Dietary Behaviours among 222,401 School-Going Adolescents from 61 Countries. *Sci. Rep.* **2023**, *13*, 1–12. [CrossRef]
88. O'Neil, A.; Quirk, S.E.; Housden, S.; Brennan, S.L.; Williams, L.J.; Pasco, J.A.; Berk, M.; Jacka, F.N. Relationship Between Diet and Mental Health in Children and Adolescents: A Systematic Review. *Am. J. Public Health* **2014**, *104*, e31. [CrossRef]
89. Kamalrathne, T.; Amaratunga, D.; Haigh, R.; Kodituwakku, L. Need for Effective Detection and Early Warnings for Epidemic and Pandemic Preparedness Planning in the Context of Multi-Hazards: Lessons from the COVID-19 Pandemic. *Int. J. Disaster Risk Reduct.* **2023**, *92*, 103724. [CrossRef]
90. Shafaati, M.; Chopra, H.; Priyanka; Khandia, R.; Choudhary, O.P.; Rodriguez-Morales, A.J. The next Pandemic Catastrophe: Can We Avert the Inevitable? *New Microbes New Infect.* **2023**, *52*, 101110. [CrossRef]

nutrients

MDPI

*Article*

# Associations Between Body Appreciation, Body Weight, Lifestyle Factors and Subjective Health Among Bachelor Students in Lithuania and Poland: Cross-Sectional Study

**Vilma Kriaučionienė [1,2,*,†], Danuta Gajewska [3,†], Asta Raskilienė [1,4], Joanna Myszkowska-Ryciak [3], Julia Ponichter [3], Lina Paulauskienė [2] and Janina Petkevičienė [1,2]**

1 Health Research Institute, Faculty of Public Health, Lithuanian University of Health Sciences, Tilžės Str. 18, 47181 Kaunas, Lithuania; asta.raskiliene@lsmu.lt (A.R.); janina.petkeviciene@lsmu.lt (J.P.)
2 Department of Preventive Medicine, Faculty of Public Health, Lithuanian University of Health Sciences, Tilžės Str. 18, 47181 Kaunas, Lithuania; lina.paulauskiene@lsmu.lt
3 Department of Dietetics, Institute of Human Nutrition Sciences, Warsaw University of Life Sciences, 159C Nowoursynowska Str., 02-776 Warsaw, Poland; danuta_gajewska@sggw.edu.pl (D.G.); joanna_myszkowska-ryciak@sggw.edu.pl (J.M.-R.); julia_ponichter@sggw.edu.pl (J.P.)
4 Institute of Biological Systems and Genetic Research, Lithuanian University of Health Sciences, Eivenių Str. 4, 44307 Kaunas, Lithuania
* Correspondence: vilma.kriaucioniene@lsmu.lt
† These authors contributed equally to this work.

**Abstract: Background/Objectives:** Positive body image is linked to improved mental and physical well-being, healthier lifestyles, and fewer unhealthy weight control behaviors. Cultural factors also play a role in influencing body appreciation. This study investigated the associations between body appreciation, body weight, lifestyle factors, and subjective health among bachelor's students in Lithuania and Poland. **Methods:** A cross-sectional online survey was conducted with 1290 students from universities in both countries. The Body Appreciation Scale-2 (BAS-2) measured body appreciation, while participants provided self-reported data on their dietary habits, physical activity, sleep, health perceptions, and body weight and height. Linear regression models explored associations between BAS-2 scores, actual and perceived body weight, lifestyle habits, and subjective health. **Results:** Gender and country-based differences in body appreciation were observed. Lithuanian female students reported a higher median BAS score of 33 compared to 32 among Polish female students ($p = 0.02$), despite having a higher median BMI (22.3 kg/m$^2$ vs. 21.1 kg/m$^2$, $p = 0.001$). Positive body appreciation was linked to healthier dietary behaviors, such as higher consumption of fruits, vegetables, fish, and regular breakfasts. Additionally, greater physical activity and sufficient sleep were associated with higher body appreciation, while higher intake of sweets, sugary drinks, and fast food correlated with lower BAS-2 scores. Both BMI and perceived weight were negatively associated with body appreciation, particularly among females. **Conclusions:** Body appreciation is closely linked to body weight, healthier lifestyle, and positive health perceptions, suggesting that promoting healthier habits may improve body appreciation.

**Keywords:** students; body appreciation; body weight; nutrition habits; physical activity; subjective health

## 1. Introduction

Body appreciation is the recognition and acceptance of one's body, regardless of its size, shape, or appearance. It means maintaining a positive attitude toward the body, focusing on its functionality and uniqueness rather than just its physical appearance or how it fits society's beauty standards. Researchers have become increasingly interested in studying body appreciation to explore its associations with lifestyle, health perception, and overall well-being. The Body Appreciation Scale-2 (BAS-2), developed by Tylka and

Wood-Barcalow, is a validated tool used to assess positive attitudes toward one's body [1]. The study has demonstrated strong internal consistency ($\alpha = 0.97$), test–retest reliability ($r = 0.90$), and construct validity (convergent, incremental, and discriminant) for the BAS-2 [1]. The scale has also been shown to be invariant across genders and was considered a reliable and valid tool for measuring the body appreciation of students [2].

Young adulthood, particularly in a university setting, is a critical period for identity formation, including self-image and health habits, which are often shaped by social, cultural, and academic environments [3]. University students face unique pressures related to academic performance, social integration, and, in recent years, the increased influence of social media on appearance ideals, which can significantly impact body appreciation and lifestyle behaviors. Additionally, university environments are structured to facilitate recruitment for large-scale surveys, as students are typically accessible through institutional channels, enabling the analysis of a relatively diverse sample across demographic groups. University students often come from varied backgrounds, including different regions, family structures, and socioeconomic statuses, offering a good possibility to analyze diverse populations [3].

Previous studies have demonstrated that higher BAS scores are associated with healthier behaviors. Instead of engaging in extreme dieting or harmful weight loss practices, individuals with high body appreciation were more likely to adopt a balanced and sustainable diet, along with moderate physical activity, respecting the body's needs [1,4–6]. Research also shows that students with greater body appreciation are engaged in healthier behaviors, such as regular physical activity and healthier eating patterns. On the contrary, those with lower body appreciation scores often display poorer dietary choices and are more prone to sedentary lifestyles [2,5,7].

Gender differences in body appreciation have been well-documented. Studies have shown that women generally report lower BAS scores compared to men, largely due to societal emphasis on thinness for women and muscularity for men [8,9]. Body dissatisfaction among young women strongly predicts disordered eating, depressive symptoms, and unhealthy weight control practices. Conversely, men's body image concerns focus more on muscularity and strength [10]. A study by Wawrzyniak et al. found that body dissatisfaction was more frequently reported among female adolescents and increased with age for both genders [11]. The literature describes the high prevalence of body weight dissatisfaction among female adolescents as "normative discontent" [12]. Furthermore, men are more likely to underestimate their body weight compared to women, even though a larger proportion of men are classified as overweight [13,14].

Social media's influence on body appreciation has received considerable attention. Social platforms often promote narrow beauty ideals, leading to body dissatisfaction, particularly among women who engage in social comparisons. Studies have shown that increased use of social media intensifies body image concerns and negatively affects body appreciation, especially in cultures that highly value thinness and physical appearance [15,16].

Cultural differences play a significant role in body appreciation. While some studies indicate that body image concerns exist worldwide, the level of body appreciation can vary based on cultural and regional norms [6,17]. For instance, a study conducted across several European countries found that Polish students reported higher levels of body dissatisfaction compared to their peers in Lithuania. This discrepancy reflects differing societal norms regarding beauty and body size [18].

This study aims to examine the associations between body appreciation, body weight, and lifestyle factors among bachelor students in Lithuania and Poland. Although these two neighboring countries share some similarities, they also exhibit differences in dietary habits and societal norms, which may influence perceptions of body image

## 2. Materials and Methods

### 2.1. Study Design and Sample

The online cross-sectional study was conducted among bachelor students at the largest Universities of Applied Sciences in Lithuania, located in Vilnius, Kaunas, Klaipėda, and Šiauliai. Additionally, it included major Polish universities such as the Warsaw University of Life Sciences, Warsaw University of Technology, University of Gdańsk, and Poznań University of Economics. The survey at the Lithuanian universities took place during the second semester of the 2021–2022 academic year. The study in Poland was conducted from February to December 2022.

The data collection process involved sending emails to all students in the randomly selected faculties, providing information about the survey, and inviting them to participate. The online survey was open for three weeks, and additional reminders were sent during the first and second weeks to encourage completion. Participation in the study was voluntary and anonymous, ensuring that respondents felt comfortable providing their answers.

A self-administered questionnaire was completed by 1290 students: 709 students (216 males and 493 females) from Lithuanian universities, and 581 (121 males and 460 females) from Polish universities.

The study protocol received ethical approval from the Bioethics Centre at the Lithuanian University of Health Sciences (protocols BEC-GVM(M)-80, BEC-GM(M)-119) and the Ethics Committee for Research with Human Participation at the Institute of Human Nutrition Sciences of the Warsaw University of Life Sciences (Resolutions No. 3/2022 from 28 January 2022). Additionally, permission to conduct the study was obtained from the administration of the participating universities, ensuring compliance with institutional regulations and policies.

### 2.2. Measurements

A standardized questionnaire developed for this study was used in both countries. Students were asked about the frequency of consumption of selected foods, their physical activity levels, harmful behaviors, self-reported weight and height, attitudes toward weight gain, healthy eating, health perception, and body appreciation. A food frequency questionnaire was employed to assess the students' dietary habits. Respondents were asked to indicate the frequency of eating breakfast and consuming 19 various food items, such as meat and meat products, poultry, fish/seafood, milk and dairy products, bread, cereal products, fresh vegetables, fruits, nuts/seeds, confectionery, sweets, soft drinks, energy drinks, fast foods, and snacks. Response options included the following categories: (1) 'never', (2) '1–4 times a month', (3) 'several times a week', (4) 'daily', and (5) 'several times a day'. Based on the reported frequency of food consumption, respondents were categorized into two groups: (1) 'At least several times a week' ('several times per day', 'daily' and 'several times a week') and (2) '1–4 times a month or never'.

Physical activity was assessed by asking two questions: (1) 'How much time do you spend sitting on a typical weekday (e.g., sitting at a desk, watching television, reading)?' and (2) 'In your leisure time, how often do you engage in physical exercise for at least 30 min that makes you at least mildly short of breath or perspire?' The possible answers were: 'every day', '4–6 times a week', '2–3 times a week', 'once a week', 'once a month', and 'never' coded from 1 to 7. Based on their responses, participants were categorized into two groups: (1) those physically active at least four times a week and (2) those physically active less often. Additionally, participants were asked to report their average daily sleep duration in hours.

Participants were also asked to report their actual weight (in kilograms) and height (in centimeters). Body mass index (BMI) was calculated by dividing weight (in kg) by the square of height (in meters). BMI values were categorized according to the WHO criteria: underweight (BMI < 18.5 kg/m$^2$), normal weight (BMI 18.5–24.9 kg/m$^2$), overweight (BMI 25–29.9 kg/m$^2$), and obesity (BMI $\geq$ 30 kg/m$^2$) [19]. Additionally, students were asked, 'Would you be concerned if you gained weight?', with possible five answers from

(1) 'never' to (5) 'very often. The question 'How important is it for you to eat healthy?' had five responses from (1) 'very important' to (5) 'not at all'. Regarding health perception, students were questioned to what extent they keep an eye on their health. Response options included: 'not at all', 'not much', 'to some extent', and 'very much'. Respondents were categorized as those who kept an eye on health answering 'to some extent' and 'very much' and the rest. Additionally, participants were asked about their health: 'How would you assess your present health status?', with the following answer categories: (1) 'good', (2) 'reasonably good', (3) 'average, (4) 'rather poor', and (5) 'poor'. Those who chose answers 'good' and 'reasonably good' were grouped as having a positive health assessment.

The Body Appreciation Scale-2 (BAS-2) was used to assess students' body appreciation [1]. This scale includes 10 items, each rated on a five-point Likert scale ranging from 1 (never) to 5 (always). In the current study, Cronbach's alpha coefficient was $\alpha = 0.962$ for the Lithuanian version and 0.945 for the Polish version, indicating a strong internal consistency of the questionnaire. The overall score was determined by adding all responses, with a minimum sum of 10 and a maximum of 50. Higher scores reflect a more positive perception of body image.

### 2.3. Statistical Analysis

The categorical variables were presented as percentages and were compared using the Pearson chi-square test and z-test with Bonferroni correction for multiple comparisons. The continuous variables were presented as median and interquartile range, as all analyzed variables did not meet the criteria for normal distribution (Kolmogorov–Smirnov test). The Mann–Whitney test was used to compare the distributions of continuous variables between countries. The associations of BAS scores with BMI and perceived weight were analyzed using Spearman correlation analysis by country and gender. Multivariable linear regression analysis was used to associate the BAS scores with lifestyle factors, health perception, and other variables. Our data satisfied the assumptions of linear regression: (1) the response variable (BAS) was continuous, while the explanatory variables were either continuous or binary (gender and country); (2) the relationship between the outcome and the explanatory variables was linear; (3) the residuals were normally distributed; and (4) there was no multicollinearity among the explanatory variables. Initially, separate models were calculated for each variable, adjusting for gender, age, country, and BMI. Subsequently, variables that were uncorrelated and contributed to an increase in R square were included in the final model.

Data analysis was conducted using the IBM SPSS Statistics software package, version 29.0 (IBM Corp.: Armonk, NY, USA, released 2022). $p$-values < 0.05 were considered as significant.

## 3. Results

The main characteristics of the study population are presented in Table 1. The median age for men was 19 years in Lithuania and 20 years in Poland ($p = 0.01$), while for women, it was 20 years in both countries ($p = 0.359$).

Body appreciation levels were comparable for males in both Lithuania and Poland, with median BAS scores of 34 and 35, respectively. In contrast, Lithuanian female students had a higher median BAS score of 33, compared to 32 for Polish female students ($p = 0.02$). While the median BMI for males was similar in both countries, Lithuanian female students had a higher median BMI of 22.3 kg/m$^2$, compared to 21.1 kg/m$^2$ for their Polish counterparts ($p = 0.001$).

Polish students reported spending more time sitting than Lithuanian students, 8 h per day and 6 h, respectively. Additionally, Polish students consumed more fruits and vegetables than Lithuanians. They also had a higher frequency of breakfast consumption. Lithuanian students ate fish, legumes, nuts, and seeds more frequently than their Polish counterparts. Concerns about weight gain were more common among Polish students, whereas Lithuanian students tended to rate their health more positively than their Polish peers.

**Table 1.** The characteristics of students in Lithuania and Poland.

| Characteristics | Male Students | | | Female Students | | |
|---|---|---|---|---|---|---|
| | Lithuania | Poland | p-Value | Lithuania | Poland | p-Value |
| Age (years), median (IR) | 19 (2) | 20 (3) | 0.01 | 20 (4) | 20 (3) | 0.359 |
| Body Appreciation Scale (scores), median (IR) | 34 (10.8) | 35 (13.5) | 0.809 | 33 (12.5) | 32 (14) | 0.020 |
| Body mass index, (kg/m$^2$), median (IR) | 23.3 (4.6) | 23.5 (4.1) | 0.834 | 22.3 (4.8) | 21.1 (4.1) | 0.001 |
| Sitting time (hours), median (IR) | 6 (4) | 8 (4) | 0.001 | 6 (4) | 8 (4) | 0.001 |
| Sleep duration (hours) | 7 (1) | 7 (2) | 0.001 | 7 (2) | 7 (2) | 0.294 |
| Vegetable portions a day, median (IR) | 1 (1) | 2 (2) | <0.001 | 1 (1) | 2 (2) | <0.001 |
| Fruit portions a day, median (IR) | 1 (1) | 1 (2) | 0.001 | 1 (1) | 2 (1) | <0.001 |
| Breakfast (days a week), median (IR) | 5 (4) | 7 (3) | 0.025 | 5 (4) | 7 (3) | <0.001 |
| *Consumption of food products at least several times a week (% of participants)* | | | | | | |
| Fish | 43.1 | 14.9 | <0.001 | 36.7 | 7.8 | <0.001 |
| Porridge/Cereals | 62.5 | 47.9 | 0.010 | 58.6 | 62.6 | 0.208 |
| Legumes | 45.4 | 22.3 | <0.001 | 42.0 | 33.0 | <0.004 |
| Nuts and seeds | 65.3 | 32.2 | <0.001 | 56.6 | 39.8 | <0.001 |
| Sweets (chocolate, candies) | 60.2 | 56.2 | 0.476 | 56.6 | 65.0 | 0.008 |
| Sugary drinks (soda) | 53.2 | 52.1 | 0.836 | 35.1 | 40.7 | 0.077 |
| Fast food | 41.7 | 25.6 | 0.003 | 21.9 | 13.7 | 0.001 |
| Snacks (chips, roasted peanuts, etc.) | 51.4 | 42.1 | 0.103 | 31.2 | 27.6 | 0.220 |
| *Perceptions and Behavior (% of participants)* | | | | | | |
| Believe it is important to eat healthy | 46.8 | 51.2 | 0.430 | 62.1 | 61.1 | 0.755 |
| Worry about weight gain | 7.4 | 15.4 | <0.001 | 40.6 | 73.7 | <0.001 |
| Take care of health | 77.8 | 65.3 | 0.013 | 86.4 | 62.8 | <0.001 |
| Evaluate health as good | 81.5 | 62.0 | <0.001 | 77.9 | 67.8 | <0.001 |

IR—interquartile range; p-Value from Mann–Whitney or Pearson chi-square tests.

Most students in both countries were classified as having a normal weight (Figure 1). A higher proportion of males in Lithuania and Poland were categorized as overweight, while a smaller proportion were classified as underweight when compared to females. The distribution of males across body weight status categories was quite similar in both countries. In contrast, Polish female students showed a slightly higher prevalence of underweight (15.7%) and a lower prevalence of overweight (10.9%) compared to Lithuanian females, who had rates of 10.5% for underweight and 18.7% for overweight.

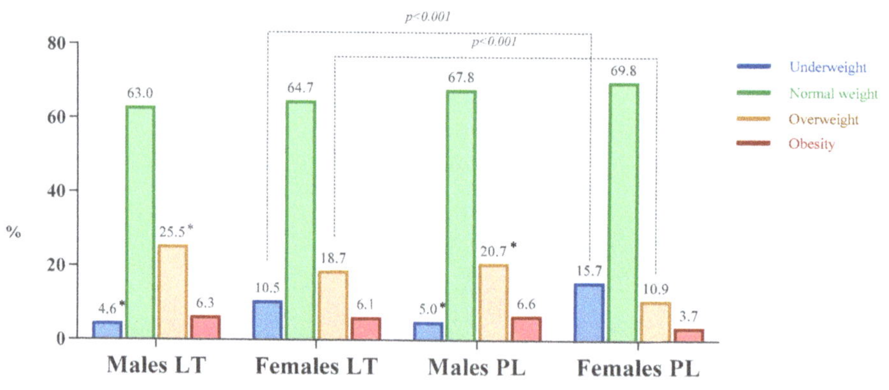


**Figure 1.** Distribution of male and female students by body weight status in Lithuania and Poland. * $p < 0.05$ compared to females in Lithuania or Poland ($\chi^2$ test with Bonferroni corrections); LT—Lithuania, PL—Poland.

A larger percentage of Lithuanian males described themselves as 'just right' compared to Polish males, while more Polish males perceived themselves as 'too fat' (Figure 2). Among females, weight perceptions were more consistent between the two countries. Additionally, more females in both countries considered themselves 'too thin' compared to their male counterparts. In Lithuania, a higher proportion of females than males perceived themselves as 'too fat', while in Poland, more females than males considered themselves 'just right'.

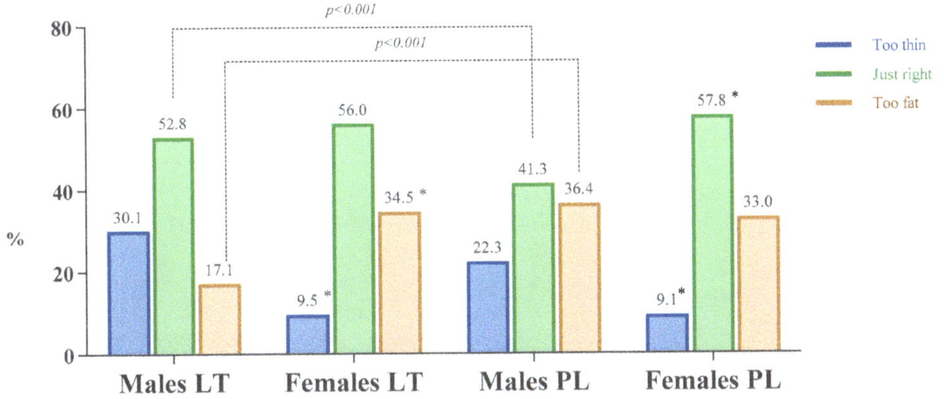


**Figure 2.** Distribution of male and female students by body weight perception in Lithuania and Poland. * $p < 0.05$ compared to males in Lithuania or Poland ($\chi^2$ test with Bonferroni corrections); LT—Lithuania, PL—Poland.

The Spearman correlation analysis indicated a significant negative association between BAS scores and BMI among female students in both Lithuania and Poland (Table 2). For male students, this negative association was statistically significant only in Poland. Additionally, perceived weight demonstrated a negative correlation with body appreciation for both genders in Poland and females in Lithuania, suggesting that students who perceive themselves as overweight tend to have lower BAS scores.

**Table 2.** Spearman correlation coefficients between body appreciation scale and BMI as well as perceived weight.

| Variable | Sex | Lithuania | | Poland | | Total | |
|---|---|---|---|---|---|---|---|
| | | r | *p*-Value | r | *p*-Value | r | *p*-Value |
| BMI | M | −0.008 | 0.912 | −0.284 | 0.002 | −0.120 | 0.028 |
| | F | −0.276 | <0.001 | −0.186 | <0.001 | −0.209 | <0.001 |
| Perceived weight | M | −0.133 | 0.051 | −0.387 | <0.001 | −0.232 | <0.001 |
| | F | −0.386 | <0.001 | −0.362 | <0.001 | −0.372 | <0.001 |

r—correlation coefficient; M—males; F—females.

The results of the multivariable linear regression analysis, which calculated separate models for each variable, adjusting for gender, age, country, and BMI, indicated that body appreciation is associated with lifestyle, health perceptions, and other analyzed factors. (Table 3). A higher intake of vegetables and fruits, along with regular breakfast consumption, was positively associated with BAS scores. Additionally, higher frequencies of consuming fish, porridge, legumes, and nuts were linked to better body appreciation. On the contrary, a high intake of sweets, sugary drinks, fast food, and unhealthy snacks was related to lower BAS scores, suggesting that limiting less nutritious foods might contribute to a more positive body image. Students who considered healthy eating important had higher BAS scores compared to those with the opposite view.

**Table 3.** Associations of body appreciation scale with lifestyle factors, health, and other variables (multivariable linear regression analysis *).

| Variable | Lithuania | | | Poland | | | Total | | |
|---|---|---|---|---|---|---|---|---|---|
| | β | CI | *p*-Value | β | CI | *p*-Value | β | CI | *p*-Value |
| Breakfast frequency | 0.63 | 0.40; 0.86 | <0.001 | 1.03 | 0.69; 1.38 | <0.001 | 0.76 | 0.57; 0.96 | <0.001 |
| Vegetable daily portions | 0.75 | 0.16; 1.34 | 0.013 | 0.61 | 0.25; 0.97 | 0.001 | 0.62 | 0.32; 0.92 | <0.001 |
| Fruit daily portions | 0.73 | 0.13; 1.33 | 0.017 | 0.57 | 0.00; 1.13 | 0.049 | 0.64 | 0.24; 1.05 | 0.002 |
| Fish consumption | 1.65 | 0.86; 2.43 | <0.001 | 1.711 | 0.46; 2.96 | 0.007 | 1.66 | 0.98; 2.33 | <0.001 |
| Porridge consumption | 1.55 | 0.93; 2.17 | <0.001 | 1.17 | 0.39; 1.96 | 0.003 | 1.32 | 0.84; 1.81 | <0.001 |
| Legumes consumption | 1.68 | 0.97; 2.39 | <0.001 | 1.11 | 0.17; 2.05 | 0.020 | 1.41 | 0.84; 1.98 | <0.001 |
| Nuts consumption | 2.50 | 1.97; 3.03 | <0.001 | 1.70 | 0.84; 2.56 | <0.001 | 2.16 | 1.69; 2.62 | <0.001 |
| Sweets consumption | −1.83 | −2.49; −1.18 | <0.001 | −0.41 | −1.33; 0.51 | 0.381 | −1.30 | −1.84; −0.76 | <0.001 |
| Sugary drinks consumption | −1.58 | −2.21; −0.96 | <0.001 | 0.25 | −0.47; 0.97 | 0.495 | −0.68 | −1.16; −0.21 | 0.005 |
| Fast food consumption | −1.93 | −2.91; −0.95 | <0.001 | −0.28 | −1.76; 1.21 | 0.713 | −1.36 | −2.19; −0.53 | 0.001 |
| Snacks consumption | −1.22 | −2.05; −0.39 | 0.004 | −0.40 | 1.50; 0.69 | 0.468 | −0.89 | −1.56; −0.22 | 0.009 |
| Sitting hours | −0.38 | −0.57; −0.18 | <0.001 | −0.28 | −0.56; 0.01 | 0.057 | −0.34 | −0.50; −0.17 | <0.001 |
| Exercises during leisure time | −1.72 | −2.07; −1.38 | <0.001 | −0.52 | −0.97; −0.07 | 0.023 | −1.18 | −1.46; −0.89 | <0.001 |
| Sleep hours | 1.65 | 1.18; 2.12 | <0.001 | 1.34 | 0.70; 1.98 | <0.001 | 1.49 | 1.11; 1.87 | <0.001 |
| Health perception | −1.70 | −2.43; −0.97 | <0.001 | −4.43 | −5.30; −3.56 | <0.001 | −2.87 | −3.43; −2.30 | <0.001 |
| Taking care of health | 3.34 | 2.48; 4.20 | <0.001 | 3.67 | 2.76; 4.58 | <0.001 | 3.52 | 2.90; 4.14 | <0.001 |
| Worrying about weight gain | −2.67 | −3.18; −2.16 | <0.001 | −3.27 | −3.80; −2.74 | <0.001 | 2.97 | 2.60; 3.34 | <0.001 |
| Importance of eating healthily | −2.92 | −3.58; −2.27 | <0.001 | −2.59 | −3.35; −1.83 | <0.001 | −2.75 | −3.24; −2.26 | <0.001 |

* Separate models were calculated for each variable, adjusting for gender, age, and BMI, as well as the country in models where combined data were used. β—unstandardized regression coefficient; CI—confidence interval.

Students who engaged more frequently in leisure-time physical activity reported higher body appreciation, while longer hours of sitting negatively affected BAS scores. Adequate sleep also played a significant role, as it was associated with better body appreciation, underscoring the importance of rest in promoting a positive body image.

Health perception was also associated with BAS. Students who rated their health more positively or took care of their health had higher BAS scores. However, those worried about potential weight gain tended to have lower BAS scores.

It is important to note that more statistically significant associations were observed among Lithuanian students compared to Polish students. In Poland, BAS scores were not statistically significantly related to the consumption of sweets, sugary drinks, fast food, snacks, or hours spent sitting, although trends were similar to those in Lithuania.

Multivariable linear regression analysis was conducted, including all uncorrelated variables that contributed to an increase in the R-squared value (Table 4). The results indicated that body appreciation was negatively correlated with several factors: BMI, the frequency of sweets consumption, a decrease in exercise frequency, a declining perception of health, concerns about weight gain, and being a Polish student. Conversely, body appreciation was positively associated with increased consumption of fish and nuts, longer sleep hours, and age. Additionally, males had higher BAS scores compared with females. Overall, this model accounted for 36.2% of the variations in BAS scores among students.

**Table 4.** Associations of body appreciation scale with analyzed variables (multivariable linear regression analysis).

| Variable | Unstandardized Coefficients | CI | *p*-Value | Standardized Coefficients |
|---|---|---|---|---|
| BMI | −0.22 | −0.33; −0.11 | <0.001 | −0.10 |
| Fish consumption | 0.90 | 0.30; 1.50 | 0.003 | 0.08 |
| Nuts consumption | 1.36 | 0.93; 1.79 | <0.001 | 0.16 |
| Sweets consumption | −0.65 | −1.12; −0.18 | 0.006 | −0.06 |
| Exercises during leisure time | −0.60 | −0.85; −0.34 | <0.001 | −0.11 |
| Sleep hours | 0.92 | 0.59; 1.25 | <0.001 | 0.13 |
| Health perception | −2.08 | −2.58; −1.58 | <0.001 | −0.19 |
| Worrying about weight gain | −2.67 | −3.00; −2.33 | <0.001 | −0.42 |
| Age | 0.23 | 0.13; 0.32 | <0.001 | 0.11 |
| Gender (males vs. females) | 1.56 | 0.56; 2.56 | 0.002 | 0.08 |
| Country (Poland vs. Lithuania) | −3.40 | −4.31; −2.50 | <0.001 | −0.19 |

R square 0.362; CI—confidence interval.

## 4. Discussion

This study examined body appreciation, BMI, lifestyle, and other factors among university students in Lithuania and Poland. Notable differences in the analyzed characteristics were observed between genders and countries. The results revealed strong associations between body appreciation and both actual and perceived body weight, as well as lifestyle choices, and subjective health.

Consistent with earlier studies, the dietary habits of Lithuanian and Polish students do not align with recommended dietary guidelines. A significant number of students consume insufficient amounts of vegetables, fruits, fish, legumes, and nuts, while frequently eating sweets, sugary drinks, fast food, and snacks. Other research has also shown that university students tend to have high-calorie diets that are rich in sugar and fat but low in essential nutrients, including fruits, vegetables, and fish [13,20–22]. Meal skipping, particularly breakfast, is common and has been linked to unhealthy patterns such as weight gain [20,22,23].

In our study, Lithuanian students reported consuming fish, legumes, nuts, and seeds more frequently than their Polish counterparts. Previous studies conducted in Poland, Germany, and Slovakia have also indicated differences in diet quality between these countries, highlighting that Polish students tend to consume a less healthy diet [18,24]. The highest level of knowledge about food and nutrition was observed among students from Poland. However, this knowledge did not correlate with a healthier diet [24]. Polish adolescents aged 13 to 19 who were dissatisfied with their body weight were less likely to meet dietary recommendations [11]. Many respondents expressed a willingness to engage in various activities aimed at increasing exercise, changing eating habits, or seeking assistance from professionals, such as personal trainers, to achieve their desired body shape. Conversely, concerns about gaining weight and losing attractiveness were commonly mentioned as factors influencing their perceptions of personal appearance [25].

It is important to note that dietary habits among Lithuanian students have improved from 2000 to 2017, with females exhibiting healthier eating patterns than males [13]. Moreover, Polish females also showed healthier food choices compared to males [11,26,27]. Other studies confirm that males are more likely to consume poorer diets, characterized by fast food and snacks, compared to females [13,28–30].

Our data indicate that the majority of students in both countries were of normal weight. Overweight was more prevalent among males, while underweight was more common in females. Polish females demonstrated slightly higher rates of underweight and lower rates of overweight compared to their Lithuanian counterparts. The COVID-19 pandemic contributed to weight gain among students, with some of these changes persisting even after the pandemic [31,32]. Overweight and obesity are increasing globally, affecting young people, including students, and influencing body image [33–35]. Students with higher BMI are more likely to experience body dissatisfaction and engage in unhealthy behaviors [31,36].

In our study, Polish females reported significantly lower BAS scores compared to Lithuanian females. This difference occurred despite Polish females having a lower prevalence of overweight. Additionally, Polish female students expressed more concern about weight gain than Lithuanian students. In terms of male perceptions, Polish students were more likely to believe they were overweight compared to Lithuanian male students, even though the prevalence of overweight and obesity was similar among males in both countries. These findings underscore the complex influence of cultural and gender-specific factors on body image perceptions in Lithuania and Poland [4,37].

Research data have shown that females are generally more dissatisfied with their weight and appearance and are more likely to engage in weight control practices compared to males [10,38]. In contrast, males often focus on muscularity, frequently viewing themselves as underweight or insufficiently muscular [39,40]. Social media has intensified body image concerns by promoting an ideal of thinness for women and muscularity for men, which contributes to body dissatisfaction and unhealthy behaviors [1,41,42]. Social platforms amplify these pressures through likes and comparisons, fostering disordered eating, especially in cultures that idealize thinness [43–46].

Our data align with previous studies showing that a higher BMI is associated with lower body appreciation, except for Lithuanian male students. Other studies also found that individuals with a higher BMI generally reported lower BAS scores, indicating greater body dissatisfaction [25,47]. This association was particularly pronounced among females for whom societal pressures and the promotion of thinness as the ideal body type by the media intensify body image concerns [10]. Males experience less pressure regarding body size and BMI compared to females [8]. This difference may explain why Lithuanian male students did not demonstrate a significant association between BMI and body appreciation.

In our study, BAS scores were negatively correlated with perceived weight in both countries and across genders. Other authors demonstrated that people who see themselves as overweight, regardless of their actual weight, tend to experience lower levels of body appreciation and higher levels of psychological distress [34]. Additionally, individuals with greater body dissatisfaction are more likely to engage in harmful weight control practices, such as restrictive dieting and excessive exercise, which can have negative effects on both their physical and mental well-being [20,46,48]. Research indicates that students feel pressure regarding their body appearance in fitness centers [47]. Thus, the increase in obesity prevalence among students, linked to physical inactivity and poor dietary habits, may heighten the risk of body dissatisfaction, particularly in societies where thinness is idealized [13,33].

Our data indicated that body appreciation is linked to the dietary habits of students. Higher consumption of vegetables, fruits, fish, porridge, legumes, nuts, and regular breakfast was associated with higher BAS scores. This is consistent with findings, which show that healthier diets are related to a positive body image [49,50]. Conversely, the intake of sweets, sugary drinks, fast food, and snacks of Lithuanian students was associated with lower BAS scores. Earlier studies have also shown that poor diets are related to negative body image and poor mental health [50,51].

The current study found a positive association between physical activity and body appreciation, which supports previous findings indicating that regular exercise enhances body image and self-esteem [7]. In contrast, sitting time was associated with lower body appreciation among Lithuanian students, reflecting research that shows inactivity negatively affects body image and overall health [52,53]. The increasing prevalence of sedentary behavior, driven by longer screen time, is a global concern.

High levels of weight importance during adolescence were found to be predictive of persistent dieting and disordered eating during young adulthood for both males and females [54].

Health perceptions demonstrated a strong association with body appreciation. Our findings indicated that students who rated their health positively or were more health-conscious reported higher BAS scores. This aligns with previous research suggesting a relationship between body image perception and better health outcomes [50].

This study has several limitations that should be acknowledged. Firstly, the cross-sectional design limits the ability to establish causal relationships between body appreciation and lifestyle factors such as diet, physical activity, and health perceptions. Secondly, the study relied on self-reported data for weight and height, which may introduce bias due to participants underreporting or overreporting these anthropometric measures, as well as their dietary habits and body weight perceptions. Social desirability bias may also affect responses, especially regarding body image and health behaviors. Additionally, the sample consisted only of bachelor students from various universities in Lithuania and Poland, which may limit the generalizability of the findings to larger populations or students in other countries. The lack of data on socioeconomic status and mental health, both of which could influence body appreciation and lifestyle behaviors, is another limitation that needs to be considered in future research. Lastly, the analysis did not include detailed measurements of social media use, which is known to significantly impact body image and should be explored further.

Despite these limitations, the study has several significant strengths. It examines a large and diverse sample from two countries, which allows for meaningful cross-cultural comparisons. This provides valuable insights into how regional and cultural differences may affect body appreciation and related lifestyle factors. The study employs the validated Body Appreciation Scale-2, ensuring accurate and consistent measurement of body image across different genders and cultures. Additionally, by including various health-related behaviors such as diet, physical activity, and sleep, along with subjective health assessments, the study offers a comprehensive overview of factors connected to body appreciation.

Future research should include longitudinal studies to explore the causal relationships between body appreciation and health behaviors over time. Expanding the sample to include students from different regions or countries could enhance the generalizability of the findings. Additionally, examining the roles of mental health, socioeconomic status, and the influence of social media on body appreciation would provide more comprehensive insights. Interventions aimed at promoting a positive body image, particularly among female students, could be developed to encourage healthier lifestyle choices and improve overall well-being.

## 5. Conclusions

This study emphasizes the strong links between body appreciation, lifestyle factors, such as diet and physical activity, and health perceptions among university students in Lithuania and Poland. Female students, especially in Poland, reported lower levels of body appreciation despite having lower rates of overweight. This suggests that cultural and gender-specific pressures may influence body image. Higher body appreciation was associated with healthier eating habits, increased physical activity, better sleep quality, and positive health perceptions. Conversely, sedentary behavior and a higher intake of sweets and fast food were linked to lower body appreciation. Promoting healthier lifestyle habits may support improved body appreciation among students.

**Author Contributions:** Conceptualization: V.K., J.P. (Janina Petkevičienė) and D.G.; Methodology: V.K., J.P. (Janina Petkevičienė), D.G. and J.M.-R.; Formal analysis: J.P. (Janina Petkevičienė), A.R. and V.K.; Writing—original draft preparation: V.K., J.P. (Janina Petkevičienė) and D.G.; Writing—review and editing: A.R., J.M.-R., D.G., J.P. (Julia Ponichter) and L.P. All authors have read and agreed to the published version of the manuscript.

**Funding:** This research received no external funding.

**Institutional Review Board Statement:** This study was conducted by the Declaration of Helsinki and approved by the Bioethics Centre at the Lithuanian University of Health Sciences (protocols BEC-GVM(M)-80 from 28 February 2022, BEC-GM(M)-119 from 13 April 2021) and the Ethics Committee for Research with Human Participation at the Institute of Human Nutrition Sciences of the Warsaw University of Life Sciences (Resolutions No. 3/2022 from 28 January 2022).

**Informed Consent Statement:** Informed consent was obtained from all subjects involved in the study.

**Data Availability Statement:** The data presented in this study are available on request from the corresponding author (This is due to ethical reasons as indicated in the Bioethics permissions).

**Conflicts of Interest:** The authors declare no conflicts of interest.

## References

1. Tylka, T.L.; Wood-Barcalow, N.L. The Body Appreciation Scale-2: Item refinement and psychometric evaluation. *Body Image* **2015**, *12*, 53–67. [CrossRef] [PubMed]
2. Baceviciene, M.; Jankauskiene, R. Associations between body appreciation and disordered eating in a large sample of adolescents. *J. Nutr.* **2020**, *12*, 752. [CrossRef]
3. Almoraie, N.M.; Alothmani, N.M.; Alomari, W.D.; Al-Amoudi, A.H. Addressing Nutritional Issues and Eating Behaviors Among University Students: A Narrative Review. *Nutr. Res. Rev.* **2024**, 1–16. [CrossRef] [PubMed]

4. Avalos, L.; Tylka, T.L.; Wood-Barcalow, N.L. The body appreciation scale: Development and psychometric evaluation. *Body Image* **2005**, *2*, 285–297. [CrossRef]
5. Abu Alwafa, R.; Badrasawi, M. Factors associated with positive body image among Palestinian university female students, cross-sectional study. *Health Psychol. Behav. Med.* **2023**, *11*, 2278289. [CrossRef] [PubMed]
6. Swami, V.; Tran, U.S.; Stieger, S.; Aavik, T.; Ranjbar, H.A.; Adebayo, S.O.; Afhami, R.; Ahmed, O.; Aimé, A.; Akel, M.; et al. Body appreciation around the world: Measurement invariance of the Body Appreciation Scale-2 (BAS-2) across 65 nations, 40 languages, gender identities, and age. *Body Image* **2023**, *46*, 449–466. [CrossRef]
7. Shi, L.; Jiang, L.; Zhou, S.; Zhou, W.; Yang, H. Self-appreciation is not enough: Exercise identity mediates body appreciation and physical activity and the role of perceived stress. *Front. Psychol.* **2024**, *15*, 1377772. [CrossRef]
8. Quittkat, H.L.; Hartmann, A.S.; Düsing, R.; Buhlmann, U.; Vocks, S. Body Dissatisfaction, Importance of Appearance, and Body Appreciation in Men and Women over the Lifespan. *Front. Psychiatry* **2019**, *10*, 864. [CrossRef]
9. Swami, V.; Ng, S.K. Factor structure and psychometric properties of the Body Appreciation Scale-2 in university students in Hong Kong. *Body Image* **2015**, *15*, 68–71. [CrossRef]
10. Neumark-Sztainer, D.; Wall, M.M.; Chen, C.; Larson, N.I.; Christoph, J.M.; Sherwood, N.E. Eating, activity, and weight-related problems from adolescence to adulthood. *Am. J. Prev. Med.* **2018**, *55*, 133–141. [CrossRef]
11. Wawrzyniak, A.; Myszkowska-Ryciak, J.; Harton, A.; Lange, E.; Laskowski, W.; Hamulka, J.; Gajewska, D. Dissatisfaction with Body Weight among Polish Adolescents Is Related to Unhealthy Dietary Behaviors. *Nutrients* **2020**, *12*, 2658. [CrossRef] [PubMed]
12. Fortes, L.S.; Cipriani, M.; Coelho, F.D.; Paes, T.P.; Ferreira, M.E.C. Does self-esteem affect body dissatisfaction levels in female adolescents? A autoestima afeta a insatisfação corporal em adolescentes do sexo feminino? *Rev. Paul. Pediatr.* **2014**, *32*, 236–240. [CrossRef] [PubMed]
13. Kriaucioniene, V.; Raskiliene, A.; Petrauskas, D.; Petkeviciene, J. Trends in Eating Habits and Body Weight Status, Perception Patterns and Management Practices among First-Year Students of Kaunas (Lithuania) Universities, 2000–2017. *Nutrients* **2021**, *13*, 1599. [CrossRef]
14. Gruszka, W.; Owczarek, A.J.; Glinianowicz, M.; Bąk-Sosnowska, M.; Chudek, J.; Olszanecka-Glinianowicz, M. Perception of body size and body dissatisfaction in adults. *Sci. Rep.* **2022**, *12*, 1159. [CrossRef]
15. Vincente-Benito, I.; Ramírez-Durán, M.D.V. Influence of Social Media Use on Body Image and Well-Being Among Adolescents and Young Adults A Systematic Review. *J. Psychosoc. Nurs. Ment. Health Serv.* **2023**, *61*, 11–18. [CrossRef] [PubMed]
16. Mills, J.S.; Musto, S.; Williams, L.; Tiggemann, M. "Selfie" Harm: Effects on Mood and Body Image in Young Women. *Body Image* **2018**, *27*, 86–92. [CrossRef]
17. Azevedo, A.; Azevedo, Â.S. Implications of Socio-Cultural Pressure for a Thin Body Image on Avoidance of Social Interaction and on Corrective, Compensatory or Compulsive Shopping Behaviour. *Int. J. Environ. Res. Public Health* **2023**, *20*, 3567. [CrossRef] [PubMed]
18. Mikolajczyk, R.T.; Ansari, W.E.; Maxwell, A.E. Food consumption frequency and perceived stress and depressive symptoms among students in three European countries. *Nutr. J.* **2009**, *8*, 31. [CrossRef]
19. World Health Organization. *Obesity: Preventing and Managing the Global Epidemic. Report on a WHO Consultation*; Report No. 894; WHO: Geneva, Switzerland, 2000.
20. Ramón-Arbués, E.; Granada-López, J.M.; Martínez-Abadía, B.; Echániz-Serrano, E.; Antón-Solanas, I.; Nieto-Ventura, A. Eating Behavior and Relationships with Stress, Anxiety, Depression, and Insomnia in University Students. *Nutrients* **2021**, *13*, 2324. [CrossRef]
21. Whatnall, M.C.; Soo, Z.M.; Patterson, A.J.; Hutchesson, M.J. University Students Purchasing Food on Campus More Frequently Consume More Energy-Dense, Nutrient-Poor Foods: A Cross-Sectional Survey. *Nutrients* **2021**, *13*, 1053. [CrossRef]
22. Arslan, N.; Aslan Ceylan, J.; Hatipoğlu, A. The relationship of fast-food consumption with sociodemographic factors, body mass index and dietary habits among university students. *Nutr. Food Sci.* **2023**, *53*, 112–123. [CrossRef]
23. Alafif, N.; Alruwaili, N.W. Sleep duration, body mass index, and dietary behaviour among KSU students. *Nutrients* **2023**, *15*, 510. [CrossRef] [PubMed]
24. Suliga, E.; Cieśla, E.; Michel, S.; Kaducakova, H.; Martin, T.; Śliwiński, G.; Braun, A.; Izova, M.; Lehotska, M.; Kozieł, D.; et al. Diet Quality Compared to the Nutritional Knowledge of Polish, German, and Slovakian University Students—Preliminary Research. *Int. J. Environ. Res. Public Health* **2020**, *17*, 9062. [CrossRef] [PubMed]
25. Platta, A.M.; Mikulec, A.T.; Radzymińska, M.; Ruszkowska, M.; Suwała, G.; Zborowski, M.; Kowalczewski, P.L.; Nowicki, M. Body image and willingness to change it—A study of university students in Poland. *PLoS ONE* **2023**, *18*, e0293617. [CrossRef] [PubMed]
26. Gil, M.; Rudy, M.; Stanisławczyk, M. Gender Differences in Dietary Habits and Nutrition Knowledge Among Polish University Students. *Int. J. Environ. Res. Public Health* **2022**, *19*, 1456. [CrossRef]
27. Kowalska, A.S.; Tarnowska, A. Eating habits of Polish university students depending on the direction of studies and gender. *Brit. Food J.* **2019**, *121*, 54–70. [CrossRef]

28. Fryar, C.D.; Hughes, J.P.; Herrick, K.A. Fast Food Consumption Among Adults in the United States, 2013–2016; NCHS Data Briefs No. 322. 2018. Available online: https://stacks.cdc.gov/view/cdc/59582?utm_source=1851franchise.com&utm_medium=article&utm_campaign=fast-food-flops-and-failures-from-major-franchise-brands-2722290 (accessed on 27 October 2024).
29. Breitenbach, Z.; Raposa, B.; Szabó, Z.; Polyák, É.; Szács, Z.; Kubányi, J.; Figler, M. Examination of Hungarian college students' eating habits, physical activity and body composition. *Eur. J. Integr. Med.* **2016**, *2*, 13–17. [CrossRef]
30. Beaudry, K.M.; Ludwa, I.A.; Thomas, A.M.; Ward, W.E.; Falk, B.; Josse, A.R. First-year university is associated with greater body weight, body composition and adverse dietary changes in males than females. *PLoS ONE* **2019**, *14*, e0218554. [CrossRef]
31. Kriaučionienė, V.; Grincaitė, M.; Raskilienė, A.; Petkevičienė, J. Changes in Nutrition, Physical Activity, and Body Weight among Lithuanian Students during and after the COVID-19 Pandemic. *Nutrients* **2023**, *15*, 4091. [CrossRef]
32. Jehi, T.; Khan, R.; Halawani, R.; Dos Santos, H. Effect of COVID-19 outbreak on the diet, body weight and food security status of students of higher education: A systematic review. *Br. J. Nutr.* **2023**, *129*, 1916–1928. [CrossRef]
33. Peltzer, K.; Pengpid, S.; Samuels, T.A.; Özcan, N.K.; Mantilla, C.; Rahamefy, O.H.; Wong, M.L.; Gasparishvili, A. Prevalence of overweight/obesity and its associated factors among university students from 22 countries. *Int. J. Environ. Res. Public Health* **2014**, *11*, 7425–7441. [CrossRef] [PubMed]
34. Whatnall, M.C.; Patterson, A.J.; Brookman, S.; Convery, P.; Swan, C.; Pease, S.; Hutchesson, M.J. Lifestyle behaviors and related health risk factors in a sample of Australian university students. *J. Am. Coll. Health* **2019**, *68*, 734–741. [CrossRef] [PubMed]
35. Kyrkou, C.; Tsakoumaki, F.; Fotiou, M.; Dimitropoulou, A.; Symeonidou, M.; Menexes, G.; Biliaderis, C.G.; Michaelidou, A.M. Changing Trends in Nutritional Behavior among University Students in Greece, between 2006 and 2016. *Nutrients* **2018**, *10*, 64. [CrossRef]
36. Hao, M.; Fang, Y.; Yan, W.; Gu, J.; Hao, Y.; Wu, C. Relationship between body dissatisfaction, insufficient physical activity, and disordered eating behaviors among university students in southern China. *BMC Public Health* **2022**, *22*, 2054. [CrossRef] [PubMed]
37. He, J.; Sun, S.; Lin, Z.; Fan, X. The association between body appreciation and body mass index among males and females: A meta-analysis. *Body Image* **2020**, *34*, 10–26. [CrossRef]
38. Yong, C.; Liu, H.; Yang, Q.; Luo, J.; Ouyang, Y.; Sun, M.; Xi, Y.; Xiang, C.; Lin, Q. The Relationship between Restrained Eating, Body Image, and Dietary Intake among University Students in China: A Cross-Sectional Study. *Nutrients* **2021**, *13*, 990. [CrossRef]
39. Daniel, S.; Bridges, S.K. The drive for muscularity in men: Media influences and objectification theory. *Body Image* **2010**, *7*, 32–38. [CrossRef]
40. Watters, S.; Higgins, A. Muscle dysmorphia: An under-recognised aspect of body dissatisfaction in men. *Br. J. Nurs.* **2024**, *33*, 584–588. [CrossRef]
41. Tiggemann, M.; Anderberg, I. Muscles and Bare Chests on Instagram: The Effect of Influencers' Fashion and Fitspiration Images on Men's Body Image. *Body Image* **2020**, *35*, 237–244. [CrossRef]
42. Rounsefell, K.; Gibson, S.; McLean, S.; Blair, M.; Molenaar, A.; Brennan, L.; Truby, H.; McCaffrey, T.A. Social Media, Body Image and Food Choices in Healthy Young Adults: A Mixed Methods Systematic Review. *Nutr. Diet.* **2020**, *77*, 19–40. [CrossRef]
43. Dias, I.; Hernâni-Eusébio, J.; Silva, R. "How Many Likes?": The Use of Social Media, Body Image Insatisfaction and Disordered Eating. *Eur. Psychiatry* **2021**, *64*, S698. [CrossRef]
44. Jin, S.; Zhang, J. The Effects of Physical and Psychological Well-Being on Suicidal Ideation. *J. Clin. Psychol.* **1998**, *54*, 401–413. [CrossRef]
45. Santoniccolo, F.; Trombetta, T.; Paradiso, M.N.; Rollè, L. Gender and Media Representations: A Review of the Literature on Gender Stereotypes, Objectification and Sexualization. *Int. J. Environ. Res. Public Health* **2023**, *20*, 5770. [CrossRef] [PubMed]
46. Merino, M.; Tornero-Aguilera, J.F.; Rubio-Zarapuz, A.; Villanueva-Tobaldo, C.V.; Martín-Rodríguez, A.; Clemente-Suárez, V.J. Body Perceptions and Psychological Well-Being: A Review of the Impact of Social Media and Physical Measurements on Self-Esteem and Mental Health with a Focus on Body Image Satisfaction and Its Relationship with Cultural and Gender Factors. *Healthcare* **2024**, *12*, 1396. [CrossRef]
47. Sundgot-Borgen, C.; Sundgot-Borgen, J.; Bratland-Sanda, S.; Kolle, E.; Torstveit, M.K.; Svantorp-Tveiten, K.M.E.; Mathisen, T.F. Body appreciation and body appearance pressure in Norwegian university students comparing exercise science students and other students. *BMC Public Health* **2021**, *21*, 532. [CrossRef]
48. Eck, K.M.; Quick, V.; Byrd-Bredbenner, C. Body Dissatisfaction, Eating Styles, Weight-Related Behaviors, and Health among Young Women in the United States. *Nutrients* **2022**, *14*, 3876. [CrossRef] [PubMed]
49. Kosendiak, A.A.; Adamczak, B.B.; Kuźnik, Z.; Makles, S. How Dietary Choices and Nutritional Knowledge Relate to Eating Disorders and Body Esteem of Medical Students? A Single-Center Cross-Sectional Study. *Nutrients* **2024**, *16*, 1414. [CrossRef]
50. Jiménez-Morcillo, J.; Ramos-Campo, D.J.; Rodríguez-Besteiro, S.; Clemente-Suárez, V.J. The Association of Body Image Perceptions with Behavioral and Health Outcomes among Young Adults. *Nutrients* **2024**, *16*, 1281. [CrossRef]
51. Jankauskiene, R.; Baceviciene, M. Body Image Concerns and Body Weight Overestimation Do not Promote Healthy Behaviour: Evidence from Adolescents in Lithuania. *Int. J. Environ. Res. Public Health* **2019**, *16*, 864. [CrossRef]
52. Davis, H.A.; Rush, M.; Smith, G.T. Reciprocal relations between body dissatisfaction and excessive exercise in college women. *J. Am. Coll. Health* **2024**, *72*, 1473–1479. [CrossRef]

53. Zhang, R.; Liu, F.; Wang, X.; Wang, S. Towards active health: A study on the relationship between physical activity and body image among college students. *Heliyon* **2024**, *10*, e38465. [CrossRef] [PubMed]
54. Loth, K.A.; MacLehose, R.; Bucchianeri, M.; Crow, S.; Neumark-Sztainer, D. Predictors of dieting and disordered eating behaviors from adolescence to young adulthood. *J. Adolesc. Health* **2014**, *55*, 705–712. [CrossRef] [PubMed]

*nutrients*

MDPI

*Article*

# Muscle Mass and Vitamin B6 Are Linked to Negative Body Image in Women with Anorexia Nervosa: A Retrospective Study

**Federica Scarpina [1,2], Stefania Cattaldo [3,*], Elisa Prina [1,3], Paolo Piterà [1,3], Federico Brusa [4], Lorenzo Priano [1,2,3], Leonardo Mendolicchio [4,5] and Alessandro Mauro [1,2,3]**

1. "Rita Levi Montalcini" Department of Neurosciences, University of Turin, 10126 Turin, Italy; f.scarpina@auxologico.it (F.S.); e.prina@auxologico.it (E.P.); p.pitera@auxologico.it (P.P.); lorenzo.priano@unito.it (L.P.); alessandro.mauro@unito.it (A.M.)
2. I.R.C.C.S. Istituto Auxologico Italiano, U.O. di Neurologia e Neuroriabilitazione, Ospedale San Giuseppe, 282824 Piancavallo, Italy
3. I.R.C.C.S. Istituto Auxologico Italiano, Laboratorio di Neurobiologia Clinica, Ospedale San Giuseppe, 282824 Piancavallo, Italy
4. I.R.C.C.S. Istituto Auxologico Italiano, Laboratorio Sperimentale di Ricerche di Neuroscienze Metaboliche, Ospedale San Giuseppe, 282824 Piancavallo, Italy; f.brusa@auxologico.it (F.B.); l.mendolicchio@auxologico.it (L.M.)
5. I.R.C.C.S. Istituto Auxologico Italiano, U.O. di Riabilitazione dei Disturbi Alimentari e della Nutrizione, Ospedale San Giuseppe, 282824 Piancavallo, Italy

* Correspondence: s.cattaldo@auxologico.it

**Abstract:** Introduction. Anorexia nervosa severely impacts the physical body and mental body (i.e., body image). In this retrospective study, we investigated the relationship between the perceived body image and body composition in women with anorexia nervosa. Specifically, we aimed to verify what components (i.e., weight, body composition, and micronutrients) may predict a higher number of symptoms of negative body image in this clinical condition. Methods. Weight status and body composition, including the expressions of vitamins, and body image concerns were measured in a sample of 112 women with anorexia nervosa (age in years M = 26.78; SD = 12; range = 14–67). Results. According to the regression analysis, a higher skeletal muscle mass and a higher concentration of vitamin B6 seemed to predict a higher number of symptoms of negative body image in our sample. Conclusions. This study pointed out muscle mass and the concentration of vitamin B6 as involved in the psychological expression of body image concerns in anorexia nervosa, especially at the beginning of the disease. Thus, we may suggest including and monitoring these parameters in routine care for anorexia nervosa.

**Keywords:** anorexia nervosa; body image; physical body; body composition; vitamins

## 1. Introduction

Negative body image is a core symptom of anorexia nervosa [1]. Affected individuals may show significant perceptual distortions in estimating physical body sizes (i.e., they tend to estimate their size as larger than is objectively true) and high levels of dissatisfaction toward bodily size, shape, and appearance [2–4]. The major expression of the disease symptoms is on the physical body: thinness and low weight (even if some people with anorexia nervosa may not look very thin) [5], blue fingers, thin and fragile hair, yellowing and dry skin, and downy hair covering the body [6,7]. Also, individuals with anorexia nervosa report high levels of weakness and tiredness; they feel dizziness or fainting [5].

Swelling of the arms and legs and stress fractures or reduced bone mass can be observed [8]. Some of these signs and symptoms mirror micronutrient deficiency, such as vitamin deficiency. In the context of eating disorders, vitamin D deficiency, together with an inadequate calcium intake, is highly associated with fractures, osteoporosis, and reduced bone mass [9]. Water-soluble vitamins like vitamin B complex and vitamin C are more prone to wash out from the body during stressful conditions, like starvation and low food intake [10]. Vitamins as micronutrients play a vital functional role in maintaining high levels of health and peak physical performance [11]. They enable those complex reactions necessary to use the potential energy in macronutrients to fuel the biological processes inherent in physical training and recovery [11,12]. Thus, in the case of reduced levels of vitamins, the physical body may be perceived as less strong and less capable, which is precisely the experience of individuals with anorexia nervosa. The lower level of strength mirrors what is observed in the clinical assessment of body composition in terms of reduction of fat mass, fat-free mass, and bone mass [13], and reduced efficiency of energy processes and proteolysis [14], as suggested by lower values of the phase angle (i.e., computed according to the resistance and reactance values obtained through bioelectric impedance analysis and directly correlated with body cell mass and cellular nutritional status). A lower phase angle reflects the impact of undernutrition, as observed in malnourished or cachectic patients [15]. Notably, phase angle values have been found to increase during nutritional recovery, along with improved body composition and fluid compartment changes [15,16]. Nevertheless, high inter-individual variability is observed in this parameter because of the role played by other factors, including individual levels of physical activity, the presence of vomiting and use of laxatives, the characteristics of the metabolism in starvation, and, possibly, on the stage of recovery.

In this retrospective study, we aimed to investigate the relationship between perceived body image and body composition (even at the level of micronutrients) in women with anorexia nervosa. Specifically, to recognize a sort of body-related biomarker of negative body image in anorexia nervosa, we verified what components (i.e., weight, body composition, and micronutrients) may predict a higher number of symptoms of negative body image in our sample.

## 2. Methods

### *2.1. Participants*

This study was approved by the Ethics Committee of the involved institution (Reference number: 2022_09_27_10). All participants were volunteers who gave informed written consent. For individuals with ages lower than 18 years, parents signed the written consent. The Italian National Sanitary System covered all hospital charges, and participants were not remunerated.

In this retrospective observational study, only in-patients consecutively recruited at their admission to the hospital between February 2021 and March 2024 were included. Participants were evaluated according to routine procedures on a day-hospital basis. Individuals were included in this study if they satisfied the Diagnostic and Statistical Manual of Mental Disorders, Fifth Edition (American Psychiatric Association, 2013) criteria for anorexia nervosa, which are as follows: (i) restriction of food intake leading to weight loss or a failure to gain weight resulting in a "significantly low body weight" of what would be expected for someone's age, sex, and height; (ii) fear of becoming fat or gaining weight; and (iii) a distorted view of themselves and their condition. Both "restricting type" and "binge eating/purging type" (American Psychiatric Association, 2013) were included in this study. Moreover, participants were included if they had not previously been treated, and none declared that they took prescribed or self-administered micronutrient supplements. Participants were excluded if other pathologies not related to anorexia nervosa were recorded (e.g., neurodegenerative diseases such as brain injury and stroke). Notably, individuals in their first recovery and those in their second or more recovery were included as the chronic nature of the disease [17], since affected individuals frequently undertake more than one treatment. A clinical assessment of the eating style through the traditional Eating Disorder Inventory™–3 (EDI-3) [18], which focuses on the symptomatology associated with eating disorders, and the Binge Eating Scale [19] to assess the presence of binge eating behavior. We administered the Symptom Checklist-90 [20], which allows the evaluation of self-reported severity of psychopathological symptoms, and the Italian version [21] of the Psychological General Well-Being Index (PGWBI) [22], which measures the self-reported level of anxiety, depressed mood, positive well-being, self-control, general health, and vitality.

### 2.2. Body Image

Our participants with anorexia nervosa filled out the Body Uneasiness Test [23], which assesses body image characteristics. This self-rating scale explores various areas of body-related psychopathology, which are dissatisfaction regarding the body and its weight; avoiding and compulsive control behavior; experience of separation and foreignness regarding the body; and specific worries for certain body parts, characteristics, or functions. The main global dimension of the Global Severity Index (i.e., GSI) is read as a marker of body image concerns, with a higher score indicating greater body uneasiness. This questionnaire has been validated in individuals with a body mass index lower than 25, showing a good internal consistency (Cronbach's alpha coefficients range between 0.69 and 0.90 > 0.7) [23].

### 2.3. Weight Status and Body Composition

Because of the clinical standard routine, individuals' weight and height were measured under standardized conditions in the morning, after a fasting period of 12 h, in light clothes without shoes. Body mass index (BMI) was calculated as body weight divided by squared height ($kg/m^2$). Body weight (kg) and body height (meters) were measured with precision to the nearest 0.1 kg and 0.5 cm, respectively. A mechanical column scale (Scale-Tronix, Wheaton, IL) and a stadiometer (Scale-Tronix, Wheaton, IL, USA) were used for these measurements. Moreover, we performed a body composition analysis with a single-frequency bioelectrical impedance analysis (BIA 101, Akern®, Pisa, Italy).

After a 12 h fasting period, blood samples were collected through an indwelling cannula inserted at the cubital vein. Tubes were kept on ice and protected from light immediately after sampling. To determine the level of concentration of vitamin C, vitamin A, and vitamin E, plasma was separated with centrifugation at 3000× $g$ for 10 min at 4 °C and stored at −80 °C until analysis. For vitamins B1 and B6, whole blood was stored at −80 °C until analysis. All vitamins were determined using high-performance liquid chromatography (HPLC) with commercially available kits from Chromsystems (Chromsystems Instruments and Chemicals GmbH, Gräfelfing, Germany), according to the manufacturer's instructions. The inter- and intra-assay coefficients of variations were ≤4.2% and ≤2.3% for vitamin C, ≤3.5% and ≤4.9% for vitamin A, ≤3.2% and ≤4.1% for vitamin E, ≤5.9% and ≤4.5% for vitamin B1, and ≤5.2% and ≤4.0% for vitamin B6.

### 2.4. Statistical Analysis

Data were initially analyzed using descriptive statistics, including means, standard deviations, frequencies, and percentages. Scores about the psychological questionnaires were computed according to the seminal articles.

We performed preliminary analyses to describe our sample of participants with anorexia nervosa in terms of clinical parameters. First, we considered the value of body mass index. We split the sample into two groups: participants with a score below the threshold of 18 (i.e., underweight) and those over this threshold (i.e., normal weight). Through an independent sample *t*-test, we verified any difference between the two groups in the score of the Global Severity Index. We followed the same rationale about the different parameters from the bioelectrical impedance analysis. However, to our knowledge, there is no specific threshold for these parameters in the literature about anorexia nervosa. Thus, we split the sample into two groups, assessing if each individual's score was below or over the within-sample median. Through an independent sample *t*-test, we verified any difference between the two groups in the score of the Global Severity Index. Concerning vitamins, we verified how many participants reported a concentration within, below, or over the normal range. The reference levels of vitamins were obtained from the datasheets of the kits used for their analysis. Thus, according to the one-way analysis of variance, we verified any difference between the three groups at the score of the main index of the Global Severity Index.

Successively, we verified any relationship between the role of age, age at the onset, and disease length, and the parameters relative to the body image and the physical body.

Finally, we used a linear regression analysis approach to investigate whether and what body image concerns (the outcome, the Global Severity Index) may be associated with weight status and body composition (the predictive factors). Indeed, this statistical model is used to quantify the strength of the relationship between the outcome and the explanatory variables. First, the correlation and directionality of the data were investigated in a preliminary analysis to formulate the statistical model using Pearson's correlation coefficient (Pearson's r). Those variables significantly associated (*p*-value 0.05) with the score relative to the main outcome were further investigated with a multiple linear regression model. Goodness-of-fit was reported as $R^2$; the significance of the model was evaluated according to the F-value and *p*-value. The relative contribution of factors included in the statistical model was verified with the independent variable. Finally, for each predictive factor included in the model, the variance inflation factor (VIF) was reported as a measure of multicollinearity.

## 3. Results

### 3.1. Participants

Data relative to 112 women with anorexia nervosa were collected (age in years M = 26.78; SD = 12; range = 14–67; age in years at the symptoms onset M = 18.02; SD = 6.36; range = 10–51; disease length in years M = 8.77; SD = 10.35; range: 1–47). Overall, the majority of the sample (92.85% N = 104) satisfied the criteria for the restrictive type, while 7.14% (n = 8) of the sample was categorized as a binge/purging type. The majority (N = 63; 56.25%) of our sample was in early adulthood (18 years old to mid-30s), while 19.64% (N = 22) were adolescents (age $\leq$ 18 years old). Moreover, 22.32% (N = 25) was in adulthood (mid-30s to mid-60s), and only 1.78% (N = 2) was in late adulthood (mid-60s). In Table 1, the results relative to the psychopathological assessment of our participants are reported.

**Table 1.** Mean, standard deviation, and range relative to the scores at the questionnaires on eating behavior (Binge Eating Scale and Eating Disorders Inventory-3), symptoms of psychopathology (Symptom Checklist-90), and overall psychological well-being (Psychological General Well-Being Index) in our sample of participants affected by anorexia nervosa.

| | Mean | Standard Deviation | Min | Max |
|---|---|---|---|---|
| | **Binge Eating Scale** | | | |
| Score | 31.27 | 10.12 | 10 | 102 |
| | **Eating Disorder Inventory™–3** | | | |
| Drive for thinness | 22.10 | 8.27 | 0 | 28 |
| Bulimia | 6.21 | 7.32 | 0 | 32 |
| Body dissatisfaction | 29.04 | 8.61 | 2 | 40 |
| Eating disorder risk | 57.35 | 18.20 | 11 | 96 |
| Low self-esteem | 16.04 | 5.98 | 1 | 24 |
| Personal alienation | 15.31 | 6.44 | 2 | 28 |
| Interpersonal insecurity | 12.79 | 6.52 | 0 | 26 |
| Interpersonal alienation | 12.19 | 5.68 | 1 | 28 |
| Interoceptive deficits | 22.58 | 16.87 | 0 | 171 |
| Emotional dysregulation | 10.93 | 7.13 | 0 | 37 |
| Perfectionism | 11.38 | 5.42 | 11 | 24 |
| Asceticism | 14.54 | 6.44 | 1 | 28 |
| Maturity fears | 17.59 | 7.78 | 0 | 32 |
| Ineffectiveness | 31.36 | 11.81 | 5 | 52 |
| Interpersonal problems | 24.98 | 11.19 | 3 | 51 |
| Affective problems | 32.04 | 14.52 | 0 | 63 |
| Overcontrol | 25.92 | 10.06 | 2 | 51 |
| General psychological maladjustment | 131.89 | 43.58 | 26 | 237 |
| | **Symptom Checklist-90** | | | |
| Somatization | 1.94 | 1.02 | 0 | 6.17 |
| Obsession-compulsion | 2.28 | 0.84 | 0 | 3.90 |
| Interpersonal sensitivity | 2.23 | 0.93 | 0.11 | 3.89 |
| Depression | 2.29 | 1 | 0 | 6.85 |
| Anxiety | 2.29 | 0.99 | 0.1 | 7.20 |
| Hostility | 1.28 | 0.97 | 0 | 3.67 |
| Phobic anxiety | 1.34 | 0.90 | 0 | 3.43 |
| Paranoid ideation | 1.9 | 0.83 | 0 | 3.67 |
| Psychoticism | 1.42 | 0.66 | 0 | 3.4 |
| Total score | 2.27 | 2.86 | 0.13 | 31.19 |
| | **Psychological General Well-Being Index** | | | |
| Anxiety | 8.31 | 5.57 | 0 | 24 |
| Depression | 5.6 | 4.27 | 0 | 14 |
| Positive well-being | 4.63 | 3.38 | 4 | 15 |
| Self-control | 6.22 | 3.23 | 0 | 14 |
| General healthy | 7.20 | 3.09 | 0 | 14 |
| Vitality | 7.02 | 4.27 | 0 | 19 |
| Total score | 38.98 | 18.14 | 4 | 92 |

### 3.2. *Body Image*

The results of the Body Uneasiness Test [23], which measures body image, are reported in Table 2.

**Table 2.** Mean, standard deviation, and range relative to the scores at the questionnaire relative to body image (Body Uneasiness Test [23]) in our sample of women affected by anorexia nervosa.

| | Mean | Standard Deviation | Min | Max |
|---|---|---|---|---|
| Global Score Index | 3.22 | 1.11 | 0.5 | 6.09 |
| Weight phobia | 3.64 | 1.19 | 0.25 | 5 |
| Body image concerns | 3.59 | 1.15 | 0.78 | 5 |
| Avoidance | 2.57 | 1.51 | 0 | 12.83 |
| Compulsive self-monitoring | 3.08 | 1.39 | 0 | 5 |
| Depersonalization | 2.88 | 1.36 | 0.2 | 6.4 |
| Positive symptom total | 22.14 | 9.8 | 1 | 37 |
| Positive Symptom Distress Index | 3.23 | 0.84 | 1.36 | 5.67 |

### 3.3. *Weight Status and Body Composition*

Table 3 reports the results for the parameters used to describe the physical body.

We did not find any difference in the score at the Global Severity Index between participants with a score below the threshold of 18 in body mass index (i.e., underweight; N = 98; M = 3.16; SD = 1.17) and those over the threshold (i.e., normal weight; N = 14; M = 3.61; SD = 0.79) [t(110) = 1.41; $p = 0.15$]. When the following rationale was used about the parameters relative to the bioelectrical impedance analysis, we observed a significant difference in the Global Severity Index when the sample was split according to the phase angle, body cell mass, fat mass, body cell mass index, skeletal muscle mass, and fat-free mass index, as reported in Table 4.

**Table 3.** Mean, standard deviation, and range relative to the scores describing the physical body in our sample of women with anorexia nervosa.

| | Mean | Standard Deviation | Min | Max |
|---|---|---|---|---|
| Body Mass Index | 15.01 | 2.41 | 9.74 | 20.89 |
| | **Body Impedance Analysis** | | | |
| Phase angle (°) | 3.88 | 1.11 | 1.03 | 6.06 |
| Fat-free mass (%) | 34.04 | 4.86 | 15.30 | 45.4 |
| Body cell mass (%) | 14.77 | 4.37 | 1.4 | 28 |
| Fat mass (%) | 13.33 | 7.52 | 4.8 | 29.7 |
| Extracellular water (%) | 56.17 | 8.6 | 38.4 | 86.7 |
| Muscle mass (%) | 47.54 | 7.67 | 18.5 | 68.8 |
| Body cell mass index | 5.55 | 1.52 | 0.7 | 9.4 |
| Hydration (%) | 73.78 | 4.36 | 62.7 | 90 |
| Skeletal muscle mass (kg) | 17.38 | 2.5 | 8.1 | 24.1 |
| Fat-free mass index | 12.85 | 1.35 | 9.20 | 16.1 |
| | **Vitamins** | | | |
| Vitamin C (mg/L) | 12.03 | 5.96 | 0.20 | 30.80 |
| Vitamin A (mg/L) | 0.51 | 0.21 | 0.11 | 1.15 |
| Vitamin E (mg/L) | 15.8 | 5.49 | 2.84 | 37.4 |
| Vitamin B1 (μg/L) | 57.61 | 18.7 | 23.8 | 129 |
| Vitamin B6 (μg/L) | 31.24 | 17.42 | 6.4 | 109.58 |

**Table 4.** For each parameter of the bioelectrical impedance analysis, sample size, mean, and standard deviation relative to the Global Severity Index from the psychological questionnaire were reported when the sample was split into two groups according to the within-sample threshold. In bold, significant results ($p < 0.05$).

| | Within-Sample Threshold | Below the Threshold | Over the Threshold | Statistical Results |
|---|---|---|---|---|
| Phase angle (°) | 4.035 | n = 61<br>M = 3<br>SD = 1.18 | n = 51<br>M = 3.48<br>SD = 0.94 | t(110) = 2.35;<br>***p* = 0.02** |
| Fat-free mass (%) | 34.5 | n = 60<br>M = 3.07<br>SD = 1.10 | n = 52<br>M = 3.39<br>SD = 1.09 | t(110) = 1.55;<br>*p* = 0.12 |
| Body cell mass (%) | 14.7 | n = 59<br>M = 2.94<br>SD = 1.13 | n = 53<br>M = 3.53<br>SD = 0.99 | t(110) = 2.9;<br>***p* = 0.004** |
| Fat mass (%) | 10.9 | n = 60<br>M = 3.01<br>SD = 1.1 | n = 52<br>M = 3.46<br>SD = 1.07 | t(110) = 2.18;<br>***p* = 0.03** |
| Extracellular water (%) | 54.3 | n = 64<br>M = 3.35<br>SD = 0.99 | n = 48<br>M = 3.04<br>SD = 1.23 | t(119) = 1.48;<br>*p* = 0.15 |
| Muscle mass (%) | 48 | n = 63<br>M = 3.06<br>SD = 1.2 | n = 49<br>M = 3.42<br>SD = 0.93 | t(110) = 1.71;<br>*p* = 0.08 |
| Body cell mass index | 5.75 | n = 61<br>M = 2.92<br>SD = 1.16 | n = 51<br>M = 3.57<br>SD = 0.93 | t(110) = 3.23;<br>***p* = 0.002** |
| Hydration (%) | 73.2 | n = 57<br>M = 3.23<br>SD = 1.01 | n = 55<br>M = 3.11<br>SD = 1.19 | t(110) = 0.99;<br>*p* = 0.32 |
| Skeletal muscle mass (kg) | 17.7 | n = 57<br>M = 2.93<br>SD = 1.11 | n = 55<br>M = 3.52<br>SD = 1.02 | t(110) = 2.92;<br>***p* = 0.004** |
| Fat-free mass index | 12.85 | n = 61<br>M = 2.93<br>SD = 1.16 | n = 51<br>M = 3.57<br>SD = 0.92 | t(110) = 3.16;<br>***p* = 0.002** |

According to the results reported in Table 5, we did not find any significant differences in the score at the Global Severity Index when our sample was split into three groups according to the expression of the vitamins (i.e., within, below, or over the normal range). The only exception was the results relative to vitamin C: participants with a concentration below the normal range reported a significantly lower score at the Global Severity Index than those within the normal range.

### 3.4. The Role of Age, Age at the Onset, and Disease Length

As shown in Table 6, more negative symptoms in almost all the components measured by the questionnaire relative to body image were significantly related to lower demographic age, age at the onset, and disease length. The only exception was the Positive Symptom Distress Index (PSDI) (i.e., the mean intensity of all disliked body parts), which was not significantly related to the demographic parameters. As observed in Table 7, older participants reported significantly a lower fat mass and a lower skeletal muscle mass, as well as lower concentrations of vitamin B6 but higher levels of extracellular water. A longer disease length was associated with lower fat mass, lower skeletal muscle mass, and lower expressions of vitamin B6. As shown in Table 8, when we considered the physical parameters of body mass index and body impedance analysis, we observed a significant relationship between multiple scores relative to the body image concerns. The sign of the coefficient was positive, meaning that a more muscular body (as suggested by higher scores for body mass index, phase angle, fat-free mass, body cell mass, fat mass, muscle mass, body cell mass index, skeletal muscle mass, and fat-free mass index) was significantly related with a more negative body image (as suggested by higher scores at the psychological questionnaire). The only exceptions were represented by the parameters relative to the parameters of extracellular water and hydration, about which we observed a negative sign of correlation. About vitamins, we observed a less pronounced pattern. The most consistent result emerged about vitamin B6: a higher expression of this vitamin was significantly associated with a higher Global Severity Index of negative body image. Moreover, it was related to higher expressions of avoidance-related behaviors and a higher score at the positive symptom total (PST) relative to body uneasiness. Finally, higher expressions of vitamin C were related to higher positive symptom total (PST) of body uneasiness, and lower expressions of vitamin E were related to more severe body image concerns (BIC).

**Table 5.** Sample size, mean, standard deviation, and range (minimum and maximum) relative to the Global Severity Index from the psychological questionnaires are reported for the sample when it was divided into three groups about the level of concentration of the vitamins (i.e., within, below or over the normal range). In bold, significant results ($p < 0.05$).

| | N | Mean | Standard Deviation | | Range Min | Range Max | Statistical Results |
|---|---|---|---|---|---|---|---|
| | | | **Vitamin C** | | | | |
| Within the normal range | 68 | 3.35 | 1.11 | 0.14 | 0.68 | 6.09 | F(2,111) = 3.54; $p = 0.032$ normal vs. below $p = 0.02$; other comparison $\geq 0.08$ |
| Below the normal range | 13 | 2.48 | 1.21 | 0.34 | 0.5 | 4.62 | |
| Over the normal range | 31 | 3.27 | 0.95 | 0.17 | 0.94 | 4.5 | |
| | | | **Vitamin A** | | | | |
| Within the normal range | 72 | 3.35 | 1.11 | 0.13 | 0.68 | 6.09 | F(2,111) = 1.44; $p = 0.23$ |
| Below the normal range | 18 | 2.95 | 1.17 | 0.28 | 0.5 | 4.85 | |
| Over the normal range | 22 | 3.02 | 1.01 | 0.22 | 1.09 | 4.5 | |
| | | | **Vitamin E** | | | | |
| Within the normal range | 90 | 3.25 | 1.07 | 0.11 | 0.5 | 4.85 | F(2,111) = 0.58; $p = 0.55$ |
| Below the normal range | 1 | 2.12 | | | 2.12 | 2.12 | |
| Over the normal range | 21 | 3.14 | 1.28 | 0.28 | 0.68 | 6.09 | |
| | | | **Vitamin B1** | | | | |
| Within the normal range | 102 | 3.22 | 1.12 | 0.11 | 0.5 | 6.09 | F(2,111) = 0.12; $p = 0.88$ |
| Below the normal range | 2 | 3.58 | 0.19 | 0.14 | 3.44 | 3.71 | |
| Over the normal range | 8 | 3.14 | 1.12 | 0.39 | 1.09 | 4.12 | |
| | | | **Vitamin B6** | | | | |
| Within the normal range | 59 | 3.07 | 1.15 | 0.15 | 0.5 | 4.85 | F(2,111) = 1.36; $p = 0.26$ |
| Below the normal range | 1 | 2.82 | – | – | 2.82 | 2.82 | |
| Over the normal range | 52 | 3.41 | 1.05 | 0.15 | 0.85 | 6.09 | |

**Table 6.** Pairwise correlation matrix among the scores reported at the questionnaire relative to body image (columns) and the parameters of demographical age, age at the onset, and disease length. In bold, $p \leq 0.05$; in bold and italic, $p \leq 0.001$.

| | Age | Age at the Onset | Disease Lenght |
|---|---|---|---|
| GSI | ***−0.580*** | ***−0.374*** | ***−0.441*** |
| WP | ***−0.466*** | ***−0.263*** | ***−0.378*** |
| BIC | ***−0.580*** | ***−0.415*** | ***−0.417*** |
| Av | ***−0.436*** | ***−0.314*** | ***−0.311*** |
| CSM | ***−0.519*** | ***−0.280*** | ***−0.428*** |
| D | ***−0.453*** | ***−0.283*** | ***−0.351*** |
| PST | ***−0.528*** | ***−0.431*** | ***−0.347*** |
| PSDI | −0.142 | −0.158 | −0.067 |

−1 … +1

**Table 7.** Pairwise correlation matrix among the scores reported at the parameters relative to the physical body (columns) and the parameters of demographical age, age at the onset, and disease length. In bold, $p \leq 0.05$; in bold and italic, $p \leq 0.001$.

| | Age | Age at the Onset | Disease Lenght |
|---|---|---|---|
| Body Mass Index | −0.139 | 0.029 | −0.180 |
| Phase Angle | −0.129 | −0.147 | −0.059 |
| Fat-Free Mass | −0.125 | −0.028 | −0.127 |
| Body Cell Mass | −0.174 | −0.098 | −0.142 |
| Fat Mass | ***−0.245*** | −0.111 | **−0.216** |
| Extra Cellular Water | ***0.209*** | 0.137 | 0.155 |
| Muscle Mass | −0.122 | −0.092 | −0.083 |
| Body Cell Mass Index | −0.153 | −0.099 | −0.116 |
| Hydration | 0.122 | 0.149 | 0.048 |
| Skeletal Muscle Mass | ***−0.391*** | −0.100 | ***−0.384*** |
| Fat-Free Mass Index | −0.033 | 0.017 | −0.049 |
| Vitamin C | −0.135 | −0.160 | −0.057 |
| Vitamin A | 0.061 | 0.158 | −0.026 |
| Vitamin E | 0.166 | 0.185 | 0.078 |
| Vitamin B1 | 0.053 | 0.125 | −0.015 |
| Vitamin B6 | ***−0.246*** | −0.158 | **−0.188** |

−1 … +1

**Table 8.** Pairwise correlation matrix among the scores reported at questionnaire relative to body image (columns) and the parameters describing the physical body (i.e., body mass index and the parameters relative to body impedance analysis) and its micronutrients (i.e., vitamins). In bold, $p \leq 0.05$; in bold and italic, $p \leq 0.001$.

| | Body Mass Index | Phase Angle | Fat-Free Mass | Body Cell Mass | Fat Mass | Extra Cellular Water | Muscle Mass | Body Cell Mass Index | Hydration | Skeletal Muscle Mass | Fat-Free Mass Index | Vitamin C | Vitamin A | Vitamin E | Vitamin B1 | Vitamin B6 |
|---|---|---|---|---|---|---|---|---|---|---|---|---|---|---|---|---|
| GSI | **0.250** | ***0.305*** | **0.241** | ***0.309*** | 0.172 | ***−0.303*** | **0.239** | ***0.312*** | −0.178 | ***0.310*** | **0.227** | 0.136 | −0.038 | −0.138 | −0.031 | **0.191** |
| WP | ***0.348*** | ***0.371*** | ***0.311*** | ***0.384*** | **0.244** | ***−0.390*** | ***0.298*** | ***0.397*** | **−0.252** | ***0.275*** | ***0.320*** | 0.101 | 0.029 | −0.122 | −0.010 | 0.118 |
| BIC | **0.244** | ***0.325*** | **0.241** | ***0.332*** | 0.173 | ***−0.309*** | **0.241** | ***0.329*** | **−0.201** | ***0.314*** | **0.217** | 0.155 | −0.118 | **−0.212** | 0.002 | 0.176 |
| Av | 0.038 | 0.080 | 0.053 | 0.080 | 0.017 | −0.081 | 0.080 | 0.079 | −0.012 | **0.203** | 0.037 | 0.056 | −0.044 | −0.037 | −0.076 | **0.214** |
| CSM | ***0.283*** | ***0.272*** | **0.253** | ***0.279*** | **0.222** | ***−0.268*** | 0.176 | ***0.273*** | −0.145 | ***0.287*** | **0.234** | 0.177 | −0.055 | −0.105 | −0.044 | 0.163 |
| D | 0.147 | **0.240** | **0.205** | ***0.289*** | 0.045 | ***−0.284*** | ***0.282*** | ***0.288*** | −0.163 | ***0.274*** | 0.183 | 0.064 | 0.115 | −0.106 | −0.001 | 0.130 |
| PST | **0.203** | 0.188 | 0.113 | 0.139 | **0.225** | −0.183 | 0.103 | 0.166 | −0.137 | ***0.264*** | 0.109 | **0.213** | −0.003 | 0.089 | 0.045 | **0.186** |
| PSDI | ***0.269*** | ***0.362*** | 0.176 | ***0.322*** | **0.205** | ***−0.322*** | **0.204** | ***0.332*** | ***−0.284*** | −0.062 | **0.218** | −0.057 | 0.013 | −0.010 | −0.065 | −0.056 |

−1 … +1

### 3.5. Regression Analysis

The Global Severity Index was included in the model as the main outcome. We introduced the parameters of body mass index (VIF = 4.32), phase angle (VIF = 5.632), extracellular water (VIF = 7.99), hydration (VIF = 6.07), skeletal muscle mass (VIF = 1.64), fat-free mass index (VIF = 4.56), and B6 (VIF = 1.02) as predictors since they were significantly related to the main outcome (Table 6). Notably, other parameters relative to bioelectrical impedance analysis emerged as significantly correlated to the main outcome. However, they showed a very high multicollinearity. Then, we selected those parameters that may be globally descriptive in the case of anorexia nervosa [14,15,24]. The model was significant [R2 = 0.237; F(7,95) = 3.903; $p$ = 0.001]. Skeletal muscle mass [B = 0.145; $p$ = 0.008] and B6 [B = 0.01; $p$ = 0.026] were significant; no other parameters were significant [$p \geq 0.17$]. Thus, higher levels of skeletal muscle mass and higher concentrations of vitamin B6 significantly predicted a higher level of body image concerns in our sample.

## 4. Discussion

Anorexia nervosa severely impacts both the physical and mental body. In this retrospective study, we aimed to identify any specific component of the physical body (even in the micronutrients) affecting the mental body. According to our results, two physical body-related components were significantly related to a negative body image: the skeletal muscle mass and the concentration of vitamin B6.

It is well known that in anorexia nervosa, we generally observe a loss of body fat. Interestingly, skeletal muscle mass is also often reduced [25], with negative side effects on the levels of vitamin D, bone mineral density, bodily strength, metabolic function, and physical performance [26]. Overall, individuals with anorexia nervosa have smaller muscle size and reduced energy expenditure when compared with healthy controls. Additionally, some studies reported that the recovery from this clinical disease was not enough to restore muscle mass. It should be observed that a muscular body, especially in women, is far from the ideal mental body in anorexia nervosa, while it is more characteristic of orthorexia [27], which is an eating disorder in which people express an obsessive focus on healthy eating, and bigorexia [28,29], which is a body dysmorphic disorder in which individuals express the desire to have less fat mass and the obsession with increasing muscle mass. Interestingly, our results suggest that a lower skeletal muscle mass is associated with older age and longer disease length, in line with the evidence that this parameter generally does not restore in the long term. Moreover, younger women at the beginning of anorexic symptoms could still have a muscular body, which may, in turn, increase the negative body image. Thus, our evidence that a higher skeletal muscle mass predicts more negative body image concerns is not surprising.

The novelty of this research is the result relative to the micronutrient component of vitamin B6. Our results suggested that it may play a role in body image in anorexia nervosa. We observed that a higher concentration of vitamin B6 was significantly related to higher negative body image. Vitamin B6 is water soluble and can be obtained from different foods and supplements. In the healthy population, its deficiency is not very common [30,31]; instead, its lower expression generally occurs in eating disorders, including anorexia nervosa, due to insufficient dietary intake, malabsorption, and use of certain medications [32,33]. Malnourished, elderly, and anorexic individuals are at higher risk of developing vitamin B6 deficiency [32,34], with several side effects such as anemia, peripheral neuropathy, seborrheic dermatitis, glossitis, cheilosis, depression, celiac disease, and seizures [35,36]. Our data seems to support this evidence since, in our sample of participants with anorexia nervosa, lower concentrations of vitamin B6 were significantly associated with older age and longer disease length. However, how do we conceive the relationship between the concentration of vitamin B6 and a negative body image? To our knowledge, this relationship was not explored in the literature. B vitamins directly impact bodily energy levels, brain function, and cell metabolism. Thus, when their expression is within the normal range, people perceive the physical body as healthy and energized. However, in individuals with anorexia nervosa, an energized body may be miscategorized as a fat body and mislabeled as the experience of "feeling fat" [37], enhancing negative thoughts, emotions, and feelings (i.e., body image). On the other hand, the decreasing levels of B vitamins, which may result in perceiving the physical body as less strong, may decrease the expressions of body image concerns. Because of the lack of evidence in the literature, our hypothesis about the predictive role of the expression of vitamin B6 on body image concerns is speculative, and we cannot further establish the mechanism underlying this association. We need future investigation to confirm our hypothesis. For example, experimental studies exploring the physiological and psychological effects of vitamin B6 supplementation in individuals with anorexia nervosa could further clarify the role of this micronutrient in body image concerns.

We underlined some limitations of this study. Our sample consisted only of female participants, since there is a higher incidence of anorexia nervosa in women than in men [38]. Moreover, it is well-established that bodily perceptions and body images are different between genders [39–41]. Consequently, the evidence from this research should not be generalized to males. Moreover, it was a monocentric-based study, limiting the generalizability of our findings. It should be noted that the sample's age ranged from 14 to 67 years old, meaning that different phases of life (from adolescence to elderly) were included in the sample. In future cross-sectional studies, the age of the participants at the time of the experimental observation should be used as an inclusion/exclusion criterion. We furnished a clinical description of the global psychological functioning of our sample as well as the characteristics of the eating disorder and the expression of psychopathological symptoms. However, because of the sample size, we did not include any of this information in the statistical model (otherwise, the statistical power would be dramatically decreased). Thus, in the case of larger samples, we may suggest verifying the role of demographical information and disease-related characteristics (including the level of severity) on the main outcome. Finally, we used a self-report measurement that is highly used in clinical and research contexts to assess the mental body. However, body perceptions, especially about physical dimensions, could be measured using behavioral instruments, which should be included in future research.

## 5. Conclusions

Our evidence may suggest the level of muscle mass and the concentration of vitamin B6 as key components in experiencing a negative body image in anorexia nervosa, especially at the beginning of the disease. Notably, the role of muscle mass has some evidence in the literature, while the role of vitamin B6 in this context represents a novel and intriguing result. Because of this observation, future research is needed to support this novel evidence. Nevertheless, we strongly suggest the monitoring of these components in the routine care for anorexia nervosa, especially before and after multidisciplinary treatments.

**Author Contributions:** Conceptualization, F.S., S.C., A.M., L.M. and L.P.; methodology, F.S., S.C., E.P., P.P. and F.B.; formal analysis, F.S., S.C., E.P. and P.P.; data curation, S.C., E.P., P.P. and F.B.; writing—original draft preparation, F.S., E.P. and P.P.; writing—review and editing, S.C.; supervision,

A.M., L.M. and L.P.; funding acquisition, A.M., L.M. and L.P. All authors have read and agreed to the published version of the manuscript.

**Funding:** This research was funded by the Italian Ministry of Health Research (Ricerca Corrente).

**Institutional Review Board Statement:** The study was conducted in accordance with the Declaration of Helsinki, and approved by the Ethics Committee of Istituto Auxologico Italiano (protocol code 2022_09_27_10 and date of approval 27 September 2022).

**Informed Consent Statement:** Informed consent was obtained from all subjects involved in the study.

**Data Availability Statement:** The data that support the findings of this study are available in Zenodo at the following link: https://doi.org/10.5281/zenodo.13923377.

**Conflicts of Interest:** The authors declare no conflicts of interest.

## References

1. American Psychiatric Association. *Diagnostic and Statistical Manual of Mental Disorders: DSM-5TM*, 5th ed.; American Psychiatric Publishing, Inc.: Arlington, VA, USA, 2013; p. xliv, 947. ISBN 978-0-89042-554-1.
2. Mölbert, S.C.; Thaler, A.; Mohler, B.J.; Streuber, S.; Romero, J.; Black, M.J.; Zipfel, S.; Karnath, H.-O.; Giel, K.E. Assessing Body Image in Anorexia Nervosa Using Biometric Self-Avatars in Virtual Reality: Attitudinal Components Rather than Visual Body Size Estimation Are Distorted. *Psychol. Med.* **2018**, *48*, 642–653. [CrossRef] [PubMed]
3. Dalhoff, A.W.; Romero Frausto, H.; Romer, G.; Wessing, I. Perceptive Body Image Distortion in Adolescent Anorexia Nervosa: Changes After Treatment. *Front. Psychiatry* **2019**, *10*, 748. [CrossRef] [PubMed]
4. Provenzano, L.; Porciello, G.; Ciccarone, S.; Lenggenhager, B.; Tieri, G.; Marucci, M.; Dazzi, F.; Loriedo, C.; Bufalari, I. Characterizing Body Image Distortion and Bodily Self-Plasticity in Anorexia Nervosa via Visuo-Tactile Stimulation in Virtual Reality. *J. Clin. Med.* **2019**, *9*, 98. [CrossRef] [PubMed]
5. Zipfel, S.; Giel, K.E.; Bulik, C.M.; Hay, P.; Schmidt, U. Anorexia Nervosa: Aetiology, Assessment, and Treatment. *Lancet Psychiatry* **2015**, *2*, 1099–1111. [CrossRef]
6. Schulze, U.M.; Pettke-Rank, C.V.; Kreienkamp, M.; Hamm, H.; Bröcker, E.B.; Wewetzer, C.; Trott, G.E.; Warnke, A. Dermatologic Findings in Anorexia and Bulimia Nervosa of Childhood and Adolescence. *Pediatr. Dermatol.* **1999**, *16*, 90–94. [CrossRef]
7. Samela, T.; Sampogna, F.; Fabbricatore, M.; Fania, L.; Papanikolaou, E.S.; Abeni, D. Skin Signs in Eating Disorders: A Literature Review. *Eat Weight Disord.* **2022**, *27*, 867–879. [CrossRef]
8. LaBan, M.M.; Wilkins, J.C.; Sackeyfio, A.H.; Taylor, R.S. Osteoporotic Stress Fractures in Anorexia Nervosa: Etiology, Diagnosis, and Review of Four Cases. *Arch. Phys. Med. Rehabil.* **1995**, *76*, 884–887. [CrossRef]
9. Voulgaridou, G.; Papadopoulou, S.K.; Detopoulou, P.; Tsoumana, D.; Giaginis, C.; Kondyli, F.S.; Lymperaki, E.; Pritsa, A. Vitamin D and Calcium in Osteoporosis, and the Role of Bone Turnover Markers: A Narrative Review of Recent Data from RCTs. *Diseases* **2023**, *11*, 29. [CrossRef]
10. Lindschinger, M.; Tatzber, F.; Schimetta, W.; Schmid, I.; Lindschinger, B.; Cvirn, G.; Stanger, O.; Lamont, E.; Wonisch, W. A Randomized Pilot Trial to Evaluate the Bioavailability of Natural versus Synthetic Vitamin B Complexes in Healthy Humans and Their Effects on Homocysteine, Oxidative Stress, and Antioxidant Levels. *Oxidative Med. Cell. Longev.* **2019**, *2019*, 6082613. [CrossRef]
11. Lukaski, H.C. Vitamin and Mineral Status: Effects on Physical Performance. *Nutrition* **2004**, *20*, 632–644. [CrossRef]
12. Volpe, S.L. Micronutrient Requirements for Athletes. *Clin. Sports Med.* **2007**, *26*, 119–130. [CrossRef] [PubMed]
13. Hübel, C.; Yilmaz, Z.; Schaumberg, K.E.; Breithaupt, L.; Hunjan, A.; Horne, E.; García-González, J.; O'Reilly, P.F.; Bulik, C.M.; Breen, G. Body Composition in Anorexia Nervosa: Meta-Analysis and Meta-Regression of Cross-Sectional and Longitudinal Studies. *Int. J. Eat. Disord.* **2019**, *52*, 1205–1223. [CrossRef] [PubMed]
14. Małecka-Massalska, T.; Popiołek, J.; Teter, M.; Homa-Mlak, I.; Dec, M.; Makarewicz, A.; Karakuła-Juchnowicz, H. Application of Phase Angle for Evaluation of the Nutrition Status of Patients with Anorexia Nervosa. *Psychiatr. Pol.* **2017**, *51*, 1121–1131. [CrossRef] [PubMed]
15. Murray Hurtado, M.; Martín Rivada, Á.; Quintero Alemán, C.; Ruiz Alcántara, M.P.; Ramallo Fariña, Y. Body Composition and Nutritional Status Changes in Adolescents with Anorexia Nervosa. *An. Pediatría* **2023**, *99*, 162–169. [CrossRef]
16. Scalfi, L.; Marra, M.; Caldara, A.; Silvestri, E.; Contaldo, F. Changes in Bioimpedance Analysis after Stable Refeeding of Undernourished Anorexic Patients. *Int. J. Obes. Relat. Metab. Disord.* **1999**, *23*, 133–137. [CrossRef]
17. Hay, P.J.; Touyz, S.; Sud, R. Treatment for Severe and Enduring Anorexia Nervosa: A Review. *Aust. N. Z. J. Psychiatry* **2012**, *46*, 1136–1144. [CrossRef]

18. Clausen, L.; Rosenvinge, J.H.; Friborg, O.; Rokkedal, K. Validating the eating disorder inventory-3 (EDI-3): A comparison between 561 female eating disorders patients and 878 females from the general population. *J. Psychopathol. Behav. Assess.* **2010**, *33*, 101–110. [CrossRef]
19. Gormally, J.; Black, S.; Daston, S.; Rardin, D. The Assessment of Binge Eating Severity among Obese Persons. *Addict. Behav.* **1982**, *7*, 47–55. [CrossRef]
20. Derogatis: Brief Symptom Inventory. Available online: https://scholar.google.com/scholar_lookup?&title=Brief%20Symptom%20Inventory-18%20(BSI-18):%20Administration,%20scoring,%20and%20procedures%20manual&publication_year=2000&author=Derogatis,LR (accessed on 15 October 2024).
21. Grossi, E.; Compare, A. Psychological General Well-Being Index (PGWB). In *Encyclopedia of Quality of Life and Well-Being Research*; Michalos, A.C., Ed.; Springer: Dordrecht, The Netherlands, 2014; pp. 5152–5156. ISBN 978-94-007-0753-5.
22. Dupuy, H.J. The psychological general well-being (PGWB) index. In *Assessment of Quality of Life in Clinical Trials of Cardiovascular Therapies*; Wenger, N.K., Mattson, M.E., Furberg, C.D., Elinson, J., Eds.; Le Jacq: New York, NY, USA, 1984; pp. 170–183.
23. Cuzzolaro, M.; Vetrone, G.; Marano, G.; Garfinkel, P.E. The Body Uneasiness Test (BUT): Development and Validation of a New Body Image Assessment Scale. *Eat. Weight Disord.* **2006**, *11*, 1–13. [CrossRef]
24. Mattar, L.; Godart, N.; Melchior, J.C.; Pichard, C. Anorexia Nervosa and Nutritional Assessment: Contribution of Body Composition Measurements. *Nutr. Res. Rev.* **2011**, *24*, 39–45. [CrossRef]
25. Essén, B.; Fohlin, L.; Thorén, C.; Saltin, B. Skeletal Muscle Fibre Types and Sizes in Anorexia Nervosa Patients. *Clin. Physiol.* **1981**, *1*, 395–403. [CrossRef] [PubMed]
26. El Ghoch, M.; Pourhassan, M.; Milanese, C.; Müller, M.J.; Calugi, S.; Bazzani, P.V.; Dalle Grave, R. Changes in Lean and Skeletal Muscle Body Mass in Adult Females with Anorexia Nervosa before and after Weight Restoration. *Clin. Nutr.* **2017**, *36*, 170–178. [CrossRef] [PubMed]
27. Oberle, C.D.; Lipschuetz, S.L. Orthorexia Symptoms Correlate with Perceived Muscularity and Body Fat, Not BMI. *Eat Weight Disord.* **2018**, *23*, 363–368. [CrossRef]
28. Levinson, C.A.; Rodebaugh, T.L.; White, E.K.; Menatti, A.R.; Weeks, J.W.; Iacovino, J.M.; Warren, C.S. Social Appearance Anxiety, Perfectionism, and Fear of Negative Evaluation: Distinct or Shared Risk Factors for Social Anxiety and Eating Disorders? *Appetite* **2013**, *67*, 125–133. [CrossRef]
29. Devrim Lanpir, A.; Bilgic, P. Bigorexia: Body dysmorphic disorder or eating disorder? *J. Health Sci.* **2018**, *27*, 64–69.
30. Ho, C.-L.; Quay, T.A.W.; Devlin, A.M.; Lamers, Y. Prevalence and Predictors of Low Vitamin B6 Status in Healthy Young Adult Women in Metro Vancouver. *Nutrients* **2016**, *8*, 538. [CrossRef]
31. Spinneker, A.; Sola, R.; Lemmen, V.; Castillo, M.J.; Pietrzik, K.; González-Gross, M. Vitamin B6 Status, Deficiency and Its Consequences—An Overview. *Nutr. Hosp.* **2007**, *22*, 7–24.
32. Rock, C.L.; Vasantharajan, S. Vitamin Status of Eating Disorder Patients: Relationship to Clinical Indices and Effect of Treatment. *Int. J. Eat. Disord.* **1995**, *18*, 257–262. [CrossRef]
33. Hanachi, M.; Dicembre, M.; Rives-Lange, C.; Ropers, J.; Bemer, P.; Zazzo, J.-F.; Poupon, J.; Dauvergne, A.; Melchior, J.-C. Micronutrients Deficiencies in 374 Severely Malnourished Anorexia Nervosa Inpatients. *Nutrients* **2019**, *11*, 792. [CrossRef]
34. Jungert, A.; Neuhäuser-Berthold, M. Determinants of Vitamin B6 Status in Community-Dwelling Older Adults: A Longitudinal Study Over a Period of 18 Years. *J. Gerontol. A Biol. Sci. Med. Sci.* **2020**, *75*, 374–379. [CrossRef]
35. Institute of Medicine (US). Standing Committee on the Scientific Evaluation of Dietary Reference Intakes and its Panel on Folate, Other B Vitamins, and Choline. In *Dietary Reference Intakes for Thiamin, Riboflavin, Niacin, Vitamin B6, Folate, Vitamin B12, Pantothenic Acid, Biotin, and Choline*; The National Academies Collection: Reports funded by National Institutes of Health; National Academies Press (US): Washington, DC, USA, 1998; ISBN 978-0-309-06411-8.
36. Office of Dietary Supplements—Vitamin B6. Available online: https://ods.od.nih.gov/factsheets/VitaminB6-HealthProfessional/ (accessed on 22 October 2024).
37. Cooper, M.J.; Deepak, K.; Grocutt, E.; Bailey, E. The Experience of "feeling Fat" in Women with Anorexia Nervosa, Dieting and Non-Dieting Women: An Exploratory Study. *Eur. Eat. Disord. Rev.* **2007**, *15*, 366–372. [CrossRef] [PubMed]
38. Martínez-González, L.; Fernández-Villa, T.; Molina, A.J.; Delgado-Rodríguez, M.; Martín, V. Incidence of Anorexia Nervosa in Women: A Systematic Review and Meta-Analysis. *Int. J. Environ. Res. Public Health* **2020**, *17*, 3824. [CrossRef] [PubMed]
39. Furnham, A.; Radley, S. Sex Differences in the Perception of Male and Female Body Shapes. *Personal. Individ. Differ.* **1989**, *10*, 653–662. [CrossRef]

40. Kiefer, I.; Leitner, B.; Bauer, R.; Rieder, A. Body Weight: The Male and Female Perception. *Soz Praventivmed* **2000**, *45*, 274–278. [CrossRef]
41. Fallon, A.E.; Rozin, P. Sex Differences in Perceptions of Desirable Body Shape. *J. Abnorm. Psychol.* **1985**, *94*, 102–105. [CrossRef]

nutrients MDPI

*Article*

# A Longitudinal Study on Body Image Perception and Size among Italian Early Adolescents: Changes over Time and Discrepancies between Genders

Emanuela Gualdi-Russo [1,*], Sabrina Masotti [1,2,*], Natascia Rinaldo [1,2], Federica De Luca [1], Stefania Toselli [3], Gianni Mazzoni [1,2], Simona Mandini [1,2] and Luciana Zaccagni [1,2]

[1] Department of Neuroscience and Rehabilitation, Faculty of Medicine, Pharmacy and Prevention, University of Ferrara, 44121 Ferrara, Italy; rnlnsc@unife.it (N.R.); federica.deluca@unife.it (F.D.L.); gianni.mazzoni@unife.it (G.M.); simona.mandini@unife.it (S.M.); luciana.zaccagni@unife.it (L.Z.)
[2] Center for Exercise Science and Sports, University of Ferrara, 44123 Ferrara, Italy
[3] Department for Life Quality Studies, University of Bologna, 47921 Rimini, Italy; stefania.toselli@unibo.it
* Correspondence: emanuela.gualdi@unife.it (E.G.-R.); sabrina.masotti@unife.it (S.M.)

**Abstract:** Background/Objectives: The discrepancy between the current perceived body image (BI) and beauty ideals leads to dissatisfaction, which is believed to be common among adolescents. This study aimed to investigate the stability in BI perceptions and dissatisfaction during early adolescence. Another aim was to highlight differences in dissatisfaction according to Body Mass Index (BMI) and inconsistencies in weight status perception. Methods: Two hundred and nine participants (mean age at first survey: 11.33 $\pm$ 0.38 years) were enrolled in this longitudinal study with three years of follow-up. Data on size and BI perceptions were collected through individual interviews. Stature and weight were directly measured. Results: Findings indicated significant changes over three years in anthropometric traits but not in weight status prevalence or BI perception and dissatisfaction, except for the ideal figure in males and weight control in females. The results also indicated a significant difference in BI perception and dissatisfaction by BMI categories. Self-reported body measurements were found to be unreliable with a weak-to-moderate agreement between self-perceived and actual weight status. Conclusions: BI perception appears quite stable across the three years considered. Dissatisfaction is similar in both genders, although a tendency toward different gender aesthetic ideals is already appreciable in early adolescence. We suggest that the lower frequency of normal-weight adolescents compared with peers in previous studies is attributable to the effects of the recent pandemic. Given the growing dissatisfaction with increasing BMI and misinterpretations of weight status, school actions to promote a healthy lifestyle and positive BI should be undertaken.

**Keywords:** perceived body image; perceived weight status; body dissatisfaction; body mass index; adolescents

## 1. Introduction

Body image (BI) perception corresponds to a subjective image of the body, which may disregard its actual appearance. Indeed, it is a composite construct that includes feelings, evaluations, and attitudes related to the body [1,2]. This perception is also influenced by feedback from other people [3]. When there are negative cognitions and feelings toward body image, we speak of body image dissatisfaction [4].

The importance of BI perception is so great that it can affect the individual's physical and mental health status. Judging one's appearance depends heavily on the ideal body image in society [4], which associates the ideal of beauty with thinness in the Western world [5,6]. BI dissatisfaction, defined as individuals' detrimental and subjective mental representation of body appearance, is widespread among adolescents [7]. At the developmental stage of adolescence, when major mental, physical, and social changes occur, the perception of BI is generally exacerbated with a tendency to develop concerns about it. At this stage, adolescents are particularly susceptible to the comments and opinions of peers and adults about their bodies. They also tend to abandon family as a reference to focus on friends and peers, such that the perceptions and evaluations of these reference groups condition their self-concept and self-esteem [8]. Dissatisfaction with BI, due to discrepancies between reality, perception of one's body, and desired ideal, is often determined in adolescents, deeply affecting their self-esteem [9].

Adolescent self-esteem is conditioned by the perception of BI and the judgment given by peers according to fixed canons that, following sexual dimorphism, impose thin and toned bodies for females and muscular and strong bodies for males [5,6]. Unlike in some non-Western countries [10], what happens in Western countries is that these body ideals end up increasing the risk of psychological and physical illnesses often accompanied by eating disorders. An eating disorder is a severe mental health condition that can develop from a poor BI and can have physical health consequences with possible cardiovascular, endocrine, gastrointestinal, and fertility complications. These eating disorders include anorexia, bulimia, and binge eating disorder [11].

During the pubertal transition, adolescents increase in stature and weight and change their body structure and composition [12,13]. In this period of pubertal development, differences between males and females increase in physical appearance and psychological profile, with various health implications [14]. Along with sexual dimorphism, there is growing awareness among adolescents of gender differences that are reflected in physical appearance. It is mainly girls who manifest problems regarding their BI and body weight, so much so that more than 70% of female adolescents wish to change their body weight or shape [15]. The onset of puberty can result, especially in girls, in a series of transformations that drive them away from the ideal of thinness, exposing them to teasing from peers [16]. These aspects have great relevance as body dissatisfaction in adolescence is believed to be a significant risk factor leading to eating disorders [17,18]. Epidemiological studies conducted in Italy have shown that a high percentage of adolescents are at risk for eating disorders [19] with increased underweight or overweight status [20].

Studies in the literature assessing changes in body dissatisfaction during adolescence have led to inconsistent results, which is probably because of the different developmental ages and contexts analyzed [21,22]. Cross-sectional studies cannot answer the questions, which are still open, concerning the possible changes in BI perception and dissatisfaction from early adolescence when growth and development occur.

Therefore, using a longitudinal design, our main purpose was to examine possible changes in body perception and dissatisfaction in girls and boys across middle school years. We hypothesized that girls are characterized by greater and increasing body dissatisfaction with age than boys because they wish to comply with an ideal of thinness.

Another purpose was to explore body dissatisfaction by BMI over three years and any inconsistencies in participants' perceptions of their weight status. In this regard, we hypothesized that dissatisfaction is greater in unhealthy weight categories and that there is a proper perception of weight status.

## 2. Materials and Methods

### *2.1. Participants and Procedures*

This longitudinal study is part of more extensive ongoing research on the physical and mental health of children and early adolescents in northern Italy [23–27].

The Bioethics Committee of the University of Bologna approved the study procedures (approval code no. 2.18).

Before this longitudinal study was undertaken, we determined the minimum sample size required to test the primary study hypothesis by the G*Power statistical program (version 3.1.9.6; Universitat Kiel, Kiel, Germany) by ANOVA for repeated measures (3 measurements) in 2 groups (males and females) for 95% power, medium effect size, and 0.05 significance level, resulting in 44 participants.

To this end, we invited all students from a middle school in Ferrara (Emilia-Romagna region, Northern Italy) to participate. The school (chosen on a convenience basis) cooperated by sending an informational letter to all students' families and requesting written, signed parental consent. Twenty students without signed consent from their parents were excluded from the surveys. The final sample included 209 schoolchildren (121 boys and 88 girls) who had signed consent and agreed to participate. Anonymity and confidentiality were assured to all participants, who had the right to withdraw from the study at any time.

The longitudinal data of each schoolchild completing grades 6–8 were collected at a 1-year interval from November 2020 to November 2023. The mean age in the first survey for participants was 11.33, with a standard deviation (SD) of 0.38.

Due to careful monitoring, no missing values were found in the body image perception questions or anthropometric measurements collected.

### 2.2. Anthropometric Survey

A simple proxy for body composition in these early adolescents was obtained through Body Mass Index (BMI), that is, the weight-to-stature ratio (kg/m$^2$). Through this index, the students' weight status can be assessed according to Cole cutoffs [28,29] as underweight (UW), normal weight (NW), overweight (OW), or obese (O). To this end, the same operator directly measured stature to the nearest 0.1 cm with an anthropometer (Magnimeter, Raven Equipment Ltd., Dunmow, Essex, UK) and weight to the nearest 0.5 kg with a mechanical scale (SECA, Basel, Switzerland) on participants dressed in light clothing, following standardized anthropometric methods [30–32].

### 2.3. Size and Body Image Perception

Before the anthropometric survey, students' face-to-face interviews were conducted within a separate setting allowing for greater privacy and concentration.

After some basic socio-demographic questions, we tested the participants regarding awareness of body size by asking the following: "How tall are you?", "How much do you weigh?" (for these first two questions, given the age of the participants, the answer "I don't know" was also scheduled), "Which weight status do you think best represents you?" (choice: underweight, normal weight, overweight/obese). We tested participants for possible dissatisfaction and weight interventions by asking the following: "Do you control your weight?" (choice: never, sometimes, often, always), "Have you tried to lose weight in the last year?" (choice: never, sometimes, often, always) and "If yes, please indicate in what manner" (choice: reducing food, increasing physical activity).

Moreover, to assess body image perception and dissatisfaction, we employed the scale of Childress et al. [33] consisting of eight drawings of female or male figures obtained by child adaptation of an adult figure scale [34]. These eight silhouettes of equal stature were sorted separately by gender in an incremental manner from a skinny figure (silhouette 1) to an obese one (silhouette 8). On this scale, participants had to mark which figure best represented their actual body image and which represented their ideal figure. This evaluation was repeated longitudinally in all surveying years. BI dissatisfaction (Feel minus Ideal Discrepancy, FID) was assessed by subtracting the selected ideal figure from the actual feel figure [35,36]. Dissatisfaction increases as the absolute value of the score rises: positive values indicate that the dimensions perceived as real are greater than ideal ones; negative values indicate that the dimensions perceived as real are smaller than ideal ones. No dissatisfaction exists when the figure perceived as real equals the ideal figure (FID score = 0).

Aware of the difference between the terms sex and gender [37], we use the former when related only to biological characteristics and the latter in all other cases (e.g., when related to cultural, psychological, or behavioral characteristics or when these factors are variously intertwined with biological factors).

### 2.4. Statistical Analysis

We calculated absolute and relative frequencies for each qualitative variable (questionnaire responses and weight status categories) and means and SD for each quantitative variable (anthropometric and BI perception traits) separately by gender. The normality of quantitative variables was assessed through the Kolmogorov–Smirnov test.

The chi-squared test was used to detect whether the observed frequencies differed from the expected ones in comparisons by gender and time. To test the effect of sex on normally distributed quantitative variables at times T0, T1, and T2, we used ANOVA for repeated measurements. A nonparametric Friedman rank-sum test was applied to variables not normally distributed, such as variables related to BI perception. In addition, in these cases, we performed a gender comparison separately at times T0, T1, and T2 with the Wilcoxon test.

A paired samples t-test was applied to compare measured anthropometric traits with the ones reported by participants.

The main effects of perceived BI variables with the different BMI categories were analyzed by the Kruskal–Wallis one-way analysis of variance with rank ANOVA.

To assess participants' ability to place themselves in the correct weight status category, we calculated Cohen's kappa coefficient, a measure of agreement for qualitative variables.

All the analyses were conducted using Statistica for Windows, Version 11.0 (StatSoft Srl, Tulsa, OK, USA), and R software version 4.4.1 for Windows.

$p$ values less than 0.05 were assumed to be statistically significant.

## 3. Results

### 3.1. Changes in Anthropometric Traits, Weight Status, BI Perception

Table 1 shows the anthropometric characteristics of participants, measured directly by us and self-reported by participants during the three years (Time 0, 1, 2), and statistical comparisons.

The ANOVA for repeated measures shows a significant effect of time and a significant interaction of time and sex on five traits (stature and weight, BMI; stature and weight, self-reported). Specifically, concerning interaction, it appears that participants increased in dimensions over the three years, but the male sex showed greater increases than the female sex, especially in stature. Even time had a statistically significant effect on BMI, as calculated from self-reported data. The main effect of sex on traits was not significant.

The comparison between direct and self-reported measurements revealed the low reliability of the reported values (pairwise $t$-test): the self-reported weight was significantly lower than that measured in boys in all three surveys (T0 $p < 0.0001$; T1 $p = 0.0014$; T2 $p = 0.0124$) and in girls at T0 ($p = 0.0002$) and T1 ($p = 0.0009$); self-reported stature was significantly higher than that measured only in girls at T2 ($p = 0.0155$).

The second part of Table 1 shows the prevalence of weight status in the sample in three years. Normal-weight participants were prevalent in both sexes in all three surveys. In boys, there was a tendency for an increase in underweight and overweight from T0 to T2 and a concomitant decrease in obesity. Conversely, in girls, underweight tended to decrease from T0 to T1/T2 while obesity tended to increase. However, chi-square comparisons show that changes from time T0 to T2 did not achieve statistical significance in males or females. In comparisons between the two sexes for the prevalence of different weight status categories, no significant differences were found at T0, T1, and T2 ($p > 0.05$).

Table 2 shows the BI perception variables and answers to questions submitted to participants.

The perception of BI did not differ among time T0, T1, and T2 except for the ideal figure in males (tending to a more robust ideal figure in T2). In separate gender comparisons at each time, there was a significant difference concerning feel figures (T0: W = 4229, $p$-value = 0.009; T1: W = 4340, $p$-value = 0.03; T2: W = 4227, $p$-value = 0.008) and ideal ones (T0: W = 3165.5, $p$-value < 0.0001; T1: W = 3617.5, $p$-value < 0.0001; T2: W = 2959.5, $p$-value < 0.0001). In particular, girls perceived themselves as thinner than their male peers and liked slimmer body figures more than boys. This trend was reflected in the FD scores showing the highest dissatisfaction in the females even without reaching the statistical significance level ($p > 0.05$).

**Table 1.** Anthropometric characteristics and comparisons by sex with ANOVA for repeated measures (total n = 209; males n = 121; females n = 88) and with chi-square for weight status frequencies over time.

| Variables | Boys | | | Girls | | | *ANOVA* | | |
|---|---|---|---|---|---|---|---|---|---|
| | T0 | T1 | T2 | T0 | T1 | T2 | *Sex* | *Measures* | *Sex * Measures* |
| | *Mean (SD)* | *Mean (SD)* | *Mean (SD)* | *Mean (SD)* | *Mean (SD)* | *Mean (SD)* | *F (p)* | *F (p)* | *F (p)* |
| Stature (cm) | 150.25 (8.48) | 156.99 (9.09) | 164.00 (9.05) | 151.31 (7.07) | 156.08 (6.31) | 158.17 (5.95) | 3.04 (0.08) | 1209.63 (<0.0001) | 142.87 (<0.0001) |
| Weight (kg) | 45.48 (11.49) | 50.97 (12.28) | 56.58 (13.21) | 45.58 (9.49) | 50.56 (10.05) | 53.47 (10.16) | 0.54 (0.46) | 491.77 (<0.0001) | 16.14 (<0.0001) |
| BMI (kg/m$^2$) | 19.94 (3.78) | 20.53 (3.92) | 20.89 (3.89) | 19.82 (3.37) | 20.72 (3.72) | 21.34 (3.65) | 0.12 (0.73) | 69.68 (<0.0001) | 3.83 (0.02) |
| Stature (cm) self-reported | 150.08 [1] (11.22) | 155.95 (11.96) | 163.75 (12.04) | 151.01 [2] (8.09) | 155.88 (6.11) | 159.08 (6.36) | 1.62 (0.20) | 216.33 (<0.0001) | 18.49 (<0.0001) |
| Weight (kg) self-reported | 43.17 [1] (10.22) | 49.45 (11.93) | 55.44 (12.77) | 43.26 [2] (7.91) | 48.27 (8.95) | 52.62 (8.95) | 1.70 (0.19) | 229.59 (<0.0001) | 6.98 (0.001) |
| | Boys | | | | | | Girls | | |
| | T0 | T1 | T2 | *Chi-square* | | | T0 | T1 | T2 |
| | *N (%)* | *N (%)* | *N (%)* | *p-value* | | *p-value* | *N (%)* | *N (%)* | *N (%)* |
| *Weight status* | | | | >0.05 | | >0.05 | | | |
| UW | 8 (6.6) | 9 (7.4) | 10 (8.3) | | | | 5 (5.7) | 2 (2.3) | 2 (2.3) |
| NW | 76 (62.8) | 73 (60.3) | 72 (59.5) | | | | 55 (62.5) | 60 (68.2) | 56 (63.6) |
| OW | 25 (20.7) | 29 (24.0) | 31 (25.6) | | | | 24 (27.3) | 21 (23.9) | 24 (27.3) |
| O | 12 (9.9) | 10 (8.3) | 8 (6.6) | | | | 4 (4.5) | 5 (5.7) | 6 (6.8) |

Note: The "I don't know" response to the self-reported dimensions resulted in a reduction of the sample as follows: [1] n = 113 in Stature, n = 109 in Weight; [2] n = 83 in Stature, n= 81 in Weight.

**Table 2.** BI perception and dissatisfaction by gender and weight control.

| BI Variables | Boys | | | *Friedman Rank-Sum Test* | | Girls | | |
|---|---|---|---|---|---|---|---|---|
| | T0 | T1 | T2 | | | T0 | T1 | T2 |
| | *Mean (SD)* | *Mean (SD)* | *Mean (SD)* | *p-Value* | *p-Value* | *Mean (SD)* | *Mean (SD)* | *Mean (SD)* |
| Feel Figure | 4.55 (1.30) | 4.59 (1.33) | 4.77 (1.26) | 0.099 | 0.17 | 4.09 (1.21) | 4.16 (1.28) | 4.31 (1.24) |
| Ideal Figure | 3.90 (0.89) | 3.89 (0.94) | 4.13 (0.68) | 0.0008 | 0.15 | 3.28 (0.79) | 3.42(0.83) | 3.45 (0.86) |
| FID (score) | 0.65 (1.25) | 0.69 (1.31) | 0.64 (1.22) | 0.73 | 0.47 | 0.81 (1.19) | 0.74 (1.20) | 0.85 (1.15) |
| | Boys | | | *Chi-square* | | Girls | | |
| | T0 | T1 | T2 | | | T0 | T1 | T2 |
| Questions | *N (%)* | *N (%)* | *N (%)* | *p-value* | *p-value* | *N (%)* | *N (%)* | *N (%)* |
| *Which weight status do you think best represents you?* | | | | >0.05 | >0.05 | | | |
| -UW | 10 (8.3) | 13 (10.7) | 13 (10.7) | | | 7 (8.0) | 7 (8.0) | 5 (5.7) |
| -NW | 80 (66.1) | 78 (64.5) | 81 (66.9) | | | 60 (68.2) | 60 (68.2) | 58 (65.9) |
| -OW/O | 31 (25.6) | 30 (24.8) | 27 (22.3) | | | 21 (23.9) | 21 (23.9) | 25 (28.4) |
| *Do you control your weight?* | | | | >0.05 | <0.001 | | | |
| -never | 36 (29.8) | 33 (27.3) | 47 (38.8) | | | 26 (29.5) | 34 (38.6) | 44 (50.0) |
| -sometimes | 68 (56.2) | 69 (57.0) | 57 (47.1) | | | 56 (63.6) | 41 (46.6) | 25 (28.4) |
| -often | 13 (10.7) | 12 (9.9) | 15 (12.4) | | | 5 (5.7) | 8 (9.1) | 14 (15.9) |
| -always | 4 (3.3) | 7 (5.8) | 2 (1.7) | | | 1 (1.1) | 5 (5.7) | 5 (5.7) |
| *Have you tried to lose weight in the last year?* | | | | >0.05 | >0.05 | | | |
| -never | 74 (61.2) | 73 (60.3) | 80 (66.1) | | | 55 (62.5) | 50 (56.8) | 43 (48.9) |
| -sometimes | 36 (29.8) | 34 (28.1) | 28 (23.1) | | | 26 (29.5) | 25 (28.4) | 34 (38.6) |
| -often | 7 (5.8) | 8 (6.6) | 12 (9.9) | | | 5 (5.7) | 11 (12.5) | 11 (12.5) |
| -always | 4 (3.3) | 6 (5.0) | 1 (0.8) | | | 2 (2.3) | 2 (2.3) | 0 (0) |

The perception of weight status did not show significant changes in the three surveys or comparisons between genders in each survey ($p > 0.05$).

Considering the answer to the question about participant weight control, most boys and girls in sixth and seventh grades said they had tried to control their weight "sometimes".

Fifty percent of eighth-grade girls answered "never". However, more than 20% of girls of the same age said they did it often or always. Notably, this percentage had doubled or tripled from previous classes. Thus, significant changes were observed in females over the three years compared to males. Gender comparison was found to have significant differences only in T2 ($p < 0.05$).

Through the other question, no significant change was observed over the three years regarding weight loss attempts in the two genders. We found that girls tried to lose weight more than boys, but the differences between the genders were close to statistical significance only at T2 ($0.10 < p > 0.05$). Among participants who reported trying to lose weight, most of them took action by both reducing food and increasing physical activity (Males: 40.4% in T0, 45.8% in T1, and 43.9% in T2; Females: 36.4% in T0, 39.4% in T1, and 42.2% in T2); a part of them merely reduced food (Males: 8.5% in T0, 18.7% in T1, and 22% in T2; Females: 27.3% in T0, 30.3% in T1, and 28.9% in T2) and another increased physical activity (Males: 29.8% in T0, 22.9% in T1, and 24.4% in T2; Females: 27.3% in T0, 39.4% in T1, and 17.8% in T2).

### 3.2. *BI Perception and Weight Status*

Table 3 compares the relevant BI perception variables in T0, T1, and T2 according to the BMI category in boys and girls. All the BI variables were significantly different among weight status categories except for the ideal figure in males (in T0, T1, and T2) and females (in T0 and T1). The mean feel figure and, to a lesser extent, the ideal figure tended to increase with BMI in both genders over the three times considered. The same trend was observed for the FID. Negative FID values characterized underweight male participants (only in T1 for underweight females), indicating their propensity toward more robust ideal figures than those with whom they identified (feel figures). In contrast, the overweight/obese participants showed positive FID values, demonstrating their leaner ideals compared to the feel figures that characterized them.

Finally, we calculated Cohen's kappa coefficient to assess the agreement between the current weight status assessed by BMI (through direct anthropometric measurements) and the perceived weight status of the participant (through the answer to the question: *Which weight status do you think best represents you?*) (Table 4).

**Table 3.** BI perceptions by BMI categories in boys and girls.

| **BI Variables** | **BMI Categories** | | | | | | | | |
|---|---|---|---|---|---|---|---|---|---|
| | **T0** | | | **T1** | | | **T2** | | |
| ***Boys*** | UW *Mean (SD)* | NW *Mean (SD)* | OW/O *Mean (SD)* | UW *Mean (SD)* | NW *Mean (SD)* | OW/O *Mean (SD)* | UW *Mean (SD)* | NW *Mean (SD)* | OW/O *Mean (SD)* |
| Feel Figure | 2.75 (0.89) | 4.16 (0.99) | 5.76 (0.95) | 3.00 (1.12) | 4.22 (1.02) | 5.66 (81.15) | 3.30 (0.95) | 4.39 (0.90) | 5.85 (1.14) |
| *Kruskal–Wallis* | *H* 54.81 | | *p* <0.0001 | *H* 41.45 | | *p* <0.0001 | *H* 47.45 | | *p* <0.0001 |
| Ideal Figure | 3.75 (1.04) | 3.82 (0.76) | 4.11 (1.07) | 3.78 (0.83) | 3.90 (0.92) | 3.89 (1.01) | 3.90 (0.74) | 4.11 (0.64) | 4.23 (0.74) |
| *Kruskal–Wallis* | *H* 2.91 | | *p* 0.23 | *H* 0.41 | | *p* 0.81 | *H* 2.48 | | *p* 0.29 |
| FID (score) | −1.00 (1.20) | 0.34 (0.90) | 1.65 (1.18) | −0.78 (0.97) | 0.32 (0.93) | 1.72 (1.30) | −0.60 (0.70) | 0.28 (0.88) | 1.62 (1.23) |
| *Kruskal–Wallis* | *H* 42.00 | | *p* <0.0001 | *H* 44.35 | | *p* <0.0001 | *H* 45.61 | | *p* <0.0001 |

**Table 3.** *Cont.*

| BI Variables | BMI Categories | | | | | | | | |
|---|---|---|---|---|---|---|---|---|---|
| | T0 | | | T1 | | | T2 | | |
| *Girls* | UW *Mean (SD)* | NW *Mean (SD)* | OW/O *Mean (SD)* | UW *Mean (SD)* | NW *Mean (SD)* | OW/O *Mean (SD)* | UW *Mean (SD)* | NW *Mean (SD)* | OW/O *Mean (SD)* |
| Feel Figure | 3.60 (1.82) | 3.65 (0.82) | 5.04 (1.23) | 2.00 (0.00) | 3.72 (1.01) | 5.35 (0.98) | 3.00 (1.41) | 3.82 (0.90) | 5.30 (1.21) |
| *Kruskal–Wallis* | *H* 22.76 | | *p* <0.0001 | *H* 37.46 | | *p* <0.0001 | *H* 27.35 | | *p* <0.0001 |
| Ideal Figure | 3.60 (1.34) | 3.29 (0.66) | 3.21 (0.92) | 2.50 (0.71) | 3.40 (0.79) | 3.54 (0.90) | 2.50 (0.71) | 3.34 (0.75) | 3.73 (0.98) |
| *Kruskal–Wallis* | *H* 0.31 | | *p* 0.86 | *H* 3.26 | | *p* 0.20 | *H* 6.94 | | *p* 0.03 |
| FID (score) | 0.00 (1.58) | 0.36 (0.73) | 1.82 (1.25) | −0.50 (0.71) | 0.32 (0.79) | 1.81 (1.33) | 0.50 (2.12) | 0.48 (0.83) | 1.57 (1.30) |
| *Kruskal–Wallis* | *H* 32.03 | | *p* <0.0001 | *H* 31.41 | | *p* <0.0001 | *H* 18.57 | | *p* 0.0001 |

**Table 4.** Consistency between current weight status and self-perceived weight status.

| Current Weight Status | Self-Perceived Weight Status | | | | | | | | |
|---|---|---|---|---|---|---|---|---|---|
| | T0 | | | T1 | | | T2 | | |
| **Boys** | UW | NW | OW/O | UW | NW | OW/O | UW | NW | OW/O |
| UW | 3 | 3 | 2 | 5 | 4 | 1 | 5 | 4 | 1 |
| NW | 7 | 62 | 7 | 7 | 57 | 8 | 7 | 61 | 4 |
| OW/O | 0 | 15 | 22 | 1 | 17 | 21 | 1 | 16 | 22 |
| *Kappa* | 0.439 | | | 0.405 | | | 0.476 | | |
| **Girls** | UW | NW | OW/O | UW | NW | OW/O | UW | NW | OW/O |
| UW | 2 | 1 | 2 | 1 | 1 | 0 | 1 | 1 | 1 |
| NW | 5 | 46 | 4 | 5 | 50 | 6 | 4 | 47 | 5 |
| OW/O | 0 | 13 | 15 | 1 | 9 | 15 | 0 | 10 | 19 |
| *Kappa* | 0.424 | | | 0.454 | | | 0.508 | | |

Based on Cohen's kappa, agreement strength appeared weak-moderate in both genders and for all repetitions [38,39]. The coefficient trend over the three years indicates a gradually increasing awareness of one's weight status as age increases, particularly in girls. It is also evident that there was a tendency for underweight participants of both genders to overestimate their weight status and underestimate it in overweight/obese participants.

## 4. Discussion

The adolescent stage constitutes, on the one hand, a period of intense growth and remarkable morpho-metric changes with the appearance of secondary sexual characters and, on the other hand, a sensitive period for mental health because of the need to adapt to a new appearance of one's body [40]. This is compounded by the beauty ideals imposed especially by the Western world, also due to social media, that exalt muscular development in males and an ectomorphic, thin body in females [5,6]. When adolescents perceive that their bodies differ from these ideals, dissatisfaction with their body image is often determined. In particular, psychological factors and pressures from peers, family, and the media to conform to these socially prescribed body ideals contribute to the onset and persistence of body dissatisfaction, explaining the associations between BI and weight status [41].

This longitudinal study examined anthropometric and body image perception changes in early adolescents over three years to assess any trends over time and any gender differences.

As expected, significant changes were observed in most anthropometric traits and indices with repeated measurements over time, as well as sex and the interaction between these two factors. Regarding weight status, we found no significant differences in category frequencies over the three years, nor were there any significant differences between sexes. Participants also had a clear awareness of their size (stature and weight): the reliability of the reported values most likely depends on whether participants remember the measures routinely taken by the family doctor. Moreover, slightly lower self-reported values were probably due to measurements taken previously in growing subjects during medical checkups.

Comparing the weight status of the examined sample with data from the recent literature for Italian adolescents, we found a lower percentage of normal-weight boys and, especially, girls compared to data collected longitudinally in the Emilia-Romagna region five or more years earlier ([23]: 11 years of age: 68.8% in males and 82.4% in females; 12 years of age: 76.2% in males and 85.7% in females; 13 years of age: 77.8% in males and 82.9% in females) and the national average data referred to the 11 and 13 years of age and collected cross-sectionally two years earlier ([42]: 11 years of age: 71.6% in males and 80.3% in females; 13 years of age: 73.1% in males and 83.1% in females). Also noteworthy is that there was twice the frequency of underweight in the male subsample examined and similar frequencies in the female subsample compared with national data ([42]: 11 years of age: 3.3% in males and 4.3% in females; 13 years of age: 1.9% in males and 2.5% in females); in the regional study [23], underweight was present with very low frequencies in 11- and 12-year-old girls (11 years of age: 1.5%; 12 years: 1.5%) while no underweight was found among boys and 13-year-old girls. The overweight/obese participants of our sample also showed remarkable differences with generally higher frequencies, particularly in girls (>10%), compared with regional data ([23]: 11 years of age: 31.3% in males and 16.2% in females; 12 years of age: 23.8% in males and 12.8% in females; 13 years of age: 22.2% in males and 17.1% in females) and national data ([42]: 11 years of age: 25.1% in males and 15.4% in females; 13 years of age: 25.0% in males and 14.4% in females). In summary, the extreme categories (underweight, overweight/obese) generally show higher frequencies than other Italian samples previously examined. A possible explanation for this pattern, in addition to the observed worldwide trend in the prevalence of obesity and overweight in children and adolescents [43], is that the sample we examined was adversely affected by the COVID-19 outbreak and the virus containment measures, foremost of which was the lockdown with the interruption of school activities in attendance, sports, and social activities. This situation is recognized in the literature as being generally responsible for the rise in BMI [44–46] due to the intake of unhealthy foods, lifestyle changes, and poor engagement in physical activities [47]. The negative influences of lockdown on eating behaviors have been shown in the Italian population [47] and other populations around the world [48,49]. While this hypothesis is acceptable to justify the origin of the phenomenon, it is interesting to note that the higher category of weight status (examined longitudinally) remained constant over time also in the post-COVID period (the state of emergency ended on 31 March 2022) [50], as comparisons in later surveys of our study show. In this regard, we cannot rule out the possibility that the bad habits acquired during the COVID isolation have persisted. Moreover, in contrast to the increased prevalence of overweight/obese described in several populations as an effect of the pandemic, we also found an increased prevalence of underweight in the examined sample of adolescents. This trend, probably attributable to the negative impact of the restrictions on mental health and possibly inadequate nutrition, was also found in other recent European studies on Polish adolescents aged 11–15 years [51] and German primary school children [52].

In this situation, observing any trends in the perception of BI may be particularly interesting. Among the quantitative variables related to BI perception, feel figures and FID in both genders and the ideal figures in females, on the other hand, remained stable over time. Meanwhile, the ideal figures varied significantly over time in boys: the progressive preference for a more robust body ideal is due to the likely misconception that the selected ideal figures correspond to greater muscularity [41,53]. The ideal figures chosen by girls were significantly thinner than in boys at all ages, and there was a tendency for a higher FID score in girls than in boys. This trend corresponds to what has been reported in the literature for adults regarding the greater BI dissatisfaction of women, the thinner BI ideal in females, and the more muscular BI ideal in males [54–57]. In particular, under the influence of Western media, the ideal male body is generally presented as muscular, athletic, and lean [5,56] and the female body thin, tall, and ectomorphic [5,56,58]. The longitudinal study allows us to appreciate how these trends are nuanced during early adolescence and evolve slowly over the period under consideration. Although the degree of BI dissatisfaction (FID) was higher in girls than in boys approaching +1 (the dissatisfaction threshold according to Mendo–Lazaro, 17 [59]), this did not reach statistical significance in the gender comparisons in the likeness of what was found in a cross-sectional study of a sample of Italian children aged 5 to 12 years examined in 2022–2023 [25]. In early adolescents, as observed in preadolescents [59], dissatisfaction with BI is similar in both genders, and the typical gender-differentiated pattern evidenced at later ages is not clearly visible [5] also if we can already distinguish the predominant ideal gender aesthetics.

According to the answers provided to the questions presented to the participants, in addition to a decline in weight control over the three years (significantly in females), the results show that most participants never attempted to lose weight, keeping this trend stable over the three years, consistent with their young age. Among those who made attempt to reduce body weight, the most common method was simultaneous diet control and exercise, followed by food reduction alone and, ultimately, by increased physical activity alone.

As previously reported, an analysis of the perceived vs. ideal figure shows that both genders prefer a leaner body. This trend is even more evident when considering the sample divided into weight status categories: FID increases with BMI. In particular, moving from underweight to overweight, boys' dissatisfaction scores rose to a greater extent but decreased with age (T0: +2.65; T1: +2.50; T2: +2.22) compared to those of girls (T0: +1.82; T1: +2.31; T2: +1.07). This analysis also reveals negative scores across the three surveys in underweight males who were found to be dissatisfied with their status and with a desire for more robustness. In contrast, scores varied slightly between complete satisfaction and mild dissatisfaction at the three times in underweight girls. In this regard, it should be pointed out that overweight/obese adolescents are often the object of teasing by their peers [60]. According to recent research conducted in Italy [61], these individuals showed not only the highest ratings of verbal victimization related directly to body shape but they were frequently subjected to being physically attacked and barred from playing sports and participating in group activities. Social relationships are generally more problematic and worse as weight status (actual or self-rated) increases.

This study demonstrates the need to use directly measured anthropometric measures because self-referred ones are not adequately reliable (especially weight). Moreover, the consistency between the weight status perceived and the actual one established by BMI was low: we observed a weak-moderate agreement between current weight status and self-perceived weight status, which means that the reliable data in the examined sample varied from 15% to 63% [39]. These interrater reliability levels can be considered low in terms of the possible need for health care. In particular, the health risks of some participants in the extreme weight categories should be emphasized: a few underweight boys and girls perceived themselves to be overweight/obese, and a few overweight/obese boys and girls perceived themselves to be underweight. These misperceptions can result, if left untreated, in BI or eating disorders [59]. Moreover, about half of the overweight/obese participants perceived their weight status as normal during the three surveys: underestimating one's weight status results in the continuation of any obesogenic behaviors, with the consequent risk of increasing overweight/obesity status [56,62]. School and local health services policy should safeguard adolescents' physical and mental health by involving families in an integrated effort aimed at a healthy lifestyle. This will also be effective against various chronic health and social problems, including peer bullying victimization.

Our study had various strengths. The most relevant of these strengths concerns the study's longitudinal design: we examined the same sample of early adolescents over three years. The second strength concerns the direct detection of anthropometric traits involved in the BMI calculation by the same expert operator. The third point concerns the face-to-face interaction between the adolescent and the operator in detecting the participants' BI perception and awareness of body size. This allowed participants to focus on the questions without fear of being judged by peers. As for the study's weaknesses, we did not examine the participants' eating and living habits or the families' socioeconomic conditions, although these may affect their nutritional status [63], nor did we assess their ethnic origins [64]. Moreover, a comparison with other studies is difficult because of the small number of research with a longitudinal design in early adolescents during and after COVID periods and because of the use of different possible methods of BI assessment (different silhouette scales, questionnaires). Finally, the standardized silhouette scale of Childress et al. (1993) [33] that we applied did not allow for the assessment of ideal muscularity in males. To the best of our knowledge, however, there are currently no scales for assessing muscularity in early adolescents, unlike adults [57,65].

We expect that further research can provide a longer perspective on longitudinal variability in weight status and BI perception in cohorts of adolescents who have experienced the effects of lockdown from COVID-19 in different parts of the world. Research would also be desirable to arrive at silhouette scales that distinguish between muscularity and fatness for a better interpretation of adolescent BI.

## 5. Conclusions

Our study demonstrates the stability of early adolescents' BI perception: only the body ideal changes over time in boys toward greater robustness. In females, a tendency toward greater dissatisfaction and a preference for leaner body ideals are already present at this young age. However, while conforming to gender ideals, preadolescent males and females show similar levels of BI dissatisfaction, although this has been shown to increase with BMI.

Concerning the weight status of the examined sample, we found an increase in the prevalence of both overweight/obesity and underweight compared with previous studies. These weight patterns can be attributed to lingering consequences of psychological and living conditions established during restrictions by the COVID-19 pandemic. The self-perception of weight status is not entirely satisfactory, showing a tendency to overestimation in underweight adolescents and underestimation in overweight/obese adolescents.

We expect that, especially in the school setting, actions will be taken to promote a healthy lifestyle and BI. In particular, it would be necessary to implement educational efforts at the school level to equip students, teachers, and parents with valuable tools such as guidelines for healthy living (from nutrition to physical activity) to control unhealthy weight status, negative BI, and their consequences. At the same time, early detection of BI perception issues should be encouraged as it can be critical in identifying at-risk adolescents.

**Author Contributions:** Conceptualization, E.G.-R. and L.Z.; methodology, E.G.-R. and L.Z.; software, S.T. and S.M. (Sabrina Masotti); formal analysis, E.G.-R., L.Z. and S.M. (Sabrina Masotti); investigation, L.Z., N.R., F.D.L. and S.M. (Sabrina Masotti); resources, G.M., S.M. (Simona Mandini), and S.T.; data curation, F.D.L., N.R. and S.T.; writing—original draft preparation, E.G.-R.; writing—review and editing, E.G.-R., L.Z., S.M. (Sabrina Masotti), N.R., F.D.L., S.T., G.M. and S.M. (Simona Mandini); visualization, G.M. and S.M. (Simona Mandini); supervision, G.M.; project administration, L.Z. and S.M. (Simona Mandini). All authors have read and agreed to the published version of the manuscript.

**Funding:** This research received no external funding.

**Institutional Review Board Statement:** The study was conducted in accordance with the Declaration of Helsinki and approved by the Ethics Committee of the Local Bioethics Committee of the University (Ethical Approval Ref. no. 2.18, Bologna, 11 July 2019).

**Informed Consent Statement:** Informed consent was obtained from all subjects involved in the study. The written consent of the parents or guardians was obtained for all participants.

**Data Availability Statement:** Data are available upon request due to ethical restrictions regarding participants' privacy. Requests for the data may be sent to the corresponding author.

**Acknowledgments:** The authors would like to thank all study participants and school team members for their time and willingness to collaborate on this project.

**Conflicts of Interest:** The authors declare no conflicts of interest.

## References

1. Hosseini, S.A.; Padhy, R.K. Body Image Distortion. In *StatPearls [Internet]*; StatPearls Publishing: Treasure Island, FL, USA, 2024.
2. Thompson, J.K.; Altabe, M.N. Psychometric qualities of the Figure Rating Scale. *Int. J. Eat. Disorder.* **1991**, *10*, 615–619. [CrossRef]
3. Maruf, F.A.; Akinpelu, A.O.; Nwankwo, M.J. Perceived body image and weight: Discrepancies and gender differences among University undergraduates. *Afr. Health Sci.* **2012**, *12*, 464–472. [CrossRef] [PubMed]
4. Grogan, S. *Body Image: Understanding Body Dissatisfaction in Men, Women, and Children*; Routledge: Oxfordshire, UK, 2021. [CrossRef]
5. Gualdi-Russo, E.; Rinaldo, N.; Masotti, S.; Bramanti, B.; Zaccagni, L. Sex Differences in Body Image Perception and Ideals: Analysis of Possible Determinants. *Int. J. Environ. Res. Public Health* **2022**, *19*, 2745. [CrossRef]
6. Suhag, K.; Rauniyar, S. Social Media Effects Regarding Eating Disorders and Body Image in Young Adolescents. *Cureus* **2024**, *16*, e58674. [CrossRef] [PubMed]
7. Cushman, G.K.; Rea, K.E.; Westbrook, A.L.; Alexander, C.; Reed, B. Body image dissatisfaction during the first year of pediatric inflammatory bowel disease diagnosis. *J. Pediatr. Gastroenterol. Nutr.* **2024**, *78*, 614–622. [CrossRef] [PubMed]
8. Fernández, P.; Labra, J.A.; Méndez, M.T.; González, C.; Coballes, S.; Souto-Gestal, A. The Moderating Effect of Sex and Age on the Pattern of Body Image by Pre-Adolescents and Adolescents and Its Relationship with the Time They Spend Doing Sports. *Sustainability* **2023**, *15*, 10061. [CrossRef]
9. Jiménez, R.; Arriscado, D.; Dalmau, J.M.; Gargallo, E. Determinantes de la satisfacción corporal en adolescentes de la Rioja. *Rev. Esp. Salud. Pública* **2021**, *95*, e1–e12.
10. Gualdi-Russo, E. Body image perception in ethnic minorities: Implications for prevention. In *Ethnic Diversities, Hypertension and Global Cardiovascular Risk*; Modesti, P.A., Cappuccio, F.P., Parati, G., Eds.; Springer International Publishing AG: Cham, Switzerland, 2018.
11. Health and Social Care Committee. The Impact of Body Image on Mental and Physical Health. Second Report of Session 2022–2023. 2 August 2022. Available online: https://publications.parliament.uk/pa/cm5803/cmselect/cmhealth/114/report.html (accessed on 1 October 2024).
12. Tanner, J.M. *Auxologia dal Feto all'Uomo: La Crescita Fisica dal Concepimento Alla Maturità*; UTET: Torino, Italy, 1981.
13. Malina, R.; Bouchard, C.; Bar-Or, O. *Growth, Maturation, and Physical Activity*; Human Kinetics: Champaign, IL, USA, 2004.
14. Hoyt, L.T.; Niu, L.; Pachucki, M.C.; Chaku, N. Timing of puberty in boys and girls: Implications for population health. *SSM Popul. Health* **2020**, *10*, 100549. [CrossRef]
15. Lacroix, E.; Smith, A.J.; Husain, I.A.; Orth, U.; von Ranson, K.M. Normative body image development: A longitudinal meta-analysis of mean-level change. *Body Image* **2023**, *45*, 238–264. [CrossRef]
16. Compian, L.J.; Gowen, L.K.; Hayward, C. The interactive effects of puberty and peer victimization on weight concerns and depression symptoms among early adolescent girls. *J. Early Adolesc.* **2009**, *29*, 357–375. [CrossRef]
17. Foster, L.; Lundh, L.G.; Daukantaité, D. Disordered eating in a 10-year perspective from adolescence to young adulthood: Stability, change, and body dissatisfaction as a predictor. *Scand. J. Psychol.* **2024**, *65*, 32–41. [CrossRef] [PubMed]
18. Galmiche, M.; Dechelotte, P.; Lambert, G.; Tavolacci, M.P. Prevalence of eating disorders over the 2000-2018 period: A systematic literature review. *Am. J. Clin. Nutr.* **2019**, *109*, 1402–1413. [CrossRef] [PubMed]
19. D'Anna, G.; Lazzeretti, M.; Castellini, G.; Ricca, V.; Cassioli, E.; Rossi, E.; Silvestri, C.; Voller, F. Risk of eating disorders in a representative sample of Italian adolescents: Prevalence and association with self-reported interpersonal factors. *Eat. Weight. Disord.* **2022**, *27*, 701–708. [CrossRef] [PubMed]
20. D'Anna, G.; Lucherini Angeletti, L.; Benvenuti, F.; Melani, G.; Ferroli, M.; Poli, F.; Villano, R.G.; Ricca, V.; Rotella, F. The association between sport type and eating/body image concerns in high school students: A cross-sectional observational study. *Eat. Weight. Disord.* **2023**, *28*, 43. [CrossRef] [PubMed]
21. de Guzman, N.S.; Nishina, A. A longitudinal study of body dissatisfaction and pubertal timing in an ethnically diverse adolescent sample. *Body Image* **2014**, *11*, 68–71. [CrossRef]
22. Bucchianeri, M.M.; Arikian, A.J.; Hannan, P.J.; Eisenberg, M.E.; Neumark-Sztainer, D. Body dissatisfaction from adolescence to young adulthood: Findings from a 10-year longitudinal study. *Body Image* **2013**, *10*, 1–7. [CrossRef]

23. Toselli, S.; Grigoletto, A.; Zaccagni, L.; Rinaldo, N.; Badicu, G.; Grosz, W.R.; Campa, F. Body image perception and body composition in early adolescents: A longitudinal study of an Italian cohort. *BMC Public Health* **2021**, *21*, 1381. [CrossRef]
24. Gualdi-Russo, E.; Rinaldo, N.; Mazzoni, G.; Mandini, S.; Masotti, S.; Toselli, S.; Zaccagni, L. Age at Menarche, Growth Velocity, and Adiposity Indices in Italian Girls Aged 10 to 14. *Children* **2022**, *9*, 1928. [CrossRef]
25. Zaccagni, L.; Rosa, L.; Toselli, S.; Gualdi-Russo, E. Sports Practice, Body Image Perception, and Factors Involved in Sporting Activity in Italian Schoolchildren. *Children* **2023**, *10*, 1850. [CrossRef]
26. Toselli, S.; Grigoletto, A.; Zaccagni, L.; Gualdi-Russo, E.; Rinaldo, N. A Pilot Study on the Association of Internet Use with Sports Practice and Sex in Italian Adolescents. *Healthcare* **2023**, *11*, 3075. [CrossRef]
27. Zaccagni, L.; De Luca, F.; Rinaldo, N.; Mazzoni, G.; Mandini, S.; Gualdi-Russo, E. The COVID-19 Experience in Adolescents: Emotional and Behavioral Recall at the End of the Pandemic. *Diseases* **2024**, *12*, 116. [CrossRef] [PubMed]
28. Cole, T.J.; Bellizzi, M.C.; Flegal, K.M.; Dietz, W.H. Establishing a standard definition for child overweight and obesity world-wide: International survey. *Br. Med. J.* **2000**, *320*, 1240–1243. [CrossRef] [PubMed]
29. Cole, T.J.; Flegal, K.M.; Nicholls, D.; Jackson, A.A. Body mass index cut offs to define thinness in children and adolescents: International survey. *Br. Med. J.* **2007**, *335*, 194–197. [CrossRef] [PubMed]
30. Lohman, T.G.; Roche, A.F.; Martorell, R. *Anthropometric Standardization Reference Manual*; Human Kinetics Books: Champaign, IL, USA, 1988.
31. Weiner, J.S.; Lourie, J.A. *Practical Human Biology*; Academic Press: Cambridge, MA, USA, 1981.
32. Gualdi Russo, E. Metodi Antropometrici: Generalità e principali caratteri antropometrici. In *Manuale di Antropologia. Evoluzione e Biodiversità Umana*; Sineo, L., Moggi Cecchi, J., Eds.; UTET: Torino, Italy, 2022; pp. 55–75.
33. Childress, A.C.; Brewerton, T.D.; Hodges, E.L.; Jarrel, M.P. The Kids' Eating Disorders Survey (KEDS): A Study of Middle School Students. *J. Am. Acad. Child. Adolesc. Psychiatry* **1993**, *32*, 843–850. [CrossRef] [PubMed]
34. Stunkard, A.J.; Sørensen, T.; Schulsinger, F. Use of the Danish Adoption Register for the study of obesity and thinness. *Res. Publ. Assoc. Res. Nerv. Ment. Dis.* **1983**, *60*, 115–120.
35. Mciza, Z.; Goedecke, J.H.; Steyn, N.P.; Charlton, K.; Puoane, T.; Meltzer, S.; Levitt, N.S.; Lambert, E.V. Development and validation of instruments measuring body image and body weight dissatisfaction in South African mothers and their daughters. *Public Health Nutr.* **2005**, *8*, 509–519. [CrossRef]
36. McHiza, Z.J.; Goedecke, J.H.; Lambert, E.V. Intra-familial and ethnic effects on attitudinal and perceptual body image: A cohort of South African mother-daughter dyads. *BMC Public Health* **2011**, *11*, 433. [CrossRef]
37. Muehlenhard, C.L.; Peterson, Z.D. Distinguishing Between Sex and Gender: History, Current Conceptualizations, and Implications. *Sex Roles* **2011**, *64*, 791–803. [CrossRef]
38. Landis, J.R.; Koch, G.G. The measurement of observer agreement for categorical data. *Biometrics* **1977**, *33*, 159–174. [CrossRef]
39. McHugh, M.L. Interrater reliability: The kappa statistic. *Biochem. Med.* **2012**, *22*, 276–282. [CrossRef] [PubMed]
40. Ramos, P.; Moreno-Maldonado, C.; Moreno, C.; Rivera, F. The Role of Body Image in Internalizing Mental Health Problems in Spanish Adolescents: An Analysis According to Sex, Age, and Socioeconomic Status. *Front. Psychol.* **2019**, *10*, 1952. [CrossRef] [PubMed]
41. Gualdi-Russo, E.; Rinaldo, N.; Zaccagni, L. Physical Activity and Body Image Perception in Adolescents: A Systematic Review. *IJERPH* **2022**, *19*, 13190. [CrossRef] [PubMed]
42. Health Behaviour in School-Aged Children (HBSC)/World Health Organization Collaborative Cross-National Survey. Alimentazione e Stato Nutrizionale. Available online: https://www.iss.it/documents/20126/0/HBSC.pdf/97b1cfee-444c-bfd4-ab58-d0b1dcb504fb?t=1603882812589 (accessed on 22 July 2024).
43. NCD Risk Factor Collaboration (NCD-RisC). Worldwide trends in underweight and obesity from 1990 to 2022: A pooled analysis of 3663 population-representative studies with 222 million children, adolescents, and adults. *Lancet* **2024**, *403*, 1027–1050. [CrossRef]
44. Salman, H.; Issi Irlayici, F.; Akçam, M. The COVID-19 Pandemic and the Prevalence of Childhood Obesity in Turkiye. *J. Pediatr. Gastroenterol. Nutr.* **2023**, *76*, e15–e19. [CrossRef] [PubMed]
45. Anderson, L.N.; Yoshida-Montezuma, Y.; Dewart, N.; Jalil, E.; Khattar, J.; De Rubeis, V.; Carsley, S.; Griffith, L.E.; Mbuagbaw, L. Obesity and weight change during the COVID-19 pandemic in children and adults: A systematic review and meta-analysis. *Obes. Rev.* **2023**, *24*, e13550. [CrossRef] [PubMed]
46. Li, H.; Song, Y.; Wang, Y.; Feng, X.; Li, C.; Peng, J.; Yu, H. Impact of the COVID-19 pandemic lockdown on Body Mass Index: A three-year follow up study in 6,156 Chinese college students. *Front. Endocrinol.* **2024**, *15*, 1387151. [CrossRef]
47. Cecchetto, C.; Aiello, M.; Gentili, C.; Ionta, S.; Osimo, S.A. Increased emotional eating during COVID-19 associated with lockdown, psychological and social distress. *Appetite* **2021**, *160*, 105122. [CrossRef]
48. Robinson, E.; Boyland, E.; Chisholm, A.; Harrold, J.; Maloney, N.G.; Marty, L.; Mead, B.R.; Noonan, R.; Hardman, C.A. Obesity, eating behavior and physical activity during COVID-19 lockdown: A study of UK adults. *Appetite* **2021**, *156*, 104853. [CrossRef]
49. Mignogna, C.; Costanzo, S.; Ghulam, A.; Cerletti, C.; Donati, M.B.; de Gaetano, G.; Iacoviello, L.; Bonaccio, M. Impact of Nationwide Lockdowns Resulting from the First Wave of the COVID-19 Pandemic on Food Intake, Eating Behaviors, and Diet Quality: A Systematic Review. *Adv. Nutr.* **2022**, *13*, 388–423. [CrossRef]

50. Italian Government Official Gazette of the Italian Republic. Decree of the President of the Council of Ministers (DPCM). Italy. 9 March 2020. Available online: https://www.gazzettaufficiale.it/eli/id/2020/03/09/20A01558/sg (accessed on 17 September 2024).
51. Artymiak, P.; Żegleń, M.; Kryst, Ł. Changes in BMI and the prevalence of overweight and obesity in children and adolescents (11–15 years) from Kraków (Poland) during COVID-19 pandemic. *Anthr. Anthropol. Anz.* **2024**. [CrossRef]
52. Samigullin, A.; Groß, G.; Gählert, J.; Buchenberger, S.; Morcos, M.; Schwertz, R.; Öste, R.; Siegel, E.; Humpert, P. Impact of Covid-19 lockdowns on the anthropometric development in primary school children in the Rhein-Neckar Region, Germany. *BMC Nutr.* **2024**, *10*, 78. [CrossRef] [PubMed]
53. Jones, D.C.; Crawford, J.K. Adolescent boys and body image: Weight and muscularity concerns as dual pathways to body dissatisfaction. *J. Youth Adolesc.* **2005**, *34*, 629–636. [CrossRef]
54. Brennan, M.A.; Lalonde, C.E.; Bain, J.L. Body image Perceptions: Do Gender Differences Exist? *Psi Chi J. Undergrad. Res.* **2010**, *15*, 1089–4136. [CrossRef]
55. Zaccagni, L.; Masotti, S.; Donati, R.; Mazzoni, G.; Gualdi-Russo, E. Body image and weight perceptions in relation to actual measurements by means of a new index and level of physical activity in Italian university students. *J. Transl. Med.* **2014**, *12*, 42. [CrossRef] [PubMed]
56. Song, H.; Cai, Y.; Cai, Q.; Luo, W.; Jiao, X.; Jiang, T.; Sun, Y.; Liao, Y. Body Image Perception and Satisfaction of Junior High School Students: Analysis of Possible Determinants. *Children* **2023**, *10*, 1060. [CrossRef]
57. Porras-Garcia, B.; Exposito-Sanz, E.; Ferrer-Garcia, M.; Castillero-Mimenza, O.; Gutiérrez-Maldonado, J. Body-Related Attentional Bias among Men with High and Low Muscularity Dissatisfaction. *J. Clin. Med.* **2020**, *9*, 1736. [CrossRef]
58. McComb, S.E.; Mills, J.S. Eating and body image characteristics of those who aspire to the slim-thick, thin, or fit ideal and their impact on state body image. *Body Image* **2022**, *42*, 375–384. [CrossRef]
59. Mendo-Lázaro, S.; Polo-Del-Río, M.I.; Amado-Alonso, D.; Iglesias-Gallego, D.; León-Del-Barco, B. Self-Concept in Childhood: The Role of Body Image and Sport Practice. *Front. Psychol.* **2017**, *8*, 853. [CrossRef]
60. Lian, Q.; Su, Q.; Li, R.; Elgar, F.J.; Liu, Z.; Zheng, D. The association between chronic bullying victimization with weight status and body self-image: A cross-national study in 39 countries. *PeerJ* **2018**, *6*, e4330. [CrossRef]
61. Bacchini, D.; Licenziati, M.R.; Garrasi, A.; Corciulo, N.; Driul, D.; Tanas, R.; Fiumani, P.M.; Di Pietro, E.; Pesce, S.; Crinò, A.; et al. Bullying and Victimization in Overweight and Obese Outpatient Children and Adolescents: An Italian Multicentric Study. *PLoS ONE* **2015**, *10*, e0142715. [CrossRef]
62. Lee, J.; Lee, Y. The association of body image distortion with weight control behaviors, diet behaviors, physical activity, sadness, and suicidal ideation among Korean high school students: A cross-sectional study. *BMC Public Health* **2016**, *16*, 39. [CrossRef] [PubMed]
63. Barakat, S.; McLean, S.A.; Bryant, E.; Le, A.; Marks, P.; National Eating Disorder Research Consortium; Touyz, S.; Maguire, S. Risk factors for eating disorders: Findings from a rapid review. *J. Eat. Disord.* **2023**, *11*, 8. [CrossRef] [PubMed]
64. Wildes, J.E.; Emery, R.E.; Simons, A.D. The roles of ethnicity and culture in the development of eating disturbance and body dissatisfaction: A meta-analytic review. *Clin. Psychol. Rev.* **2001**, *21*, 521–551. [CrossRef]
65. Frederick, D.A.; Buchanan, G.M.; Sadehgi-Azar, L.; Peplau, L.A.; Haselton, M.G.; Berezovskaya, A.; Lipinski, R.E. Desiring the muscular ideal: Men's body satisfaction in the United States, Ukraine, and Ghana. *Psychol. Men. Masculinity* **2007**, *8*, 103–117. [CrossRef]

nutrients

MDPI

*Article*

# What Motivates Men to Improve Their Health? Understanding the Roles of Self-Esteem and Influential Others in Behaviour Change

**Lisa S. McNeill**

Otago Business School, University of Otago, P.O. Box 56, Dunedin 9054, New Zealand; lisa.mcneill@otago.ac.nz; Tel.: +64-3-4795758

**Abstract:** The aim of this study was to examine men's body image confidence, social reflectivity, body image perceptions and external information sources influence regarding body assessments. Data were collected via a cross-sectional survey and found that men have a low motivation toward physical health behaviour changes such as food, alcohol and exercise evaluation, and have generally positive views of their bodies overall. Relationship status, rather than age, defined behavioural and attitudinal differences within the men studied here. Men in this study were largely uninfluenced by celebrities or fashion in developing their own body image perceptions; single males were more likely to turn to friends, the female population generally and societal norms when evaluating themselves. Males in relationships however, weighted their partner's opinion as the greatest influence, followed by their involvement in sport. This study offers an insight into the role of body confidence in male perspectives of the self, which is important for its intrinsic connection to motivations for health behaviours such as body weight management. This has implications for increasing the effectiveness of health-related product and service messaging, and public health messages regarding body weight management for men.

**Keywords:** self-esteem; male body image; health behaviours

## 1. Introduction

Self-identity, the way in which one defines oneself, relative to others, has long been examined from the perspective of women. As traditional notions of gender in self-image blur, men have begun to receive attention from researchers concerned with consumption behaviour as an extension of the self [1,2]. Much of this focus has, however, centred on products embraced by male consumers in a subversion of traditionally female categories of consumption (such as fashion and cosmetics). Where gaps still lie is in understanding men's perceptions of themselves, and how notions of self-identity might motivate men toward image management behaviours, including physical health-related behavioural change. There is evidence that men's body image satisfaction is influenced strongly by spouses and similar familial relationships, however. This study addresses those gaps by examining the impact of relationship status on men's body image perceptions and body confidence, as well as the impact of personal, social and external influence sources in body maintenance behaviours.

Now that physical health behaviours are increasingly recognised as identity related, research that examines the relationship between body image, identity and lifestyle is useful in that it focuses on the notion that positive self-perception is fundamentally linked to an individual's satisfaction with their own body image [3–5]. Where prior research has linked negative self-perceptions of women to motivations for physical change, little work has examined men's perceptions of their bodies in relation to self-perception, or motivations for men to engage in physical health-change behaviours. This study examines the notion of male body image satisfaction, the extent to which one is happy with one's body, as it relates to motivation toward identity management behaviours, such as body modification through diet and exercise.

### 1.1. Self Esteem and Body Satisfaction

Satisfaction with one's body is said to be driven by several social factors, including media, society, friends and family, with individuals utilising social comparison theory to compare their body to that of those around them [6]. Regarding health, there is conflicting evidence surrounding the role of body image associations and lifestyles of healthy or unhealthy behaviour. For example, some studies find that positive self-esteem and body image beliefs generally encourage more positive fitness and health-related behaviours (in college age students of both sexes) [7]. Others find that a high body dissatisfaction (in adult women) is associated with stronger fitness and health motivations toward appearance and weight [8]. Some recent research suggests that body dissatisfaction (in adolescents of both sexes) is associated with negative body management behaviour, including unhealthy weight loss strategies, lower physical activity, dysfunctional exercise, and increased computer usage [9]. While prior research has examined women, children and teenagers/adolescents in respect of their body image and self-esteem, little work has examined the adult male (beyond contrasting them to women). Further, there are gaps in understanding differences in body image perceptions and self-esteem between adult male subjects.

### 1.2. Male Body Image and Weight

Where perceptions of body image relate to health management actions, prior research confirms that weight evaluation plays a critical part in the relationship between body image and physical activity [9]. Gillison et al. [10] found that negative body weight perceptions endorsed extrinsic motivations for physical activity (e.g., exercise for weight loss), thus negatively predicting self-determined motivations for exercise behaviour. However, evaluation of physical activity is often absent from research that examines body image [9]. Those studies that do examine physical activity alongside body image report inconsistences between samples. For example, some studies show that participation in physical activity by adolescent girls is not associated with body image satisfaction [11]. Others claim that exercise helps to improve body image in adolescents of all sexes [12].

One explanation for many of the difference observed when examining the impact of physical activity on body image is that body image satisfaction in these contexts is driven by a self-evaluation of body weight, which then raises questions as to the relative accuracy of self-assessments of the body and the resultant satisfaction with the self, between individuals. Regarding males, recent studies in developed countries show that boys and men are more likely to underestimate their body weight [9,13] than women are. However, a larger proportion of men than women are seen as overweight (based on BMI) [13]. Those who overestimate body weight are deemed more likely to engage in weight management behaviours (such as reducing intake of dairy, fats, and oils), yet, male adolescents, who are more likely to underestimate their weight than females, are also more likely to consume higher volumes of foods associated with weight gain [9].

### 1.3. Lifestyle and Male Health Behaviours

The psychological quality of life cannot be separated from physical quality regarding body image [14], thus recent research calls for further study of the relationship between individual lifestyle factors and body image [13]. There is a growing body of research that examines body image and lifestyle influences in childhood and adolescent body maintenance motivation, yet gaps remain for work that focuses on adults [15]. Further, the exact links between body image and many individual lifestyle factors (beyond age) are still to be determined. Of those studies that have focused on adult lifestyles, many have considered body image as a female concern, adding to the misunderstanding of outcomes of male body image perceptions that tends to dominate such work [16]. It is accepted, however, that body image is a multifactorial construct that is affected by a group of psychological, physiological and social factors, each with different levels of influence on health behaviours at different lifestyle stages [13].

### *1.4. Relationships and Body Image Satisfaction*

Self-reporting happiness studies indicate that those with strong connections to others are happier than those without. In addition, particular interpersonal relationships, such as a romantic attachment, account for the greatest difference in satisfaction with the self [17,18]. If the singular lifestyle factor of relationship status is then examined further, it has been found that relationship status relates strongly to health-related quality of life perceptions, with those in relationships less worried about their health overall. Further, men in relationships appear to be the least worried of all groups, with single men and women more likely to be occupied with food control as a health mechanism related to body image perceptions [3].

Some research suggests that men in romantic relationships are said to be happier overall than women in these partnerships, with males more likely to benefit from the emotional support and gratification provided by a partner than vice versa [17]. When the influence of these interpersonal relationships on body image satisfaction is considered, it appears that peer reflexivity impacts the body weight assessments (and subsequent body satisfaction) of women but not of men [19], suggesting a difference in the role of social forces on male assessments of their own body. Indeed, other research suggests that women are more likely than their male counterparts to correctly evaluate their own weight in relation to the general population [20] and that married men are more likely to be 'happily' obese than their never married or previously married counterparts [21], i.e., acknowledging being overweight but not feeling negative or motivated toward change, regarding this. Thus, following calls for further study of adults, men and individual lifestyle factors, this study examines relationship status in adult males, as related to body image satisfaction, by proposing the following:

**H1.** *That males in relationships have more positive body esteem than do single males;*

and

**H2.** *That males in relationships are likely to have more positive body appearance perceptions than do single males.*

Social identity theory notes that self-evaluative or self-conceptual outcomes rely heavily on social context, irrespective of the cognitive categorisation of attributes (in terms of the recognition of normative beliefs or behaviours) [22]. Where body image dissatisfaction in women occurs, it is often in relation to comparison to a standard of social beauty perceived to be above that of the consumer's own [23,24]. Further, females tend to be influenced more by outside factors, such as the media and society more generally, leading to an assessment of body image related to one's 'value' in the social world [4,24]. In contrast, male body image satisfaction, when influenced by others, is said to be evaluated based on the opinions of close friends (in the case of young men) and spouses or family members (in the case of older males) [25].

One could propose that males in relationships are more likely to assess their body image based on the opinions of those closest to them, and less likely to self-categorise based on external sources of body image perceptions. Adding further emphasis to the proposed body satisfaction of men in relationships are findings such as those which state that, after the end of a relationship, males are likely to revise any weight gain and maintain this weight loss (where any female weight loss in the same instance tends to be short-lived) [21]. These findings suggest that, irrespective of actual weight or appearance, men in relationships are generally happy with their body, when compared to women, and, when single, are prompted to revise their body assessment (evidenced by the weight loss seen in newly separated males in the Sobal et al. [15] study). Thus:

**H3.** *That males in relationships are more likely to emphasise personal over social body image perception antecedents than are single males;*

and

**H4.** *That males in relationships are less likely to be impacted by external influence sources than are single males.*

## 2. Materials and Methods

### *2.1. Survey and Scale Items*

The research utilised a quantitative approach and an online survey (hosted via Qualtrics) to collect data. This study used a market research company (Research First) to access respondents, from a total database of 1 million adult contacts. Male, NZ members of the research panel were e-mailed the link to the survey by Research First and asked to complete qualifying questions on their age and sexual orientation. Respondents were recruited in a quota format in four banded age groups (18–24 years; 25–34 years; 35–44 years and 45–65 years). Respondents in this study were limited to heterosexual males, to allow for between-participant comparison, in relation to questions regarding sexual attractiveness and the influence of women, and of romantic partners, on body image perceptions, thus reducing the potential for multiple moderating variable influences. It is reasonable to propose that body image perceptions and goals may differ between sexual orientations, hence the decision to limit the focus of this study to a single sexual orientation. Survey questions examined self-perceptions of the participant's bodies, the influence of others on these perceptions and overall evaluations of personal physical appearance in relation to the body. Participants were also asked a series of behavioural health-related questions, examining participants' purposeful moderation of food, alcohol, fat, snacks/treats and increased exercise or participation in sport.

The scales employed were drawn or developed from existing studies of body and appearance perceptions [24–26]. Personal body image views, as per H1, were examined using a version of Rosenberg's [25] self-esteem scale, modified to the body context ("myself" in the original scale replaced with "my body"). The reliability of this scale has been tested since its inception, with some psychologists criticising its one-dimensional nature [27]. In this study, the scale yielded a Cronbach's alpha of 0.595, which is not ideal. Some prior studies have yielded an alpha coefficient of 0.86 using the scale, indicating good internal reliability; however, these studies have largely been undertaken with small samples (less than one hundred participants) [28]. Nevertheless, the Rosenberg scale has been accepted as an appropriate measure of self-esteem over the past 55 years.

To counter the low alpha of the Rosenberg self–esteem scale, and to examine H2, a shortened version of Netemeyer et al.'s [26] vanity scale was employed, with the 'Physical-Concern' and 'Physical-View' item sets used as a single scale to measure appearance perceptions. The Physical-Concern item set comprises five reflexive, self-determined appearance perceptions (e.g., 'the way I look is extremely important to me') and the Physical-View item set comprises six response-oriented appearance perceptions (e.g., "my looks are very appealing to others"). In this study, the shortened Netemeyer et al.'s [26] scale yielded a Cronbach's alpha of 0.914.

The last two scales employed were developed for this study from McNeill and Firman's [25] qualitative study of ideal body image factors for men. The first of the scales developed for this study, examining H3, consisted of twelve items that examined significant male body image perception factors, including muscularity; leanness; masculinity; age; fitness goals; sexual attractiveness; body satisfaction; friend opinions; family opinions; health; lifestyle; stomach fat; and sporting ability (Cronbach's alpha: 0.891). The second scale developed for this study, concerned with H4, examined external and internal influence factors on men's personal body image perceptions, with ten items, including celebrity influence; family; friends; women (generally); romantic partners; society; fashion; age; health; and sport (Cronbach's alpha: 0.889).

Participants were asked to rate survey items on a seven-point Likert scales as follows: 1 = strongly agree/extremely important; 2 = agree/important; 3 = somewhat agree/somewhat important; 4 = neither agree or disagree/neither important nor unimportant; 5 = somewhat disagree/somewhat unimportant; 6 = disagree/unimportant; and 7 = strongly disagree/not important at all. As the data were collected, responses were checked for completion time and central tendency, with concerning responses excluded. The survey contained two attention check questions, as well as some reverse items, which allowed for further data checking.

### 2.2. Sample

A total of 355 complete responses were collected. Respondent ages were roughly equally split between four age bands (18–24; 25–34; 35–44; 45–65 years) and with the ethnicity of participants representing proportionally the major ethnic groups of New Zealand (53.1% Caucasian). A total of 69.9% of respondents had a secondary school-level qualification or higher. A total of 31.3% of respondents participated in exercise at least weekly, 48.9% monthly or less and 19.8% did not exercise at all. A total of 67.8% of respondents were in a romantic relationship and 32.2% were single. Regarding relationship status and age, 18–24-year-old participants were the least likely to be in a relationship (48% of this group), while approximately one-third of each of the other three age groups of participants were in relationships.

### 2.3. Analysis

Survey responses were exported to IBM Statistical Package for Social Sciences (SPSS) 29 for analysis. To validate the measures, a factor analysis was run on the scales. All scales had Kaiser–Meyer–Olkin (KMO) values of above the threshold of 0.6 and the Eigenvalues indicated there was one dominant factor in all scales. The principal factor analyses further confirmed that all values loaded onto one component, and as such, the scales can be seen to be reliable. Results were obtained by comparing means and via ANOVA.

## 3. Results

### 3.1. Male Health-Related Behaviours

Of the age bands represented, participation in sport was relatively equally represented across all age bands, with no indication that one age group had a significantly greater or lesser participation overall. When age was examined as an initial predictor of body image attitudes (per scale), no significant differences in mean responses between age groups were found. Regarding health-related behaviours, participants did not exhibit a strong motivation toward health behavioural changes, with reductions in food and alcohol or increases in sport and exercise relatively moderate across the group (Table 1). The only significant difference in health-related behaviour was seen in relation to increasing sport participation, where single males were less likely to attempt to increase their sports participation than were those in relationships.

**Table 1.** Health-related behaviour.

| On the Scale Below, Indicate How Regularly You Engage in the Following Activities? | Do You Have a Partner or Significant Other | | | | | |
|---|---|---|---|---|---|---|
| | Yes (n = 240) | | No (n = 355) | | Total (n = 355) | |
| | Mean | SD | Mean | SD | Mean | SD |
| Watch the amount of fat I consume | 3.65 | 1.583 | 3.64 | 1.585 | 3.65 | 1.581 |
| Reduce my alcohol intake | 3.24 | 1.736 | 3.22 | 1.964 | 3.23 | 1.810 |
| Moderate my food intake | 3.52 | 1.360 | 3.55 | 1.416 | 3.53 | 1.376 |
| Cut back on snacks and treats | 3.47 | 1.477 | 3.50 | 1.495 | 3.48 | 1.481 |

**Table 1.** *Cont.*

| On the Scale Below, Indicate How Regularly You Engage in the Following Activities? | Do You Have a Partner or Significant Other | | | | | |
|---|---|---|---|---|---|---|
| | Yes (n = 240) | | No (n = 355) | | Total (n = 355) | |
| | Mean | SD | Mean | SD | Mean | SD |
| Increase my exercise | 3.39 | 1.374 | 3.52 | 1.489 | 3.43 | 1.411 |
| * Attempt to increase my participation in sport | 3.77 | 1.698 | 4.44 | 1.846 | 3.99 | 1.773 |

* Significant at 0.05, 1 = always; 7 = never.

### 3.2. Men's Body Assessments and Self-Esteem

All the men in this study were generally positive about their own body but remained neutral in response to statements indicating extreme negativity, rather than disagreeing outright with statements such as "I am inclined to think that I am a failure in relation to my body" or "I feel I do not have much to be proud of in relation to my body". However, where significant differences in the group means were seen, males in relationships were more positive than single males overall, with an emphasis on affirmative comparison of their body to others and satisfaction with their body generally (refer to Table 2). This supports H1.

**Table 2.** Men's body esteem.

| To What Extent Do You Agree with the Following Statements about Your Body? | Do You Have a Partner or Significant Other | | | | | | $p$-Value |
|---|---|---|---|---|---|---|---|
| | Yes (n = 240) | | No (n = 115) | | Total (n = 355) | | |
| | Mean | SD | Mean | SD | Mean | SD | |
| On the whole, I am satisfied with my body | 3.34 | 1.325 | 3.86 | 1.567 | 3.51 | 1.427 | 0.001 * |
| At times, I think my body is no good at all | 4.02 | 1.574 | 3.87 | 1.565 | 3.97 | 1.570 | 0.410 |
| I feel that my body has a number of good attributes | 2.87 | 1.010 | 3.10 | 1.068 | 2.94 | 1.033 | 0.051 |
| I am able to do things as well as most other people | 2.75 | 1.103 | 3.02 | 1.256 | 2.84 | 1.160 | 0.042 * |
| I feel I do not have much to be proud of in relation to my body | 4.29 | 1.477 | 4.14 | 1.388 | 4.24 | 1.449 | 0.354 |
| I certainly feel useless at times | 4.49 | 1.664 | 4.07 | 1.599 | 4.35 | 1.653 | 0.024 * |
| I feel that I am at least on an equal plane with others | 3.23 | 1.136 | 3.52 | 1.180 | 3.32 | 1.157 | 0.026 * |
| I wish I could have more respect for my body | 3.73 | 1.513 | 3.61 | 1.336 | 3.69 | 1.457 | 0.451 |
| I am inclined to feel that I am a failure in relation to my body | 4.65 | 1.598 | 4.17 | 1.574 | 4.50 | 1.604 | 0.009 * |
| I take a positive attitude towards my body | 3.16 | 1.171 | 3.39 | 1.219 | 3.23 | 1.190 | 0.084 |

* significant at 0.05, 1 = strongly agree; 7 = strongly disagree.

### 3.3. Male Body Confidence

All the men in this study were moderately confident about their own body and remained neutral in response to statements indicating extreme body confidence, rather than disagreeing outright (such as "people are envious of my good looks" and "I have the type of body that people want to look at"). Where significant differences were seen between groups, males in relationships again exhibited greater confidence than did single males, rating their looks and sexual attractiveness higher. This supports H2. Single males, however, were more likely to see a need for a continued effort in looking their best and an awareness of how others might perceive their appearance, indicating the potential for greater reflexivity in their assessments of self (refer Table 3).

**Table 3.** Appearance perceptions.

| To What Extent Do You Agree with the Following Statements about Yourself? | Do You Have a Partner or Significant Other | | | | | | *p*-Value |
|---|---|---|---|---|---|---|---|
| | Yes (n = 240) | | No (n = 115) | | Total (n = 355) | | |
| | Mean | SD | Mean | SD | Mean | SD | |
| The way I look is extremely important to me | 3.37 | 1.437 | 3.16 | 1.328 | 3.30 | 1.405 | 0.187 |
| I am very concerned about my appearance | 3.59 | 1.441 | 3.32 | 1.386 | 3.50 | 1.427 | 0.101 |
| * I would feel embarrassed if I was around people and did not look my best | 3.97 | 1.510 | 3.50 | 1.471 | 3.82 | 1.511 | 0.006 * |
| * Looking my best is worth the effort | 3.43 | 1.327 | 3.12 | 1.278 | 3.33 | 1.318 | 0.037 * |
| It is important that I always look good | 3.68 | 1.372 | 3.53 | 1.410 | 3.63 | 1.384 | 0.331 |
| People notice how attractive I am | 3.99 | 1.366 | 4.07 | 1.497 | 4.02 | 1.408 | 0.626 |
| * My looks are very appealing to others | 3.93 | 1.295 | 4.34 | 1.450 | 4.06 | 1.359 | 0.008 * |
| * People are envious of my good looks | 4.40 | 1.488 | 4.77 | 1.358 | 4.52 | 1.456 | 0.022 * |
| * I am a very good looking individual | 3.89 | 1.360 | 4.40 | 1.407 | 4.05 | 1.394 | 0.001 * |
| * My body is sexually appealing | 3.98 | 1.466 | 4.43 | 1.475 | 4.13 | 1.482 | 0.008 * |
| * I have the type of body that people want to look at | 4.29 | 1.413 | 4.63 | 1.416 | 4.40 | 1.421 | 0.031 * |

* significant at 0.05, 1 = strongly agree; 7 = strongly disagree.

### *3.4. Male Body Perceptions–Influences*

Both single males and those in a relationship indicated their health as the most influential factor on how they thought about their body, and celebrities and fashion as least important in influencing their body image. Where differences between the groups were significant, single males were more likely to be influenced by their friends, by women and by society generally, than were males in relationships. Males who were part of a couple tended to be influenced more by their partner, and by their enjoyment of sports, than did single males (refer Table 4). These findings give support to H3 and H4.

**Table 4.** Importance of influential factors on male body image.

| How Important Are the Following in Influencing the Way You Think about Your Own Body? | Do You Have a Partner or Significant Other | | | | | | *p*-Value |
|---|---|---|---|---|---|---|---|
| | Yes (n = 240) | | No (n = 115) | | Total (n = 355) | | |
| | Mean | SD | Mean | SD | Mean | SD | |
| Celebrities | 5.25 | 1.542 | 5.04 | 1.613 | 5.18 | 1.566 | 0.255 |
| Family | 3.63 | 1.522 | 3.55 | 1.540 | 3.60 | 1.527 | 0.639 |
| * Friends | 3.70 | 1.487 | 3.17 | 1.213 | 3.52 | 1.425 | 0.001 * |
| * Women | 3.55 | 1.497 | 3.09 | 1.472 | 3.40 | 1.503 | 0.006 * |
| * Your partner/loved one | 2.54 | 1.206 | 3.25 | 1.611 | 2.77 | 1.389 | 0.000 * |
| * Society, generally | 3.86 | 1.482 | 3.45 | 1.384 | 3.73 | 1.462 | 0.014 * |
| Fashions | 4.45 | 1.560 | 4.22 | 1.497 | 4.38 | 1.542 | 0.176 |
| Age | 3.73 | 1.500 | 3.50 | 1.360 | 3.66 | 1.458 | 0.174 |
| Health | 2.80 | 1.367 | 2.84 | 1.225 | 2.81 | 1.321 | 0.772 |
| * Sport | 3.62 | 1.523 | 4.06 | 1.693 | 3.76 | 1.591 | 0.015 * |

* significant at 0.05, 1 = extremely important; 7 = not important at all.

### *3.5. Physical Appearance Factors–Key Indicators*

Single males weighted most motivating factors as more important than did their non-single counterparts, and where this difference was significant, single males in the study were more likely to emphasise socially reflective factors, such as being lean, what friends think of their appearance and having a visibly flat stomach, than were males in relationships (refer to Table 5). This supports H3 and H4. Men in relationships in this study emphasised the importance of being healthy, having a good lifestyle balance and being able to participate in sports when compared to single participants, but the difference was not significant. Both groups indicated health, lifestyle and body satisfaction as the most important factors in motivating their overall body image perceptions.

**Table 5.** Key indicators of desired physical appearance.

| How Important Is the Following to You? | Do You Have a Partner or Significant Other | | | | | | $p$-Value |
|---|---|---|---|---|---|---|---|
| | Yes (n = 240) | | No (n = 115) | | Total (n = 355) | | |
| | Mean | SD | Mean | SD | Mean | SD | |
| Being muscular | 3.76 | 1.344 | 3.70 | 1.364 | 3.74 | 1.349 | 0.725 |
| * Being lean | 3.40 | 1.247 | 3.00 | 1.068 | 3.27 | 1.206 | 0.003 * |
| Being seen as very masculine | 3.93 | 1.427 | 3.82 | 1.302 | 3.90 | 1.387 | 0.462 |
| Looking better than other males in my age group | 3.95 | 1.447 | 3.77 | 1.465 | 3.89 | 1.454 | 0.252 |
| Reaching my personal fitness goals | 3.31 | 1.390 | 3.28 | 1.335 | 3.30 | 1.370 | 0.826 |
| Attracting a partner/being attractive to a partner | 3.03 | 1.229 | 2.90 | 1.370 | 2.99 | 1.276 | 0.389 |
| Feeling satisfied with my body | 2.78 | 1.065 | 2.65 | .983 | 2.74 | 1.040 | 0.282 |
| * What my friends think of my physical appearance | 3.85 | 1.419 | 3.50 | 1.280 | 3.73 | 1.383 | 0.025 * |
| What my family thinks of my physical appearance | 3.82 | 1.444 | 3.77 | 1.366 | 3.81 | 1.418 | 0.711 |
| Being healthy, rather than just looking good | 2.49 | 1.153 | 2.54 | .985 | 2.50 | 1.101 | 0.680 |
| Having a good lifestyle balance | 2.47 | 1.035 | 2.57 | 1.101 | 2.50 | 1.056 | 0.390 |
| * Being in the right physical shape for the sports I enjoy | 3.05 | 1.345 | 3.36 | 1.488 | 3.15 | 1.398 | 0.056 * |
| * Having a flat stomach | 3.36 | 1.273 | 3.06 | 1.223 | 3.26 | 1.263 | 0.038 * |

* Significant at 0.05, 1 = extremely important; 7 = not important at all.

## 4. Discussion

For health-related products and services and public health messages regarding body weight management to be effective, research that explores differentiating factors within specific gender groups is required. The aim of this study was to examine how men's body image confidence and perceptions, along with their social reflectivity, influenced their personal body assessments. The association between physical health and body image perceptions has been established in prior research, particularly among adult populations. Further, prior research confirms the intrinsic connections between body image, self-esteem and motivations for health behaviours such as body weight management [3]. Those who are more preoccupied with their body not fitting a perceived ideal are more likely to be motivated towards change regarding physical exercise or dieting [3]. However, the benefits of this association are reliant on critically accurate body assessments—essentially, that those who need to modify their lifestyle or health behaviours recognise the need for change in their own body [9]. The differences apparent in how individuals evaluate their body, and how this motivates body-related health behaviours such as diet and exercise, highlight a need for research that examines specific demographic and sociographic groups. While previous research has examined women, children and teenagers/adolescents in respect of their body image and self-esteem, little work has examined the adult male (beyond contrasting them to women). Further, there are gaps in understanding differences in body image perceptions and self-esteem between adult male subjects. This study contributes to our understanding of the heterosexual adult male and their body image evaluations and motivations toward body health-related behaviour.

The prior literature has indicated a relationship between body image positivity and motivation toward exercise in adolescents, but studies of adults indicate that body satisfaction and health-promoting lifestyles are not strongly related [13]. This study finds that adult males are generally very positive about their bodies, with satisfaction overall in how their body compares to others. This was particularly evident amongst adult males in romantic relationships, who rated themselves as better looking and more sexually attractive than did single men. In the literature concerning adolescent perceptions of the body, body image concerns are not said to promote healthy behaviours; rather, lower body esteem promotes unhealthy behaviour regarding nutrition and exercise avoidance. In contrast, the single males in this study, who exhibited lower body confidence than those males in relationships, were more motivated toward a continued effort in looking their best (based on an awareness of how others might perceive their physical appearance). In this study, age was not a significant predictor of body image perceptions or health behaviours.

Low body dissatisfaction is known to have a significant impact on lifestyle behaviours, particularly those related to fitness and health motivations [3] However, while female perceptions of their body and related body image concerns have long been a subject of research interest, consideration of males has largely been limited to contrast between them and women, regarding women's proposed body negativity and unhealthy behaviour [9]. The male participants of this study, despite representing four separate age cohorts from early adulthood to 65 years, did not indicate significant motivations toward health-directed body management behaviours, including monitoring their fat and alcohol intake, moderating food intake for health or increasing their exercise. This is a useful finding, as it illustrates a widespread issue in male health campaigning—the lack of strong motivation for change across all ages when considering easy-to-implement physical health actions. The current research suggests that social reflexivity may be a better motivator for behaviour change amongst these men, with key external influences for both single men and men in relationships identified. Further, having a positive body image may be linked to more sustainable, long-term health-promoting behaviours rather than drastic, short-term changes such as dieting or extreme exercise programs. Limitations of this study include the single country sample (New Zealand) and the cross-sectional nature of the data collection. A study that includes physical health data by participant, collected over a period, would add weight to our understanding of the research questions proposed here. Further, where relationships have been the focus of this study, it is recommended that future research extends further than heterosexual males. Other recommendations for future research would include the examination of other lifestyle factors, to better understand the relative weight of influence of the single/relationship construct. The poor reliability of the Rosenberg self-esteem scale indicates the need for further exploration of this construct, potentially via the development of a new scale, in future research.

## 5. Conclusions

Public health promotion messages and health-related products and services targeting men are often focused on the issue of body weight management. This study indicates the lack of a common set of drivers toward body maintenance and weight management across all males, but it does indicate different foci for males in relationships compared to those who are single. Age was not found to be a significant influence for male views of the body; rather, body image satisfaction tended to be influenced by other aspects of life satisfaction, as indicated by relationship status. This is an important and useful finding for those developing health/weight messages targeting men in New Zealand. Health promotion that is effective for each of the two cohorts examined here should activate key motivators of body image satisfaction in any messaging, such as the view of friends (for single males) and partners (for those in relationships), rather than focusing on personal health management drivers, including negative assessments of the self.

**Funding:** This research received no external funding.

**Institutional Review Board Statement:** This study was conducted in accordance with the Declaration of Helsinki and approved by the Ethics Committee of the University of Otago, approval #L148 20 November 2016.

**Informed Consent Statement:** Informed consent was obtained from all subjects involved in the study.

**Data Availability Statement:** Data are unavailable due to privacy or ethical restrictions.

**Conflicts of Interest:** The author declares no conflicts of interest.

## References

1. Barry, B. (Re)Fashioning Masculinity: Social Identity and Context in Men's Hybrid Masculinities through Dress. *Gend. Soc.* **2018**, *32*, 638–662. [CrossRef]
2. Barry, B. Expanding the male ideal: The need for diversity in men's fashion advertisements. *Crit. Stud. Men's Fash.* **2014**, *1*, 275–293. [CrossRef] [PubMed]

3. Zanon, A.; Tomassoni, R.; Gargano, M.; Granai, M. Body image and health behaviors: Is there a relationship between lifestyles and positive body image? *Clin. Ter.* **2016**, *167*, 63–69.
4. Rieke, S.E.; Fowler, D.; Chang, H.J.; Velikova, N. Exploration of factors influencing body image satisfaction and purchase intent: Millennial females. *J. Fash. Mark. Manag.* **2016**, *20*, 208–229. [CrossRef]
5. McNeill, L.; Douglas, K. Retailing masculinity: Gender expectations and social image of male grooming products in New Zealand. *J. Retail. Consum. Serv.* **2011**, *18*, 448–454. [CrossRef]
6. E Lowery, S.; Kurpius, S.E.R.; Befort, C.; Blanks, E.H.; Sollenberger, S.; Nicpon, M.F.; Huser, L. Body Image, Self-Esteem, and Health-Related Behaviors Among Male and Female First Year College Students. *J. Coll. Stud. Dev.* **2005**, *46*, 612–623. [CrossRef]
7. LePage, M.L.; Crowther, J.H. The effects of exercise on body satisfaction and affect. *Body Image* **2010**, *7*, 124–130. [CrossRef] [PubMed]
8. Jankauskiene, R.; Baceviciene, M.; Pajaujiene, S.; Badau, D. Are Adolescent Body Image Concerns Associated with Health-Compromising Physical Activity Behaviours? *Int. J. Environ. Res. Public Health* **2019**, *16*, 1225. [CrossRef] [PubMed]
9. Gillison, F.; Standage, M.; Skevington, S. Relationships among adolescents' weight perceptions, exercise goals, exercise motivation, quality of life and leisure-time exercise behaviour: A self-determination theory approach. *Health Educ. Res.* **2006**, *21*, 836–847. [CrossRef]
10. McIntosh-Dalmedo, S.; Devonport, T.J.; Nicholls, W.; Friesen, A.P. Examining the Effects of Sport and Exercise Interventions on Body Image Among Adolescent Girls: A Systematic Review. *J. Sport Behav.* **2018**, *41*, 245–269.
11. Kantanista, A.; Osiński, W.; Borowiec, J.; Tomczak, M.; Król-Zielińska, M. Body image, BMI, and physical activity in girls and boys aged 14–16 years. *Body Image* **2015**, *15*, 40–43. [CrossRef] [PubMed]
12. Moustafa, S.; Tawbe, Z.; Sleiman, F.; El Daouk, S.; Koubar, M.; Hoteit, M. Body Image Perception in Association with Healthy Lifestyle Behaviour's in Lebanese Men and Women. *Int. J. Sch. Cogn. Psychol.* **2017**, *4*, 201–207. [CrossRef]
13. Penedo, F.J.; Dahn, J.R. Exercise and well-being: A review of mental and physical health benefits associated with physical activity. *Curr. Opin. Psychiatry* **2005**, *18*, 189–193. [CrossRef] [PubMed]
14. Baker, L.; Gringart, E. Body image and self-esteem in older adulthood. *Ageing Soc.* **2009**, *29*, 977–995. [CrossRef]
15. El Ansari, W.; Clausen, S.V.; Mabhala, A.; Stock, C. How do I look? Body image perceptions among university students from England and Denmark. *Int. J. Environ. Res. Public Health* **2010**, *7*, 583–595. [CrossRef] [PubMed]
16. Coombs, R.H. Marital status and personal well-being: A literature review. *Fam. Relat. Interdiscip. J. Appl. Fam. Stud.* **1991**, *40*, 97–102. [CrossRef]
17. Stack, S.; Eshleman, J.R. Marital Status and Happiness: A 17-Nation Study. *J. Marriage Fam.* **1998**, *60*, 527. [CrossRef]
18. Renna, F.; Grafova, I.B.; Thakur, N. The effect of friends on adolescent body weight. *Econ. Hum. Biol.* **2008**, *6*, 377–387. [CrossRef] [PubMed]
19. Paeratakul, S.; White, M.; Williamson, D.; Ryan, D.; Bray, G. Sex, race/ethnicity, socioeconomic status, and BMI in relation to self-perception of overweight. *Obes. Res.* **2002**, *10*, 345–350. [CrossRef]
20. Sobal, J.; Rauschenbach, B.S.; Frongillo, E.A., Jr. Marital status, fatness and obesity. *Soc. Sci. Med.* **1992**, *35*, 915–923. [CrossRef]
21. Hogg, M.A.; Abrams, D. *Social Identifications: A Social Psychology of Intergroup Relations and Group Processes*; Taylor & Frances/Routledge: Florence, KY, USA, 1988.
22. D'Alessandro, S.; Chitty, B. Real or relevant beauty? Body shape and endorser effects on brand attitude and body image. *Psychol. Mark.* **2011**, *28*, 843–878. [CrossRef]
23. Martin, M.C.; Peters, C.O. Exploring adolescent girls' identification of beauty types through consumer collages. *J. Fash. Mark. Manag. Int. J.* **2005**, *9*, 391–406. [CrossRef]
24. McNeill, L.S.; Firman, J.L. Ideal Body Image: A Male Perspective on Self. *Australas. Mark. J.* **2014**, *22*, 136–143. [CrossRef]
25. Rosenberg, M. *Society and the Adolescent Self-Image*; Princeton: Princeton, NJ, USA, 1965.
26. Netemeyer, R.; Burton, S.; Lichtenstein, D. Trait aspects of vanity: Measurement and relevance to consumer behaviour. *J. Consum. Res.* **1995**, *21*, 612–626. [CrossRef]
27. Shelvin, M.E.; Bunting, B.P.; Lewis, C.A. Confirmatory factor analysis of the Rosenberg's Self-Esteem Scale. *Psychol. Rep.* **1995**, *76*, 711–714.
28. Vermillion, M.; Dodder, R.A. An examination of the Rosenberg self-esteem scale using collegiate wheelchair basketball student athletes. *Percept. Mot. Ski.* **2007**, *104*, 416–418. [CrossRef]

MDPI AG
Grosspeteranlage 5
4052 Basel
Switzerland
Tel.: +41 61 683 77 34

*Nutrients* Editorial Office
E-mail: nutrients@mdpi.com
www.mdpi.com/journal/nutrients

www.ingramcontent.com/pod-product-compliance
Lightning Source LLC
Chambersburg PA
CBHW061459210926
51850CB00008B/1764
*9783725863426*